Studies in Big Data

Volume 111

Series Editor

Janusz Kacprzyk, Polish Academy of Sciences, Warsaw, Poland

The series "Studies in Big Data" (SBD) publishes new developments and advances in the various areas of Big Data- quickly and with a high quality. The intent is to cover the theory, research, development, and applications of Big Data, as embedded in the fields of engineering, computer science, physics, economics and life sciences. The books of the series refer to the analysis and understanding of large, complex, and/or distributed data sets generated from recent digital sources coming from sensors or other physical instruments as well as simulations, crowd sourcing, social networks or other internet transactions, such as emails or video click streams and other. The series contains monographs, lecture notes and edited volumes in Big Data spanning the areas of computational intelligence including neural networks, evolutionary computation, soft computing, fuzzy systems, as well as artificial intelligence, data mining, modern statistics and Operations research, as well as self-organizing systems. Of particular value to both the contributors and the readership are the short publication timeframe and the world-wide distribution, which enable both wide and rapid dissemination of research output.

The books of this series are reviewed in a single blind peer review process.

Indexed by SCOPUS, EI Compendex, SCIMAGO and zbMATH.

All books published in the series are submitted for consideration in Web of Science.

Mariya Ouaissa · Zakaria Boulouard ·
Mariyam Ouaissa · Inam Ullah Khan ·
Mohammed Kaosar
Editors

Big Data Analytics and Computational Intelligence for Cybersecurity

 Springer

Editors
Mariya Ouaissa 🆔
Moulay Ismail University
Meknes, Morocco

Mariyam Ouaissa 🆔
Moulay Ismail University
Meknes, Morocco

Mohammed Kaosar 🆔
Murdoch University
Murdoch, Australia

Zakaria Boulouard 🆔
University of Hassan II Casablanca
El Mansouria, Morocco

Inam Ullah Khan 🆔
Islamabad Campus
Isra University, SEAS
Islamabad, Pakistan

ISSN 2197-6503   ISSN 2197-6511 (electronic)
Studies in Big Data
ISBN 978-3-031-05754-0   ISBN 978-3-031-05752-6 (eBook)
https://doi.org/10.1007/978-3-031-05752-6

This Springer imprint is published by the registered company Springer Nature Switzerland AG
The registered company address is: Gewerbestrasse 11, 6330 Cham, Switzerland

Preface

Over the past few years, big data has introduced itself as an important driver of our everyday life. New applications and systems, such as social media, wearable devices, drones, and the Internet of Things (IoT), continue to emerge and generate even more data. With the COVID-19 pandemic, the need to stay online and exchange data has become even more crucial, as most of the fields, would they be industrial, educational, economic, or service-oriented, had to go online as best as they can.

This growth in data exchange also comes with an increase in cyber-attacks and intrusion threats. Detecting cyber-attacks becomes a challenge, not only because of the sophistication of attacks but also because of the large scale and complex nature of today's IT infrastructures.

Combining the full potentiality of Artificial Intelligence and Big Data Analytics is a practice that enables analysts and decision-makers examine large amounts of data to uncover hidden patterns, correlations, and insights. This can help organizations identify new opportunities and can steer them to more efficient strategic moves. That being said, what if the strategies to optimization were actually about securing the organization's own data?

Big Data along with automated analysis brings network activity into clear focus to detect and stop threats, as well as shorten the time to remedy when attacks occur. The ability to accumulate large amounts of data provides the opportunity to examine, observe, and notice irregularities to detect network issues.

The core purpose of fitting AI and Big Data into Cybersecurity is to improve the detection of potential Cyber Threats with a more sophisticated approach. The detection in any system needs to be fast in order to detect the major and minor changes in the system.

This book will present a collection of state-of-the-art Artificial Intelligence and Big Data Analytics approaches to cybersecurity intelligence. It will illustrate the latest trends in AI and Machine Learning-based strategic defense mechanisms against malware, vulnerabilities, cyber threats, and provide solutions for the development of proactive countermeasures. It will also introduce other trending technologies, such as Blockchain, SDN and IoT, and discuss their possible impact on improving security.

The first section of this book starts with an overview over the latest advancements in Cybersecurity (Chapter "New Advancements in Cybersecurity: A Comprehensive Survey"), and then, it moves up the communication networks different layers and discusses the impact these advancements in each of these layers. For instance, the authors in chapter "CPSs Communication Using 5G Network in the Light of Security" present a new approach for communication between 5G-based cyber-physical systems, while in chapter "A Survey on Security Aspects in RPL Protocol Over IoT Networks", the authors investigate different security aspects in the RPL protocol. Chapter "Analysis of Cybersecurity Risks and Their Mitigation for Work-From-Home Tools and Techniques" analyzes cybersecurity risks and their mitigation for work-from-home tools and techniques, while chapter "A Systemic Security and Privacy Review: Attacks and Prevention Mechanisms Over IoT Layers" provides a systemic security and privacy review that goes through different attacks and prevention mechanisms over IoT layers. Chapter "Software-Defined Networking Security: A Comprehensive Review" presents a comprehensive survey on the concept of Software-Defined Networks and their security, while chapter "Detection of Security Attacks Using Intrusion Detection System for UAV Networks: A Survey" investigates different Intrusion Detection System-based solutions that are able to detect security attacks on drones.

The second section of this book starts by discussing the role of computational intelligence on cybersecurity (Chapter "Role of Computational Intelligence in Cybersecurity") in general, before moving to more specific use cases such as Intrusion Detection Systems in cyberspace (Chapter "Computational Intelligence Techniques for Cyberspace Intrusion Detection System"), or in IoT-based networks (Chapter "A Comparative Analysis of Intrusion Detection in IoT Network Using Machine Learning"). The authors in chapter "Blockchain Enabled Artificial Intelligence for Cybersecurity Systems" investigate the idea of implementing Blockchain-based Artificial Intelligence in cybersecurity systems, while the authors in chapter "Approaches for Visualizing Cybersecurity Dataset Using Social Network Analysis" use social network analysis techniques for cybersecurity purposes.

The last section of this book covers different aspects of Big Data Analytics and their applications. It starts by covering different Big Data Analytics-based approaches that can solve cybersecurity issues such as data footprinting (Chapter "Data Footprinting in Big Data"), or surveillance drone videos analysis (Chapter "An Investigation of Unmanned Aerial Vehicle Surveillance Data Processing with Big Data Analytics"). The authors of chapter "Big Data Mining Using K-Means and DBSCAN Clustering Techniques" investigate the role of DBSCAN and K-Means clustering in Big Data Analytics. Chapters "IoT Security in Smart University Systems"–"Transformation in Health-Care Services Using Internet of Things (IoT): Review" discuss the role of Big Data, IoT and their security in education and healthcare, while the authors in chaps. "A Survey of Deep Learning Methods for Fruit and Vegetable Detection and Yield Estimation", and "Bird Calls Identification in Soundscape Recordings Using Deep Convolutional Neural Network" provide some approaches to analyze agriculture and zoology data.

Meknes, Morocco   Mariya Ouaissa
El Mansouria, Morocco   Zakaria Boulouard
Meknes, Morocco   Mariyam Ouaissa
Islamabad, Pakistan   Inam Ullah Khan
Murdoch, Australia   Mohammed Kaosar

Contents

About the Editors

Dr. Mariya Ouaissa is currently a Professor at Institute Specializing in New Information and Communication Technologies, Researcher Associate and practitioner with industry and academic experience. She is a Ph.D. graduated in 2019 in Computer Science and Networks, at the Laboratory of Modelization of Mathematics and Computer Science from ENSAM-Moulay Ismail University, Meknes, Morocco. She is a Networks and Telecoms Engineer, graduated in 2013 from National School of Applied Sciences Khouribga, Morocco. She is a Co-Founder and IT Consultant at IT Support and Consulting Center. She was working for School of Technology of Meknes Morocco as a Visiting Professor from 2013 to 2021. She is member of the International Association of Engineers and International Association of Online Engineering, and since 2021, she is an "ACM Professional Member". She is Expert Reviewer with Academic Exchange Information Centre (AEIC) and Brand Ambassador with Bentham Science. She has served and continues to serve on technical program and organizer committees of several conferences and events and has organized many Symposiums/Workshops/Conferences as a General Chair also as a reviewer of numerous international journals. Dr. Ouaissa has made contributions in the fields of information security and privacy, Internet of Things security, and wireless and constrained networks security. Her main research topics are IoT, M2M, D2D, WSN, Cellular Networks, and Vehicular Networks. She has published over 20 papers (book chapters, international journals, and conferences/workshops), 8 edited books, and 5 special issue as guest editor.

Dr. Zakaria Boulouard is currently a Professor at Department of Computer Sciences at the "Faculty of Sciences and Techniques Mohammedia, Hassan II University, Casablanca, Morocco". In 2018, he joined the "Advanced Smart Systems" Research Team at the "Computer Sciences Laboratory of Mohammedia". He received his Ph.D. degree in 2018 from "Ibn Zohr University, Morocco" and his Engineering Degree in 2013 from the "National School of Applied Sciences, Khouribga, Morocco". His research interests include Artificial Intelligence, Big Data Visualization and Analytics, Optimization and Competitive Intelligence. Since 2017, he is a member of "Draa-Tafilalet Foundation of Experts and Researchers", and since

2020, he is an "ACM Professional Member". He has served on Program Committees and Organizing Committees of several conferences and events and has organized many Symposiums/Workshops/Conferences as a General Chair. He has served and continues to serve as a reviewer of numerous international conferences and journals. He has published several research papers. This includes book chapters, peer-reviewed journal articles, peer-reviewed conference manuscripts, edited books and special issue journals.

Dr. Mariyam Ouaissa is currently a Professor at Institute specializing in new information and communication technologies, Researcher Associate and Consultant Trainer in Computer Science and Networks. She received her Ph.D. degree in 2019 from National Graduate School of Arts and Crafts, Meknes, Morocco and her Engineering Degree in 2013 from the National School of Applied Sciences, Khouribga, Morocco. She is a communication and networking researcher and practitioner with industry and academic experience. Dr. Ouaissa's research is multidisciplinary that focuses on Internet of Things, M2M, WSN, vehicular communications and cellular networks, security networks, congestion overload problem and the resource allocation management and access control. She is serving as a reviewer for international journals and conferences including as IEEE Access, Wireless Communications and Mobile Computing. Since 2020, she is a member of "International Association of Engineers IAENG" and "International Association of Online Engineering", and since 2021, she is an "ACM Professional Member". She has published more than 20 research papers (this includes book chapters, peer-reviewed journal articles, and peer-reviewed conference manuscripts), 8 edited books, and 5 special issue as guest editor. She has served on Program Committees and Organizing Committees of several conferences and events and has organized many Symposiums/Workshops/Conferences as a General Chair.

Dr. Inam Ullah Khan was a Lecturer at different universities in Pakistan which include Center for Emerging Sciences Engineering & Technology (CESET), Islamabad, Abdul Wali Khan University, Garden and Timergara Campus and University of Swat. Recently, he is selected as a visiting researcher at king's college London, UK. He did his Ph.D. in Electronics Engineering from Department of Electronic Engineering, Isra University, Islamabad Campus, School of Engineering & Applied Sciences (SEAS). He had completed his M.S. degree in Electronic Engineering at Department of Electronic Engineering, Isra University, Islamabad Campus, School of Engineering & Applied Sciences (SEAS). He had done undergraduate degree in Bachelor of Computer Science from Abdul Wali Khan University Mardan, Pakistan.Apart from that his Master's thesis is published as a book on topic "Route Optimization with Ant Colony Optimization (ACO)" in Germany which is available on Amazon. He is a research scholar; he has published some research papers at international level. More interestingly he recently introduced a novel routing protocol E-ANTHOCNET in the area of flying ad hoc networks. His research interest includes Network System Security, Intrusion Detection, Intrusion Prevention, cryptography, Optimization techniques, WSN, IoT, UAV's, Mobile Ad Hoc Networks (MANETS),

Flying Ad Hoc Networks, and Machine Learning. He has served international conferences as Technical program committee member which include, EAI International Conference on Future Intelligent Vehicular Technologies, Islamabad, Pakistan and 2nd International Conference on Future Networks and Distributed Systems, Amman, Jordan, June 26–27, 2018 and now recently working on the same level at International Workshop on Computational Intelligence and Cybersecurity in Emergent Networks (CICEN'21) that will be held in conjunction with the 12th International Conference on Ambient Systems, Networks and Technologies (EUSPN 2021) which is co-organized in November 1–4, 2021, Leuven, Belgium.

Dr. Mohammed Kaosar is currently working as a Senior Lecturer in the Discipline of IT, Media and Communications, Murdoch University, Australia. Prior to that he has worked in several universities including RMIT University, Victoria University, Effat University and Charles Sturt University. Dr. Kaosar has worked in number of national and international research projects and grants. He has published many research papers in reputable journals and conferences including - IEEE Transaction on Knowledge and Data Engineering (TKDE), Data and Knowledge Engineering (DKE), IEEE International Conference on Data Engineering (ICDE), Computer Communication (ComCom), ACM SIGCOMM. He has been supervising many postgraduate students, mentoring junior colleagues, and collaborating with many national and international researchers. Dr. Kaosar is an active member of various professional organizations including – IEEE (Senior Member), Australian Computer Society (ACS), European Alliance for Innovation (EAI), The Institute for Computer Sciences, Social Informatics and Telecommunications Engineering (ICST), South Pacific Competitive Programming Association, Industrial Engineering and Operation Management (IEOM). Dr. Kaosar contributes by organizing conferences and performing editorial responsibility in some journals.

Cybersecurity in Communication Networks: Challenges and Opportunities

New Advancements in Cybersecurity: A Comprehensive Survey

Muhammad Abul Hassan, Sher Ali, Muhammad Imad, and Shaista Bibi

Abstract World is now considered as global village because of interconnected networks. Smart phones and large computing devices exchange millions of information each day. Information or data privacy is first priority for any tech company. Information security get attention both from academia and industry sectors for the purpose of data prevention, integrity and data modification. Traditional and mathematical security models are implemented to address information related issues but does not provide full proof data privacy. Computational Intelligence is power technique inspired from biological development and act as intelligent agent which detects treats in real and complex environments. Computational Intelligence is further sub divided in to Fuzzy Logic, Evaluation Computation, Artificial Neural Networks and hybrid approach. In this research study each branch of Computational Intelligence is studied from cybersecurity point of view with its merits and demerits.

Keywords Information security · Cybersecurity · Computational intelligence · Intrusion detection system

1 Introduction

Now a days securing a network is very challenging task for every tech Companies. After emergence of computer networks, integrity and data security are essential parameters of every network. Alan Turning in 1941 decipher Enigma Machine and still same type of cyber security treats are present for every network e.g. in 1950

M. A. Hassan (✉) · M. Imad
Department of Computing and Technology, Abasyn University, Peshawar 25000, Pakistan
e-mail: abulhassan900@gmail.com

S. Ali
Department of Electrical and Computer Engineering, Air University, Islamabad, Pakistan

M. A. Hassan · S. Bibi
Department of Computer Science, Federal Government Girl Degree College, Nowshera 24100, Pakistan

© The Author(s), under exclusive license to Springer Nature Switzerland AG 2022
M. Ouaissa et al. (eds.), *Big Data Analytics and Computational Intelligence for Cybersecurity*, Studies in Big Data 111,
https://doi.org/10.1007/978-3-031-05752-6_1

3

Table 1 Cyber Attacks

Name of cyber attack	Year	Impact of destruction
Operation Shady RAT	2006	Major destruction has been noticed, more than 71 organizations data were compromised
Dwell time attack	2018	Passive attack, mostly being used for stealing of important files, spying purpose and Denial of service activities. Undetected for 71 days in USA, Europe, the Middle East and Africa (EMEA) for 177 days, Asia–Pacific (APAC) for 204 days
DigiNotar's	2011	More than 500 Fake DigiNotar's certificates were found for stealing web browser data
Duqu	2011	Malware attack and provide services to the attackers
PLA Unit 61398	2014	Stole confidential business data
TAO	1998	Fake Finger printers were loaded in applications or figure ID machines triggered by USA for Stealing sensitive data
WHO	2001	Compromised several Websites and other corporation services to monitor high profile targets and exploit private conversation
Red October	2012	Collects personal data of diplomates their private communication with other higher authorities
Wanna Cry Ransomware Attack	2017	Attack computers having Microsoft windows operating system and demand for payment (amount)
Cyber-attacks on Ukraine	2017	Petya malware was injected in the Ukraine organizations, banks, electricity grids and caused major destruction

Massachusetts Institute of Technology (MIT) was compromised, in 2014 yahoo accounts data were stolen. Similarly, Table 1 represents classification of cyber-attacks and impact of destructions. More than 4.66 billion people send and receive data over internet and security of such data is very important [1, 2]. In year 2017 total 16 GB of data is collected per capita and will be increase up to 50 GB by 2022 [3]. Evaluation and deployment of Intrusion Detection System (IDS) in industries and academia shows the importance of cybersecurity systems. Intrusion Detection System shields network from active and passive attacks. At first IDS gather data then apply security measurements to prevent network from all kinds of attacks, moreover IDS send data to human network analyst system to infallible network security. Diverse algorithms are embedded in the system, these are (i) Rule-based (ii) Statistics-based (iii) Machine learning (ML) based [4–6]. Rule based systems are very fast and reliable but difficult to get updates on regular bases. To fill gaps in rule-based systems statistics methods are applied in intrusion detection system to provide processing of imprecise information but unable to handle large amount of data at a time. Machine Leaning based algorithms are now mostly embed in IDS to provide complex detection support on large amount of data [7].

In cyber security applications, filtering of data can be performed in to two ways (i) system logs and (ii) Network traffic. System logs is performed on host-based operating systems. Payload and packet headers are extracted by TCP/IP stack in network traffic [8]. Packet Capture and Netflow protocols are used in the network traffic approach [9]. Packet Capture protocol provides over all extraction of data packets which are currently being transmitted [10]. In addition, information collected from data packets comprise of packet size, flags, source to destination IP and port address and Header. Trade off must be establish to maintain integrity of private conversation and to prevent R2L and U2R non-flooding attacks. In Fig. 1 Packet Capture method is used to detect suspicious bit in a stream of incoming and outgoing bits.

NetFlow contains IP flow, Command Line Interface, NetFlow Collector, NetFlow Cache. In IP flow, Same type of IP packets are group together to examine attributes, Source and Destination IP and port address, forwarded within a router or layer-3 switch. NetFlow Cache holds data for NetFlow after packets are been examined. Command Line Interface provide over all view of data which is useful in troubleshooting process. NetFlow Collector exports data to NetFlow Collector which behave like a reporting server. Netflow are further categorized in to two types (i) Hardware-based and (ii) Software-based collector while software-based collector is mostly used for data collection [11]. In Fig. 2 flow-enabled network devices are shown.

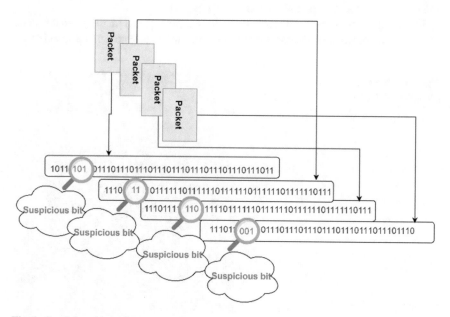

Fig. 1 Suspicious bit analyzer

Fig. 2 Network containing
flow-enabled devices

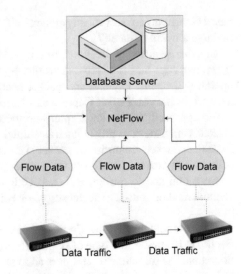

Computational Intelligence act as intelligent agent which detects treats in real and complex environments [12–14]. Natural Inspired algorithms are adopted to overcome data-driven complex problems which are not fully handled by traditional or mathematical approaches. Computational Intelligence is further divided in to four core approaches as shown in Fig. 3. These methods are adopted to overcome various problems such as optimal solution [15, 16], data hiding [17–19] and classifying normal and abnormal data in Intrusion detection system [20].

From last decades numerous techniques for Computational intelligence has been developed to overcome cybersecurity issues and according to our knowledge no

Fig. 3 Computation
intelligence categories

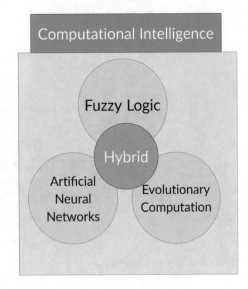

comprehensive study is conducted to discuss details about these techniques. Major contributions of this research study are (i) Brief study on various Computational techniques (ii) highlighting challenges faced by Information security systems.

2 Background Study

Intent of Information Security is prevention of information from unauthorized access, modification, destruction and replication activities. In addition, to maintain confidentiality, integrity, and availability (CIA triad) of any network. Security system is classified in to four types (i) Surveillance (ii) Prevention, Response (iii) Deception and (iv) Detection [21]. Figure 4 represents different domains of the cyber security.

2.1 Prevention

This strategy is used to prevent network from attackers or attacks to ensure integrity and security of data. Authentication techniques are utilized in prevention strategy.

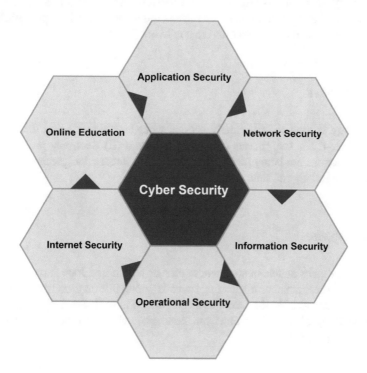

Fig. 4 Domains of cyber security

Different Software's are used to monitor user activities without side world information. Encrypted data is spread all over the network for CIA purpose. Firewalls, signatures and anomaly detections are also used to protect information but Problem associated with this strategy is small tolerance with unauthorized individuals [22, 23].

2.2 Surveillance

This strategy is used to perceive fast changing environments and attacks or threats to network. Security threats are automatically identified and take timely action to encounter such activity. Problem associated with this strategy it is very difficult to monitor both area (Technical and non-Technical), logical and physical access restrictions [24].

2.3 Detection

This strategy is designed to monitor specific kind of events occurred or trigged in a network. Detection is embedded in a network in various forms e.g. intrusion detection, anomaly detection and network scanning which aim is to pin pointing of malicious or unusual events occurred in a network [25–27].

2.4 Response

This strategy is used to take timely actions when network is being under attack. Furthermore, it has two phases (i) Treat Phase and (ii) Recovery phase. At first appreciate actions are being taken when network is under attack secondly network is restored to its original state [28].

2.5 Deception

This strategy takes attackers somewhere else or take it in a loop server to prevent data. It deceives attacker or misleading towards wrong information while preventing genuine information. Furthermore, it is categorized in to (i) Passive and (ii) Active deception. Passive hides everything from the attacker while active focused on network performance [23, 29].

2.6 Computational Intelligence for Cyber Security

Computation Intelligence was first coined by Institute of Electrical and Electronics Engineers Neural Networks Council in year 1990 and later on in 1993 Bezdek set proper definition of Computation intelligence. According to the Bezdek computational intelligence is based on numeric or low level and artificial intelligence is symbolic or knowledge base approach. Problem solving approach of both AI and CI is different. CI is based on in bottom-up approach and AI give solution is top-down manner.

2.6.1 Fuzzy Logic Approach in Information Security

Fuzzy logic was pitched in year 1965, a proposal of fuzzy logic set by Lotfi Zadeh [29–35]. Multi-value reasoning logic which aim is to get desired and suitable outcomes. Fuzzy set Theory employ membership function to identify the desired set for each element. Membership is ranging from 0 to 1 for each object mathematically shown in Eq. (1), let U be any ordinary set and fuzzy set of U. Where μ U (x) map elements in x on the real number [0,1].

$$U = \{(x, \, \mu \, U \, (x))|x \in X\} \text{ with } \mu \, U : X \rightarrow [0, 1] \qquad (1)$$

Authors in [31] conduct research study on automatic management technique which consists of three tasks. (i) Static (ii) Dynamic resource management and (iii) Service agreement. Static resource management avoids system Idle or over-burden state which comes from dynamic resource management. Auto globe and fuzzy logics are involved in second phase. Auto globe facilitates system with current situation of load while fuzzy logic generate output based on these load situations. Service agreement phase intelligently manage overall system load or particular changing demands. Problem faced by this technique is, it does not handle sudden load variations.

Authors in [33] studied Adaptive and Automatically Tuning the Intrusion Detection System (AATIDS) where network attacks are identified by Fuzzy logic. Prediction Filter forward doubtful prediction to system operator for verification. After identification of false prediction system automatically restore to its original state or early state. Problem faced by this technique is, delay in tuning caused degradation in system performance.

Authors in [36] proposed Fuzzy-IAS in which agents were divided in to different groups or societies. Ambassador agents were utilized to build communication channel between different agents. High degree of likelihood or correlation between agents were informed by ambassador agent for association. These agents have self-aware capabilities and adjust themselves accordingly. Advantage of this technique is number of associations and interconnections between agents are reduced in multi-agent group.

Authors in [37] conducted research study on Fuzzy Adaptive Buffering (FAB). For efficient utilization of mobile client's, Mobile client pull model is implemented. Two Thread model and fuzzy log inferences are used to improve pre-data fetch and sleeping cycle of data. Advantages of this technique is it successfully address buffer related issue e.g. Overflow and Underflow with excellent user experience.

Authors in [38] conducted research study on fuzzy logic bargaining system for information markets. Fuzzy logics were utilized for negotiation with buyer. During negotiation process buyer can make final decision based on these factors (i) Proposed Prices (ii) Deadline (iii) Remaining time for negotiation (iv) product relevance. But problem with this study is it does not consider buyer attitudes in their account. This problem is addressed in [39] and consider buyer preferences in to account.

Authors in [39] proposed extractor system based fuzzy logic system. In which finger prints and bio-metrics are combined for authentication. Vertex-indexed Triangulation (ViT) is used for verification of all fingerprints. Shift-free and rotation minutia local models are utilized by ViT for authentication. Later on, FE-SViT was developed to achieve high performance and accuracy. Fuzzy Rule interpolation (FIR) minimize complexity of fuzzy systems and make interfaces in sparse systems (rule based) moreover it generates maximum number of interpolation rules in reasoning process of interpolation. This technique is adopted in [40] in which dynamic fuzzy rule interpolation method is embedded in the intrusion detection system to improve overall coverage and efficiency of the system. Figure 5 represents different Cyber Security applications.

Fuzzy logic is sub branch of computation intelligence which is mainly utilized in cybersecurity. Normal and abnormal behavior detection of system is smoothly identified by fuzzy logics but main problem associated with fuzzy logics are (i) control system of fuzzy logics are still human dependent (ii) Verification and data validation is required.

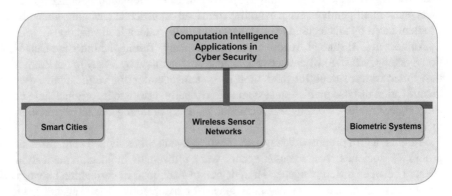

Fig. 5 Applications of computation intelligence in cyber security

2.6.2 Artificial Neural Network (ANN) Approach in Information Security

Artificial Neural Network (ANN) is collection of multiple neurons. These neurons are interconnected to accomplish difficult and different tasks like humans. Similarly, these neurons applied in data-driven applications as shown in Fig. 6.

Every network required security from unwanted events. To address these issues in the systems Artificial Neural Networks techniques are mostly applied for information security. Authors is [41] conducted research study on Generalized Regression Neural Network (GRNN) to enhance network security by detecting different type of attacks on network alert system timely during security breach.

Manual detection of malware required hours or even weeks to fix network. To address such problem CNN is proposed in [42] to extract malware effected packet bits sequences among regular bits, after that sequences of these effected bits are converted in to image. This image is further supplied to attention-based CNN where all malware and their types are predefined in to network and identify malware type.

Another unlawful activity is detected by Steganographer method where attacker steal data from user through steganography attack. Core responsibility of Steganographer is to identify convicted user with suspicious image. These suspicious images have secret information. Authors in [43] come up with solution to identify such type of attack by implementing Multi-class Deep Neural Networks based Steganographer Detection method. Six type of different embedding payloads is classified in training phase. In second phase (testing), features are extracted from each individuals (images) and identify convicted one (user).

Authors in [44] conducted research on Embedding Model based on Feed Forward Neural Networks where incoming data of a network is converted in to functions, geographical location, temporal information into dense numerical vectors for creating model locations, establishing different communities and users. Advantage of this model is it can be used in many real-world applications such as location tracing,

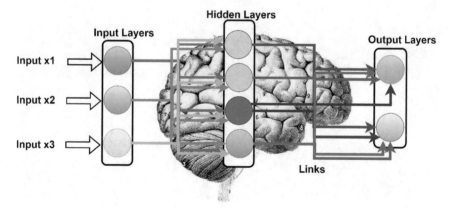

Fig. 6 Artificial neural network layers

recommendation, prediction of crime. Moreover, system is more robust compare to other models.

Artificial Neural Network (ANN) has ability to deal with noisy, incomplete data and improve processing speed. Furthermore, for any reasons if one or more ANN cells are corrupted it does not affect performance of the system, more importantly it has the ability to learn from different events. Main problem faced in ANN models are (i) Hardware dependence, (ii) Undiscovered behavior (iii) Unknown time Duration. ANN required efficient hardware and parallel processing power in accordance to the proposed structure, moreover it does not come up with proper answer (how and why) and final result or product of ANN model is mostly not optimum.

2.6.3 Evolutionary Computation Approach in Information Security

Biological Evolution is a series of process and only fittest will be survived. Evolutionary Computation (EC) is purely inspired by biological evolution theory and create methodologies to solve problems and implement them by computers [45]. EC is further divided in to (i) Ant-Colony Optimization Algorithms (ii) Genetic Programming (iii) Particle Swarm Optimization (iv) Genetic Algorithm as shown in Fig. 7.

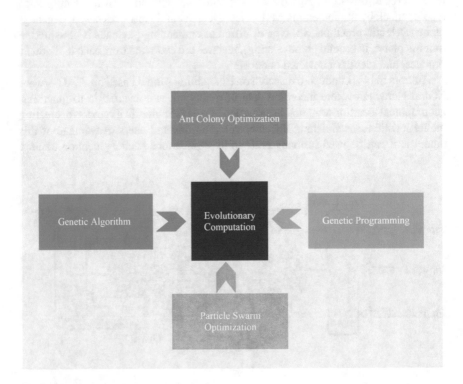

Fig. 7 Branches of evolutionary computation

Moreover, all EC algorithms have common properties e.g. (i) Recursive Algorithms approach (ii) Inherent parameters (iii) stochastic search.

Natural selection of rules are adopted in Genetic Algorithm (GA) which allow to select different set of rules by group of population. Genetic Programming (GP) is inspired by biological process or evaluation. GP is based on automatic process of evaluation and learning. Particle Swarm Optimization is Population based Stochastic Optimization Algorithm (PSOA). Which aim is to improve overall social behavior of candidate solution with respect to quality. Ant Colony Optimization is multi agent system, which mimics ant behaviors in finding food, shortest path by dropping pheromones. Graphical based representation of overall possible solution presented by using population. Components of solutions are defined by adjacent optimal solutions.

Authors in [46] conducted research study on compressed color images using loss less method for block truncation but does not achieved efficient results because of low quality of images and small payload capacity. Chaos Embedded-GA is proposed in [47] for color images. Multiple Maps Chaos are utilized in GA randomly while Peak Signal to Noise Ratio is taken as fitness function.

Authors in [48] proposed a genetic programming-based searching and repairing bugs in software's. Abnormal behavior of software is identified in first phase and then program variants are triggered until software retain its core functionality and avoid further defects. Authors in [49–51] also contributed in cybersecurity and proposed different methodologies to address cyber related issues.

Authors in [52] conducted research study on Security Risk Analysis Model (SRAM), a combination on ant colony optimization and Bayesian networks. Vulnerability Propagation path is estimated by ant colony optimization technique and provide possible risk-free planes.

Authors in [53] embed artificial Immune system in Intrusion detection in which unsupervised learning is utilized without any previous tannings and categorize network traffic in to self and non-self. Advantage of this system is it can provide both real time and online training for adaptive immune systems. Moreover, work presented in the [54–59] can also be utilized in evolutionary computation.

Evolutionary Computation played key role in addressing security related issues. Optimal Search solution is achieved by genetic algorithm. Genetic Programming techniques are utilized in normal and abnormal occurrences in Intrusion detection system. Hiding of data in images are done by particle swarm optimization. Problem faced by Evolutionary computation is it is not feasible for the real-world problems because of large number of solutions for specific problem.

3 Conclusion

Over past few decades, usage of internet is doubled. People send and receive millions of data over internet. Alan Turning in 1941 decipher enigma machine and still time to time different kind of attacks were lunched to steal sensitive and private data. Information security gained attention both from academia and industry to provide security and privacy of data. Traditional and mathematical models were implemented to identify or prevent incoming attacks but does not provide any reasonable solution. After that, Computation intelligence-based Security models were implemented and they provide accurate solution from security point of view. In this research study we have studied all branches of Computational intelligence with its advantages and disadvantages from cybersecurity point of view. During literature review we have concluded that computational intelligence is very power technique but still some work needs to be done to fill the gaps. For future direction, we will study other related models and try to find out its outcomes in a real environment.

References

1. S. Dorbala, R. Bhadoria, Analysis for security attacks in cyberphysical systems, in *Cyber-Physical Systems: A Computational Perspective* (Chapman & Hall, London, U.K., 2015), pp. 395–414
2. S.K. Khaitan, J.D. McCalley, Design techniques and applications of cyberphysical systems: A survey. IEEE Syst. J. **9**(2), 350–365 (2015)
3. R. Kabir, A.R. Onik, T. Samad, A network intrusion detection framework based on bayesian network using wrapper approach. Int. J. Comput. Appl. **166**, 13–17 (2017)
4. V. Kumar, D. Sinha, A.K. Das, S.C. Pandey, R.T. Goswami, An integrated rule-based intrusion detection system: analysis on UNSW-NB15 data set and the real time online dataset. Clust. Comput. **23**(2), 1397–1418 (2020)
5. X. Larriva-Novo, C. Sánchez-Zas, V.A. Villagrá, M. Vega-Barbas, D. Rivera, An approach for the application of a dynamic multi-class classifier for network intrusion detection systems. Electronics **9**(11), 1759 (2020)
6. H. Alqahtani, I.H. Sarker, A. Kalim, S.M.M. Hossain, S. Ikhlaq, S. Hossain, Cyber intrusion detection using machine learning classification techniques, in *International Conference on Computing Science, Communication and Security* (Springer, Singapore, 2020), pp. 121–131
7. D. Gümüşbaş, T. Yıldırım, A. Genovese, F. Scotti, A comprehensive survey of databases and deep learning methods for cybersecurity and intrusion detection systems. IEEE Syst. J. (2020)
8. O. Savas, J. Deng, *Big Data Analytics in Cybersecurity* (Auerbach, New York, NY, USA, 2017)
9. A.F. de Retana, A. Miranda-García, Á.M. Guerrero, C. Fernández-Llamas, Attacks detection on sampled netflow traffic through image analysis with convolutional neural networks (CNN), in *Computational Intelligence in Security for Information Systems Conference* (Springer, Cham, 2021), pp. 33–40
10. C. Kemp, C. Calvert, T.M. Khoshgoftaar, Detection methods of slow read DoS using full packet capture data, in *2020 IEEE 21st International Conference on Information Reuse and Integration for Data Science (IRI)* (IEEE, 2020), pp. 9–16
11. M. Fejrskov, J.M. Pedersen, E. Vasilomanolakis, Using NetFlow to measure the impact of deploying DNS-based blacklists, in *International Conference on Security and Privacy in Communication Systems* (Springer, Cham, 2021), pp. 476–496
12. R. Iqbal, F. Doctor, B. More, S. Mahmud, U. Yousuf, Big data analytics: computational intelligence techniques and application areas. Technol. Forecast. Soc. Chang. 1–13 (2018). https://doi.org/10.1016/j.techfore.2018.03.024
13. R.V. Kulkarni, A. Forster, G.K. Venayagamoorthy, Computational intelligence in wireless sensor networks: a survey. IEEE Commun. Surv. Tutor. **13**(1), 68–96 (2011)
14. E.M. El-Alfy, W.S. Awad, Computational intelligence paradigms: an overview, in *Proc. Improving Inf. Secur. Pract. Through Comput. Intell.* (2016), pp. 1–27
15. S. Fazli, M. Kiamini, A high performance steganographic method using JPEG and PSO algorithm, in *Proc. Int. Conf. IEEE Multitopic* (2008), pp. 100–105

16. F.H. Rabevohitra, J. Sang, High capacity steganographic scheme for JPEG compression using particle swarm optimization. Adv. Mater. Res. **433**, 5118–5122 (2012)
17. C. Chang, C. Lin, Y. Fan, Lossless data hiding for color images based on block truncation coding. Pattern Recognit. **41**(7), 2347–2357 (2008)
18. K. Bhowal, A.J. Pal, G.S. Tomar, P.P. Sarkar, Audio steganography using GA, in *Proc. Int. Conf. Comput. Intell. Commun. Netw.* (2010), pp. 449–453
19. M. Khodaei, K. Faez, Image hiding by using genetic algorithm and LSB substitution, in *Proc. Int. Conf. Image Signal Process* (Berlin, Germany, 2010), pp. 404–411
20. S. Sonawane, S. Karsoliya, P. Saurabh, B. Verma, Self configuring intrusion detection system, in *Proc. 4th Int. Conf. Comput. Intell. Commun. Netw.* (2012), pp. 757–761
21. A. Ahmad, S.B. Maynard, S. Park, Information security strategies: towards an organizational multi-strategy perspective. J. Intell. Manuf. **25**(2), 357–370 (2014)
22. N. Siddique, H. Adeli, Computational intelligence: synergies of fuzzy logic, in *Neural Networks and Evolutionary Computing* (Wiley, New York, NY, USA, 2013)
23. S. Liu, J. Sullivan, J. Ormaner, A practical approach to enterprise IT security. IT Prof. **5**, 35–42 (2001)
24. H. Kumar, Computational intelligence approach for flow shop scheduling problem, in *Handbook of Research on Emergent Applications of Optimization Algorithms* (IGI Global, Hershey, PA, USA, 2018), pp. 298–313
25. I. Onat, A. Miri, An intrusion detection system for wireless sensor networks, in *Proc. Int. Conf. IEEE Wireless Mobile Comput. Netw. Commun.*, vol. 3 (2005), pp. 253–259
26. W. Lee, D. Xiang, Information-theoretic measures for anomaly detection, in *Proc. Int. Symp. IEEE Secur. Privacy* (2001), pp. 130–143
27. H. Cavusoglu, B. Mishra, S. Raghunathan, The value of intrusion detection systems in information technology security architecture. Inf. Syst. Res. **16**(1), 28–46 (2005)
28. J. D'Arcy, T. Herath, M.K. Shoss, Understanding employee responses to stressful information security requirements: a coping perspective. J. Manag. Inf. Syst. **31**(2), 285–318 (2014)
29. L. Jobson, A. Stanbury, P.E. Langdon, The self-and other-deception questionnaires-intellectual disabilities (SDQ-ID and ODQ-ID): component analysis and reliability. Res. Dev. Disabil. **34**(10), 3576–3582 (2013)
30. L.A. Zadeh, Fuzzy logic, neural networks, and soft computing, in *Fuzzy Sets, Fuzzy Logic, and Fuzzy System* (World Scientific, Singapore, 1996), pp. 775–782
31. D. Gmach, S. Krompass, A. Scholz, M. Wimmer, A. Kemper, Adaptive quality of service management for enterprise services. ACM Trans. Web. **2**(1), 8 (2008)
32. S. Seltzsam, D.L. Gmach, S. Krompass, A. Kemper, Autoglobe: an automatic administration concept for service-oriented database applications, in *Proc. 22nd Int. Conf. Data Eng.* (2006), p. 90
33. L.A. Zadeh, Fuzzy logic computing with words. IEEE Trans. Fuzzy Syst. **4**(2), 103–111 (1996)
34. J. Lu, R. Wang, An enhanced fuzzy linear regression model with more flexible spreads. Fuzzy Sets Syst. **160**(17), 2505–2523 (2009)
35. Y. Zeng, M. Zhou, R. Wang, Similarity measure based on nonlinear compensatory model and fuzzy logic inference, in *Proc. Int. Conf. IEEE Granular Comput.* (2005), pp. 342–345
36. H. Duman, H. Hagras, V. Callaghan, A multi-society-based intelligent association discovery and selection for ambient intelligence environments. ACM Trans. Auton. Adapt. Syst. **5**(2), 7 (2010)
37. S. Bagchi, A fuzzy algorithm for dynamically adaptive multimedia streaming. ACM Trans. Multimed. Comput. Commun. Appl. **7**(2), 11 (2011)
38. K. Kolomvatsos, C. Anagnostopoulos, S. Hadjiefthymiades, A fuzzy logic system for bargaining in information markets. ACM Trans. Intell. Syst. Technol. **3**(2), 32 (2012)
39. J. Zhan, X. Luo, K.M. Sim, C. Feng, Y. Zhang, A fuzzy logic-based model of a bargaining game, in *Proc. Int. Conf. Knowl. Sci., Eng. Manage.* (Berlin, Germany, 2013), pp. 387–403
40. N. Naik, R. Diao, Q. Shen, Dynamic fuzzy rule interpolation and its application to intrusion detection. IEEE Trans. Fuzzy Syst. **26**(4), 1878–1892 (2018)

41. A.K. Choudhary, A. Swarup, Neural network approach for intrusion detection, in *Proc. 2nd Int. Conf. Interact. Sci. Inf. Technol. Culture Human* (2009), pp. 1297–1301

42. H. Yakura, S. Shinozaki, R. Nishimura, Y. Oyama, J. Sakuma, Malware analysis of imaged binary samples by convolutional neural network with attention mechanism, in *Proc. 8th ACM Conf. Data Appl. Secur. Privacy* (2018), pp. 127–134

43. M. Zheng, S. Zhong, S. Wu, J. Jiang, Steganographer detection based on multiclass dilated residual networks, in *Proc. Int. Conf. ACM Multimedia Retrieval* (2018), pp. 300–308

44. J. Yang, C. Eickhoff, Unsupervised learning of parsimonious general-purpose embeddings for user and location modeling. ACM Trans. Inf. Syst. **36**(3), 1–33 (2018)

45. R. Wang, W. Ji, Computational intelligence for information security: a survey. IEEE Trans. Emerg. Top. Comput. Intell. **4**(5), 616–629 (2020)

46. N.N. El-Emam, R.A.S. Al-Zubidy, New steganography algorithm to conceal a large amount of secret message using hybrid adaptive neural networks with modified adaptive genetic algorithm. J. Syst. Softw. **86**(6), 1465–1481 (2013)

47. S. Doğan, A new data hiding method based on chaos embedded genetic algorithm for color image. Artif. Intell. Rev. **46**(1), 129–143 (2016)

48. W. Weimer, T.V. Nguyen, C.L. Goues, S. Forrest, Automatically finding patches using genetic programming, in *Proc. 31st Int. Conf. Softw. Eng.* (2009), pp. 364–374

49. G. Folino, F.S. Pisani, Combining ensemble of classifiers by using genetic programming for cyber security applications, in *Proc. Eur. Conf. Appl. Evol. Comput.* (2015), pp. 54–66

50. G. Folino, F.S. Pisani, P. Sabatino, An incremental ensemble evolved by using genetic programming to efficiently detect drifts in cyber security datasets, in *Proc. Int. Conf. Genetic Evol. Comput.* (2016), pp. 1103–1110

51. S. Malhotra, V. Bali, K.K. Paliwal, Genetic programming and K-nearest neighbor classifier-based intrusion detection model, in *Proc. 7th Int. Conf. Cloud Comput., Data Sci. Eng.* (2017), pp. 42–46

52. N. Feng, H.J. Wang, M. Li, A security risk analysis model for information systems: causal relationships of risk factors and vulnerability propagation analysis. Inf. Sci. **256**, 57–73 (2014)

53. F. Hosseinpour, P.V. Amoli, F. Farahnakian, J. Plosila, T. Hämäläinen, Artificial immune system based intrusion detection: innate immunity using an unsupervised learning approach. Int. J. Digit. Content Technol. Appl. **8**(5), 1–12 (2014)

54. T. Ahmad, S. Ali, S.B.H. Shah et al., Joint mode selection and user association in D2D enabled multitier C-RAN. Clust. Comput (2022). https://doi.org/10.1007/s10586-021-03456-4

55. S. Ali, M. Sohail, S.B.H. Shah, D. Koundal, M.A. Hassan, A. Abdollahi, I.U. Khan, New trends and advancement in next generation mobile wireless communication (6G): a survey. Wirel. Commun. Mob. Comput. **2021**, (2021)

56. I.U. Khan, M.A. Hassan, M. Fayaz, J. Gwak, M.A. Aziz, Improved sequencing heuristic DSDV protocol using nomadic mobility model for FANETS. Comput. Mater. Contin. **70**(2), 3653–3666 (2022)

57. I.U. Khan, M.A. Hassan, M.D. Alshehri, M.A. Ikram, H.J. Alyamani, R. Alturki, V.T. Hoang, Monitoring system-based flying IoT in public health and sports using ant-enabled energy-aware routing. J. Healthc. Eng. **2021**, (2021)

58. M.A. Hassan, S.I. Ullah, A. Salam, A.W. Ullah, M. Imad, F. Ullah, Energy efficient hierarchical based fish eye state routing protocol for flying ad-hoc networks. Indones. J. Electr. Eng. Comput. Sci. **21**(1), 465–471 (2021)

59. M.A. Hassan, S.I. Ullah, I.U. Khan, S.B. Hussain Shah, A. Salam, A.W. Ullah Khan, Unmanned aerial vehicles routing formation using fisheye state routing for flying ad-hoc networks, in *The 4th International Conference on Future Networks and Distributed Systems (ICFNDS)* (2020), pp. 1–7

60. H. Chen, et al., Coplink center: social network analysis and identity deception detection for law enforcement and homeland security intelligence and security informatics: a crime data mining approach to developing border safe research, in *Proc. Nat. Conf. Digital Government Res.* (2005), pp. 112–113

61. G. Acampora, M. Gaeta, V. Loia, A.V. Vasilakos, Interoperable and adaptive fuzzy services for ambient intelligence applications. ACM Trans. Auton. Adapt. Syst. **5**(2), 8 (2010)
62. K. Xi, J. Hu, B.V.K. Kumar, FE-SViT: a SViT-based fuzzy extractor framework. ACM Trans. Embed. Comput. Syst. **15**(4), 78 (2016)

CPSs Communication Using 5G Network in the Light of Security

Shahbaz Ali Imran, Aftab Alam Janisar, Fahad Naveed, and Imran Fida

Abstract Irrespective of the manner that, there has been a rapid improvement in the audit, advancement, and planning of cyber-bodily structures in recent years. Nonetheless, security concerns are consistently present, yet amounts of new shortcomings, new forms of assaults, and distinct systems' compromising implantations solicitation to look at greater regarding CPSs in the light of wellness (Humayed et al., Cyber-physical systems security—a survey. IEEE Internet Things J. **4**(6), (2017); Dong et al., A security and safety framework for cyber physical system, in *International Conference on Control and Automation* (2014). In this evaluation, we talk about the prosperity and protection of cyber-physical structures and secure the correspondence of cps, the use of 5G Ultra-Reliable Low Latency Communications (URLLC), and Massive Machine Type Communications (MMTC) community shows. Instead of a cyber-bodily machine, use the time period 'CPS' for now for consistency. This study used a water supply SCADA gadget to emulate the proposed shape about the prosperity and security of the cyber-physical gadget. This research brought liveness in our proposed construction to attest that the set-off message has proven up at its target with essentially no % setback. As its progression is unnecessarily tough, so for the development, the research used node-crimson a reenactment device to impersonate the safety framework and for language, Node-JS is used, because cps improvement is particularly essential, So, in this study, arranged the movement of its correspondence by placing cps at higher elevations, which moves the water from one tank to another. The water circulation trade is sparked by the URLLC, MMTC display, and a short time later, it is thoroughly analyses the precision of the proposed structure. Because of its liveness quality, this cps safety arrangement is resistant to a wide spectrum of superior assaults. This new enhancement is heavily influenced by specific comfy ports. The research also used LOIC to thoroughly examine the insusceptibility of the proposed shape.

S. A. Imran (✉) · A. A. Janisar · I. Fida
Department of Software Engineering, Bahria University, Islamabad Campus, Islamabad, Pakistan
e-mail: Shahbazimran31@gmail.com

F. Naveed
Federal Urdu University of Arts Science and Technology, Islamabad, Pakistan

© The Author(s), under exclusive license to Springer Nature Switzerland AG 2022
M. Ouaissa et al. (eds.), *Big Data Analytics and Computational Intelligence for Cybersecurity*, Studies in Big Data 111,
https://doi.org/10.1007/978-3-031-05752-6_2

19

Keywords Security · CPS · MQTT · Cyber physical systems · 5G networks

1 Introduction

The presentation of Cyber-Physical Systems (CPSs) has improved dramatically during the last decade. CPSs are thought to be a critical component of the Industrial Internet of Things (IIoTs), as they are crucial in the fourth industrial revolution [1]. Smart applications based on cyber and physical systems are enabled by CPSs. They share many forms of information and data in real time [1]. Given the important financial potential of CPS and Industry 4.0, it is anticipated that the German industry will benefit by 267 billion Euros by 2025 [2].

CPS is made up of a number of networked systems that can monitor, process, and control authentic IoT-related devices. Aggregators, actuators, and sensors are the three major components of CPS. The CPS is capable of sensing and controlling the physical world, as well as adapting to it [1]. CPSs are now employed in a variety of well-known domains such as smart grids, autonomous systems, medical devices, and military fields (Atomic Reactor). As illustrated in Fig. 1.

There is an interdependency between their physical components and cyber operations since numerous separate CPSs are interconnected with secure communication

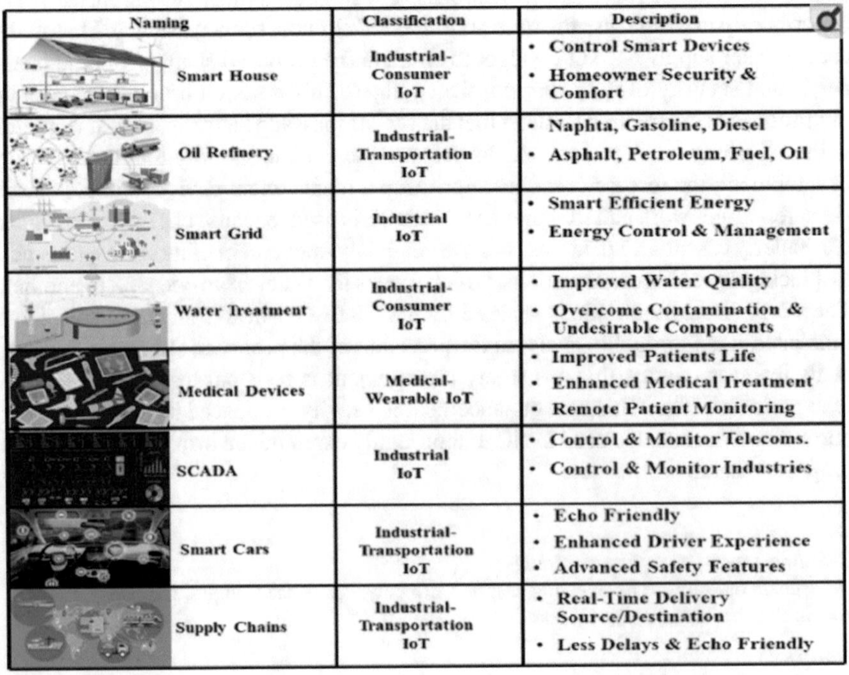

Naming	Classification	Description
Smart House	Industrial-Consumer IoT	• Control Smart Devices • Homeowner Security & Comfort
Oil Refinery	Industrial-Transportation IoT	• Naphta, Gasoline, Diesel • Asphalt, Petroleum, Fuel, Oil
Smart Grid	Industrial IoT	• Smart Efficient Energy • Energy Control & Management
Water Treatment	Industrial-Consumer IoT	• Improved Water Quality • Overcome Contamination & Undesirable Components
Medical Devices	Medical-Wearable IoT	• Improved Patients Life • Enhanced Medical Treatment • Remote Patient Monitoring
SCADA	Industrial IoT	• Control & Monitor Telecoms. • Control & Monitor Industries
Smart Cars	Industrial-Transportation IoT	• Echo Friendly • Enhanced Driver Experience • Advanced Safety Features
Supply Chains	Industrial-Transportation IoT	• Real-Time Delivery Source/Destination • Less Delays & Echo Friendly

Fig. 1 Use of CPS in the different fields [1]

and are located in different cities (e.g. Smart Grids, SCADA and Health Care, etc.) [3]. As a result, communication between two CPSs is considered to be completely safe, but this is not the case in practice [4].

Ad hoc networks have recently been a hot topic in the field of communication [5]. Cyber-attacks and technology are closely related to each other. Despite CPS's advantages, innovations, and pivotal role in the Industry 4.0 revolution, there is a significant security risk. Because of their varied behavior, CPSs are vulnerable to cyberattacks [1, 6, 7]. Because traditional IOT devices have limited resources, they are exposed to various cyber security concerns [8]. As the CPSs are now being utilized in crucial domains, thus they must be secure [3, 9]. Unmanned aerial vehicle (UAV) has gained relevance in recent years due to its highly effective and efficient structure [10]. To counter these ad hoc network security concerns, a platform is utilized to collect and monitor data from IoT-based networks [11].

The biggest security threat is Distributed Denial of Service attacks (DDoS). The first occurrence of DDoS was in 1996 [12, 13]. DDoS is the form of Denial of Service (DoS). The main difference between DoS and DDoS is "The attack source in DoS is single but in DDoS [14, 15], The attack source is multiple" where several devices are compromised and destroy the targeted system. It's a very lethal and most commonly used attack by attackers. In the year 2020 *Cisco has released a report* in which they declared that till 2018 the recorded count of DDoS is 7.9 million but it will double to over 15 million by 2023 as the number of internet users is increasing globally [12]. There are many techniques and tools to mitigate DDoS such as null routing technique, redirection techniques, load balancer infrastructure, and HoneyPots, etc. [16]. Honeypots are now used widely against DDoS. HoneyPots first formed in 1991. It acts as a dummy server and detects and mitigates the attack in other words it distinguishes the real traffic and fake traffic. It's a very healthy approach to use against DDoS [17].

Researchers have proposed different frameworks to secure the communication of CPSs using different simulation tools and presented different results. These methodologies function admirably in their considered situation but lack in their detection of DDoS and message integrity. In this research, we will propose a novel security frame for CPSs against DDoS using HoneyPots. We will secure the communication between the two CPSs. Supposedly, no framework exists in the literature that addresses the detection and prevention of attacks in the live communication between two CPSs. In order to full fill this research gap we propose the use of HoneyPots in the security framework which is used in different detection scenarios by researchers [18–21].

In this research, we will use HoneyPots to detect and mitigate DDoS. 5G is the latest iteration in cellular network technology. It promises to be about 10 times faster than 4G, which will allow for less buffering, quicker loading times, and high quality streaming. The networks are set to roll out by 2020.

There are two types of 5G networks; the first uses the current 4G LTE spectrum while the other type provides a wider bandwidth with low latency. 5G networks use beamforming to make sure that your data gets sent directly to your device without interference from any other devices on the same frequency. 5G is making headlines across industries because it not only promises faster speeds but also more reliability and security. Companies like Verizon, AT&T, and T-Mobile are investing in 5G technology to improve their wireless offerings for consumers and enterprises alike.

5G networks are the next generation of cellular networks. With these new networks, we will be able to connect more devices and enjoy faster speeds. But as we make the leap to this new network, how will its cybersecurity measure up? With 5G networks, hackers can now access a vast array of new opportunities for attack and with that come more vulnerabilities [22]. The cybersecurity risks associated with this new technology is a major concern and it is important to prepare for them ahead of time. The URLLC architecture is built with the latest 5G network technology that provides ultra-reliable low latency communications.

This new architecture will be capable of supporting applications such as remote surgery, autonomous cars, and remote control robots. Moreover, URLLC is able to provide low latency communication which makes it possible for automobiles to transmit data from one car to another and avoid crashes. The 5G network will have a significant impact on the future of communications. The more the number of connections, the more powerful a network becomes. This is why a lot of service providers are looking into developing applications that use MMTC to provide new and innovative services.

This new technology can be used for solving some problems that seem unsolvable today. It makes it possible to download huge amounts of data in just a matter of seconds and enables faster streaming with less buffering. This protocol is used in the proposed framework and shows the effective communication and secure packet transferring between CPSs.

2   Literature Review

CPSs can face a large no of complex and deadly attacks which can lead to privacy leakage, system failure, and system fault. Therefore, it is necessary to detect the attack in CPSs [23, 24].

Imran in [4] has presented a framework to secure the communication between two CPSs. The proposed framework which uses Message Queue Telemetry Transport MQTT protocol for the communication between the CPSs. The researcher simulated the security framework on a tool called Node-Red. In the research shows the flow of information in the SCADA system. In [4, 25] placed 2 CPSs in two different locations/Cities and then sends and receive the information between these two CPSs furthermore during the time of the exchange in information the researcher applied a DDoS attack using LOIC (Low Orbit Ion Cannon) and show the normal flow of information during the attack. The framework is pasted below in Fig. 2.

In another research by Utku Ozgar in [2], identical work has been done where they placed CPSs far away from each other and simulated the security framework in MATLAB Simulink and exchange the information and check the integrity of the information between the two CPSs. They have also shared the results while there's an

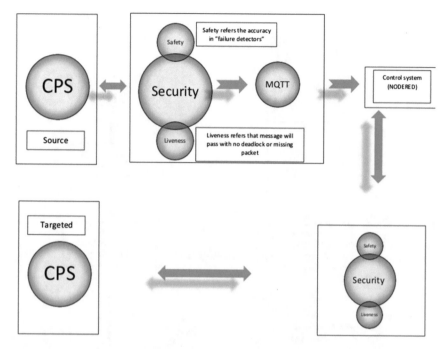

Fig. 2 SA. Imran framework for safe and secure CPSs [1]

attack the security techniques are working absolutely fine. Their presented framework is shown in the below figure.

In the above-mentioned framework in Fig. 3, it has been tested on different simulation tools on different CPSs industries in [4, 26] SCADA is used and in [2, 27] smart grid examples are used. But the research gap has emerged as they have not used any detection techniques. Their results show the information but is that information real? During the DDoS attack, information also flows normally but the rate of hits jeopardizes the system for that in our framework we will use the novel approach to secure the CPSs communication by detecting the mitigating the DDoS using HoneyPots and ensuring the flow of information. The Proposed framework will work on all kinds of CPSs communication flow. In simple words, it's a generalized communication framework.

3 Methodology

The design consists of siphons and two tanks. The primary siphon pulls water from the well within the central tank, the following siphon actions water from the first to the ensuing tank. This method has four momentous districts; every locale has Modbus RTUs or PLCs. The point of convergence of this SCADA framework may

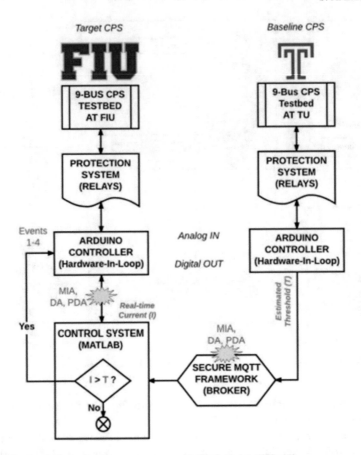

Fig. 3 Utku Ozgar frame for the secure communication between CPSs [4]

be node-pink as a way to inspect values from the tanks, start and forestall siphons, give worrying, and an HMI. The water properly/switches siphon dashboard regards show the siphon mode and standing. Tank stage set centers may be altered to change when the direction in its programming mode in this manner begins and prevents. In guide mode, the siphon may be provided and halted the toughest way possible. The tank regards have a line plan and a reference graph guide to handily see the level and the verifiable putting of the level. The alert set groups are hence set to be +15 and −15 across the siphon begin and stop set focuses, this may without a certainly amazing stretch be modified to paintings unexpectedly. At long closing, there may be a substance-organized reputation that shows inside the occasion that the alarm is high, low, or clear. A substance to speak focuses in like way broadcasts the equipped status. For liveness exceptional tests and bounds are used so its miles certified that the message may be exceeded with near no impedance. URLLC agent is used at the beginning button to move the directing of water from water nicely. URLLC and MMTCs display is used to maintain water from the water properly via tapping at

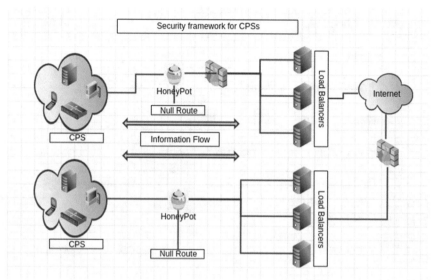

Fig. 4 Proposed security framework for CPSs

the prevent button. The Fig. 4 discern under shows the general route of the proposed framework and shops the income report.

4 Results and Discussion

To compare the immunity of this framework, implemented the framework that was implemented (without Liveness) on our NODE-RED tool and then apply the DDOS (Distributed Denial of service) where results show the successful message passing between two CPSs when the attack is applied on the framework.

Figure 5 show the immunity of our applied framework. Where the value of the start point is greater than or equal to 150 and the stop point is 200.

The Fig. 6 above shows the DDOS attack during the communication using LOIC (Low Orbit Ion Cannon) software.

The equation that is used for the graphical representation in Fig. 7 is

$$x + y = 1 \tag{1}$$

where x represents the source CPS and y denotes the targeted CPSs. During the attack the bar intersects the x and y-axis which is

$$x * 8 = 1 \tag{2}$$

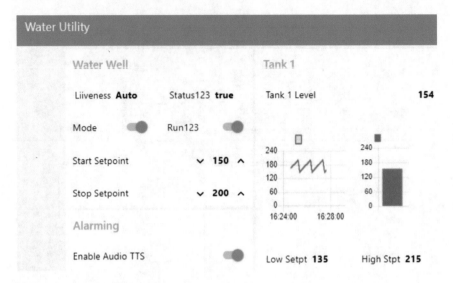

Fig. 5 Node-red simulation results

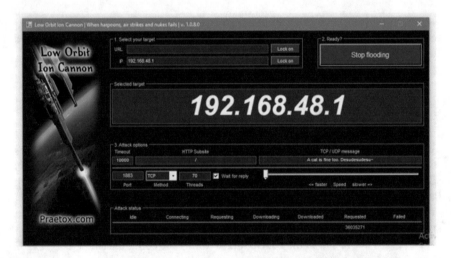

Fig. 6 LOIC throwing attack

where "8" denotes the attack frequency. Tested our result from a certified organization named "visme" and calibrated the accuracy of our framework and results are shown in the below mentioned figure.

Figure 8 illustrates the sent information, DDOS ratio, Message Integrity information received and total loss in the form of graph. Alarm alert is working fine in the form of audio on the other hand. (Figs. 3, 4 and 5), Implemented the framework defined in [28] and applied DDOS to check the immunity of their framework but the

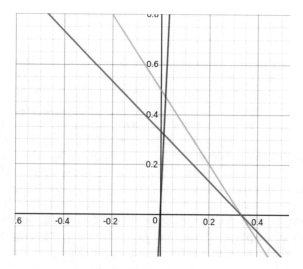

Fig. 7 Graphical representation of the packet loss

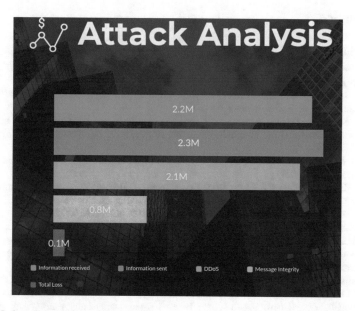

Fig. 8 Node-red simulation results

system got stuck. This study proved the importance of 5G protocols URLLC and MMTCs by implementing and securing our proposed framework.

5 Conclusion

This study cultivated the design to understand and direct the DDoS assault, the usage of honeypots and granted via intriguing indicates of 5G URLLC and MMTCs. In this study invalid guided the poisonous visitors of DDoS, and watching for the assault is begun from the internal or out of doors supply, removed that element from the device if the attack starts from the internal or external supply. Cutting-edge buildings don't ensure safety or they may be restricted [29, 30]. Special techniques have been used to construct the development for the 5G engaged business enterprise assurance framework, regardless, the research advocated a water SCADA system that movements the water from source cps to the assigned cps, that is arranged in higher places. Similarly, with liveness, it was able to obtain the connection between those two CPSs. Also executed a DDoS attack on the finished product in order to test the invulnerability of the proposed security system. Similarly, the study demonstrated the immunity of the prosperity and safety shape for CPSs.

References

1. J.A. Yaacoub, O. Salman, H.N. Noura, N. Kaaniche, A. Chehab, M. Malli, Cyber-physical systems security: limitations, issues and future trends. Microprocess Microsyst. **77**, 103201 (2020). https://doi.org/10.1016/j.micpro.2020.103201
2. S.U. Rehman, V. Gruhn, An effective security requirements engineering framework for cyber-physical systems. Technologies **6**, 65 (2018)
3. V.K. Sehgal, A. Patrick, A. Soni, L. Rajput, Smart human security framework using internet of things, cloud and fog computing, in *Proc. of Intelligent Distributed Computing* (2015), pp. 251–263
4. A. Humayed, J. Lin, F. Li, B. Luo, Cyber-physical systems security—a survey. IEEE Internet Things J. **4**(6), (2017)
5. I.U. Khan, I.M. Qureshi, M.A. Aziz, T.A. Cheema, S.B.H. Shah, Smart IoT control-based nature inspired energy efficient routing protocol for flying Ad Hoc network (FANET). IEEE Access **8**, 56371–56378 (2020). https://doi.org/10.1109/ACCESS.2020.2981531
6. Y. Wang, Z. Lin, X. Liang, W. Xu, Q. Yang, G. Yan, On modeling of electrical cyber-physical systems considering cyber security. Front. Inf. Technol. Electron. Eng. **17**(5), 465–478 (2016)
7. L. Tianbo, J. Zhao, L. Zhao, Y. Li, X. Zhang, Towards a framework for assuring cyber physical system security. Int. J. Secur. Appl. **9**(3), 25–40 (2015)
8. A. Abdollahi, M. Fathi, An intrusion detection system on ping of death attacks in IoT networks. Wirel Pers Commun **112**(4), 2057–2070 (2020). https://doi.org/10.1007/s11277-020-07139-y
9. A. Sanjab, W. Saad, On bounded rationality in cyber-physical systems security: game-theoretic analysis with application to smart grid protection (2016), http://arxiv.org/abs/1610.02110
10. I.U. Khan et al., RSSI-controlled long-range communication in secured IoT-enabled unmanned aerial vehicles. Mob. Inf. Syst. **2021**, e5523553 (2021). https://doi.org/10.1155/2021/5523553
11. I.U. Khan et al., Intelligent detection system enabled attack probability using Markov chain in aerial networks. Wirel. Commun. Mob. Comput. **2021**, e1542657 (2021). https://doi.org/10.1155/2021/1542657
12. U. Ozgur, H.T. Nair, A. Sundararajan, K. Akkaya, A.I. Sarwat, An efficient MQTT framework for control and protection of networked cyber-physical systems, in *IEEE Conference on Communications and Network Security (CNS)* (9–11 Oct 2017)

13. X. Yu, Y. Xue, Smart grids: a cyber-physical systems perspective. Proc. IEEE **104**(5), 1058–1070 (2016)
14. Y. Zacchia Lun, A. D'Innocenzo, I. Malavolta, M.D. Di Benedetto. Cyber-physical systems security: a systematic mapping study. CoRR, abs/1605.09641 (2016)
15. M.H. Cintuglu, O.A. Mohammed, K. Akkaya, A.S. Uluagac, A survey on smart grid cyber-physical system testbeds. IEEE Commun. Surv. http://ieeexplore.ieee.org/abstract/document/7740849/
16. I. Dumitrache, I.S. Sacala, M.A. Moisescu, S.I. Caramina, A conceptual framework for modeling and design of cyber-physical systems. Stud Inform Control **26**(3), 325–334 (2017)
17. D. DiMase, Z.A. Collier, K. Heffner, et al., Environ. Syst. Decis. **35**, 291 (2015). https://doi.org/10.1007/s10669-015-9540-y
18. A. Hahn, R.K. Thomas, I. Lozano, A. Cardenas, A multi-layered and kill-chain based security analysis framework for cyber-physical systems. https://doi.org/10.1016/j.ijcip.2015.08.003
19. T. Lu, J. Zhao, L. Zhao, Y. Li, X. Zhang, Towards a framework for assuring cyber physical system security. Int J Secur Appl **9**(3), 25–40 (2015)
20. A. Khalid, P. Kirisci, Z.H. Khan, Z. Ghrairi, K.-D. Thoben, J. Pannek, Security framework for industrial collaborative robotic cyber-physical systems. Comput. Ind. **97**, 123–145 (2018). https://doi.org/10.1016/j.compind.2018.02.009
21. D. Ding, Q.-L. Han, Y. Xiang, X. Ge, X.-M. Zhang, A survey on security control and attack detection for industrial cyber-physical systems. Neurocomputing (to be published). https://doi.org/10.1016/j.neucom.2017.10.009
22. O. Younis, N. Moayeri, Cyber-physical systems: a framework for dynamic traffic light control at road intersections. IEEE Wirel. Commun. Netw. Conf. **4**(6), 1–6 (2016)
23. S.F. Ochoa, G. Fortino, G. Di Fatta, Cyber-physical systems, internet of things and big data. Futur. Gener. Comput. Syst. **75**, 82–84 (2017)
24. A. Chattopadhyay, A. Prakash, M. Shafique, Secure cyber-physical systems: current trends, tools, and open research problems, in *Design, 27 Automation Test in Europe Conference Exhibition (DATE)* (2017), pp. 1104–1109
25. C. Schmittner, Z. Ma, E. Schoitsch, T. Gruber, A case study of fmvea and chassis as safety and security co-analysis method for automotive cyber-physical systems, in *Proceedings of the 1st ACM Workshop on Cyber-Physical System Security* (ACM, 2015), pp. 69–80
26. T. Lu, J. Lin, L. Zhao, Y. Li, Y. Peng (2015) A security architecture in cyber-physical systems: security theories, analysis, simulation and application fields. Int. J. Secur. Appl. **9**(7), 1–16 (2015)
27. S.R. Chhetri, J. Wan, M.A. Al Faruque, Cross-domain security of cyber-physical systems, in *Design Automation Conference (ASP-DAC), 2017 22nd Asia and South Pacific* (IEEE, 2017), pp. 200–205
28. P. Dong, Y. Han, X. Guo, F. Xie, A security and safety framework for cyber physical system, in *International Conference on Control and Automation* (2014)
29. K. Paridari, A.E.-D. Mady, S. La Porta, R. Chabukswar, J. Blanco, A. Teixeira, H. Sandberg, M. Boubekeur, Cyber-physical-security framework for building energy management system, in *7th International Conference of Cyber-Physical Systems (ICCPS)* (2016)
30. H. He, C. Maple, T. Watson, A. Tiwari, J. Mehnen, Y. Jin, B. Gabrys, The security challenges in the IoT enabled cyber-physical systems and opportunities for evolutionary computing & other computational intelligence, in *2016 IEEE Congress on Evolutionary Computation, CEC 2016* 7743900 (2016), pp. 1015–1021

A Survey on Security Aspects in RPL Protocol Over IoT Networks

Soukayna Riffi Boualam, Mariya Ouaissa, Mariyam Ouaissa, and Abdellatif Ezzouhairi

Abstract The IETF Routing Over Low-power and Lossy Networks (IETF ROLL) working group has developed a RPL protocol. It is intended for connected objects that communicate in Low Power and Lossy Networks (LLN) conforming to the 6LoWPAN protocol. The IPv6 Routing Protocol for LLN (RPL) is intended for Wireless Sensor Networks (WSN) to meet the needs of the industrial and scientific communities. The RPL protocol has drawn the attention of several researchers, it raises a growing interest in the research community. RPL relies on one or more metrics to trace the route of forwarded packets that will be forwarded to the root. However, the IETF ROLL group has not specified metrics to apply when using RPL. For this, the latter can be adapted according to specific environments. The purpose of this article is to provide an introduction and summary of what has been achieved in scientific research to develop the RPL protocol. During this article, we will present the relevant efforts of researchers to ameliorate the performance of the RPL protocol. Moreover, we present an evaluation of the performance of the RPL by modifying the number of network metrics taken into account. The goal is to involve the impact of using protocol metrics on network performance in terms of energy, time, packet loss and delay. At the end, we set up the challenges that are still open on the construction of RPL protocol.

Keywords IoT · LLN · 6LoWPAN · RPL · Security

S. R. Boualam · A. Ezzouhairi
Engineering, Systems and Applications Laboratory, National School of Applied Sciences, Sidi Mohamed Ben Abdellah University, Fez, Morocco
e-mail: abdellatif.ezzouhairi@polymtl.ca

M. Ouaissa (✉) · M. Ouaissa
Moulay Ismail University, Meknes, Morocco
e-mail: mariya.ouaissa@edu.umi.ac.ma

M. Ouaissa
e-mail: mariyam.ouaissa@edu.umi.ac.ma

© The Author(s), under exclusive license to Springer Nature Switzerland AG 2022
M. Ouaissa et al. (eds.), *Big Data Analytics and Computational Intelligence for Cybersecurity*, Studies in Big Data 111,
https://doi.org/10.1007/978-3-031-05752-6_3

1   Introduction

The Internet of Things (IoT) is a new paradigm shift in computing. The IoT is coined from the two words are Internet and object. The Internet is defined as a universal system that interconnects computer networks. It uses the TCP/IP protocols to meet the needs of users. The networks are linked by wireless, electronic and optical routing technologies. In addition, connected objects communicate with each other in Low Power and Lossy Networks (LLN). This type of network is limited in terms of processing capacity, battery life, and transmission range and transmission link quality. For this, several researches are carried out to choose the best protocol that allows efficient communication and route management thus meets these exigencies. The Routing Over Low-power and Lossy (ROLL) group of the IETF has standardized a new protocol for networks 6LoWPAN [1], IPv6 Routing Protocol for LLN (RPL) is the name of this new routing protocol.

RPL is a routing protocol that uses a gradient routing technique [2]. This technique organizes the Wireless Sensor Networks (WSN) in the form of a Direct Acyclic Graph (DAG) rooted at the nodes. It uses objective functions either standard or innovated in recent years, the goal is to reach the root from any node that is in the scenario in an optimal way.

For connected objects, the RPL protocol has become the routing protocol that allows interconnection between different networks. It is considered a proactive protocol based on objective functions. Indeed, the Objective Functions (OF) make it possible to define the rules and the constraints to select the best paths by taking into account the various routing metrics. In the definition of the RPL protocol, it does not impose the use of a single metric such as the number of hops, but it is based on the use of the objective function which is based either on a single metric or on several metrics. The choice is made according to the scenario or the need of the application. The use of OFs in the core protocol specification allows RPL to adapt with the change and optimization of its architecture easily. At this point, the RPL protocol has become flexible in its use and satisfies network designs and application exigencies.

The RPL protocol is relevant for applications used worldwide. Therefore, it is important to analyze its performance in different scenarios in order to improve its use. For this, we propose this article which illustrates the efficiency of the RPL protocol and its Quality of Service (QoS) which can provide to monitoring applications. The simulations carried out by the researchers are carried out using the Contiki COOJA simulator [3] available on [4]. Several OFs are compared by assessing their energy consumption, quality of service, network throughput, and end-to-end packet delays. The simulation results show that the RPL protocol is a powerful routing algorithm; all that remains is to select the best objective function that is compatible with the proposed scenario. It enables very fast network configuration with minimal delays [5].

2   RPL Overview

RPL is an IPv6 distance vector routing protocol designed for low-power wireless networks with limited power and bandwidth. RPL allows these types of devices to be connected through the Internet. This routing protocol can adapt in time with the change of metrics regardless of the node or link metric, the purpose of this change of metric is that it is adaptable and adequate in different scenarios.

Using a proactive approach, the RPL routing protocol constructs nodes into a Destination-Oriented Acyclic Directed Graph (DODAG). In this topology, each router targets a set of stable parents, one of these parents will be a next hop to arrive at the root of the DODAG.

DAG employs a tree to choose routes between nodes by default. Each node can have manifold parents. The DODAG is built using an OF, which determines the routing architecture based on the metrics and restrictions chosen throughout the creation process. The use of an OF makes it possible to choose the optimal route based on one or more metrics. These metrics can be node metrics like energy and hop count or link metrics like Expected Transmit Count (ETX) and Link Quality Level (LQL). Nodes in the network may have different dockets depending on the graph. The root of the graph is represented by the receiver node, which is connected to the Internet or any other external infrastructure. The router nodes in the graph's intermediate places allow traffic to be routed to the root. Leaf nodes are found at the graph's edges and are unable to forward traffic. Except for the root, each node is set up as either a router or a leaf node.

In a DODAG, RPL defines a set of ICMPv6 control messages. These authorize the exchange of information associated with DODAG. RPL identifies three control messages to exchange traffic (Fig. 1):

- **The DODAG Information Object (DIO)**: It is a control message sent in multicast by one node and forwarded by others to their neighbors (always in multicast), allowing a node to locate an RPL instance and join it using ascending route information.
- **The DODAG Information Solicitation (DIS)**: The message sent in multicast by a node when it joins a network to request information about the DODAG, or to request a DIO message, is known as a solicitation of DODAG information.
- **The Destination Advertising Object (DAO)**: This is a unicast message sent by the nodes to the DODAG to disseminate destination information. Which means that each node sends to all its parents all its known descending routes. The nodes update their routing table each time a DAO is received.

To build the network topology, the process to be followed allows each router to associate with a suitable parent based on an OF. The objective function translates metrics into rank levels. The objective function allows to specify how routing metrics are transformed into rank. It is responsible for defining how the node selects the best parent from its list of potential parents. The OF is used to select a DODAG to join, determine the rank of each node in the DODAG, generate an ordered list of DODAG

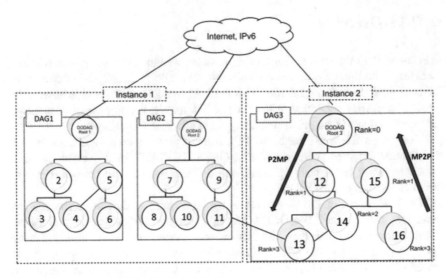

Fig. 1 RPL network with 3 DODAGs in 2 instances

parents, and select and optimize routes. The IETF has defined only two RPL goal functions are OF and MRHF. Several metrics are provided to work with the protocol; however, researchers are free in constructing the objective function. Moreover, no recommendation is made on how to combine them to take into account the QoS or to use them according to the specificities related to the application.

The OF identifies the method by which nodes designate and optimize routes within an RPL instance. The OF has several functionalities: it allows the selection of a DODAG to join, it has a role in identifying the rank of each node in the DODAG according to several criteria. This OF targets an ordered list of parents in the DODAG in order to route the suitable and optimal route to the root. The Contiki OS simulator takes into account the OF used to simulate and plot the route to the destination.

3   Related Work

In the design of the RPL protocol, the latter follows strategies to select a parent [6–9] optimally in the network construction process, this is achieved based on one or more combined metrics. Also, the RPL protocol uses an alternative path selection [10–12] and rate adjustment mechanism [13–15], congested nodes reduce the data packet transmission rate. If this task is impossible, then the nodes find another path to transmit the data packets. Indeed, for the transmission of data packets it is necessary to choose the appropriate routing metric. In recent years, the ROLL group of the IETF has suggested several metrics to construct an Objective Function for LLNs. He suggested two standard OFs are the Zero Objective Function (OF0) [6] and the Minimum Rank Objective Function with Hysteresis (MRHOF) [7].

For OF0, it is an objective function based on the minimum value of the node to reach the root node. OF0 is designed to search for the nearest root. This objective function applies if the rank of a node is very close to an abstract function of its distance from the root node. This operation is carried out by calculating the rank of each node in relation to the node value towards the root. Indeed, as the number of nodes decreases, the link to the root takes precedence and is chosen by OF0. Namely, the rank value on the parent node is always less than that of the child node. In the COOJA simulator, each link addition will increase the node. OF0 selects a closest preferred parent and a potential backup successor if one exists. The preferred parent forwards all upstream traffic without any attempt at load balancing. In the event that link conditions do not allow an upstream packet to pass through the preferred parent, then the data packet is forwarded to the potential backup successor.

For the MRHOH: It is a standard objective function based on the minimum value of the number of ETX to select the parent node to transmit the error-free data packet to its destination.

The amount of published research attests to the growing interest of the research communities in the RPL protocol. Several researches are done to analyze the performance of RPL under different contexts and platforms. We find [16] which presents a new objective function that relies on the residual energy as the main routing metric to choose the best route to its destination. The authors in this article propose an objective function which is only interested in the remaining energy, it is the only metric on which this OF is based. This new OF uses a node metric to select the next hop to the sink. The implementation of this OF uses a well-known theoretical battery model [16] from which it is necessary to take into account at runtime the battery life of the node for routing. The new OF is based on an online estimation model of the battery level in real time. The OF used for RPL takes into consideration the node's remaining energy as a metric. The research carried out reveals that the implementation of the suggested objective function increases the lifetime of the network and distributes the energy in a balanced manner between the nodes without an appreciable lack of transmission precision.

OF-FL [17] is an objective function for RPL based on fuzzy logic. This OF surmounts the limitations of the standard RPL (OF0 and MRHOF). It permits to choose the better routes to reach the root. The OF-FL objective function is defined as an objective function that permit to select the best freight forwarder candidate based on 4 criteria. The 4 metrics used are: hop count, node energy, link quality, and end-to-end delay. The combination of these metrics is achieved using fuzzy logic.

In [18], researchers realize studies in order to optimize the RPL protocol. The latter takes into account a context-awareness known as Context-Aware Objective Function (CAOF). The optimization takes into account the sensor nodes' temporal changes as well as their limited resources. The proposed CAOF aims to optimize the exploitation of power as a critical resource by making routing decisions based on battery level. Experiment results show that the proposed objective function increases lifetime by up to 44% when compared to standard non-contextual OFs. Furthermore, CAOF ensures battery operation for different nodes than non-contextual RPL OFs and improves the delivery ratio.

The authors in [19] believed that the use of a single metric was limited. As a result, a combination of link and node metrics is used. To make this combination feasible, the fuzzy logic method is applied, it allows to combine several routing metrics based on fuzzy logic rules. This combination yields an objective function known as OF-EC. This objective function aids in the selection of the best path based on link and node metrics.

SCAOF, a new objective function based on Scalable Context-Aware, is proposed by the authors in [20]. This OF enables the RPL to be tailored to Agricultural Network environmental monitoring (A-LLNs). According to the composite routing metrics approach, the actors perform the adaptation based on a combination of reliability, energy, robustness, and resources. The RPAL's performance is evaluated using simulations and field tests. In a variety of simulation scenarios and hardware test beds, the results show that SCAOF can provide network lifetime extension, high efficiency, and reliability.

Co-RPL [21] uses the corona mechanism to improve the RPL for MSN. The network area is divided into crowns by the Corona architecture. Each DODAG root has a set of crowns, which are distinguished by the number of hops required to reach the relevant route. For example, a node is in corona 3 if it is three hops from the root. This type of protocol helps guarantee QoS specifically end-to-end delay and reliability and extend better mobility support to RPL. Co-RPL also focuses on reducing energy consumption by using repair mechanisms in mobile scenarios.

For nodes to use Co-RPL and RPL in the same network, then the authors preserve backward compatibility for the coexistence of Co-RPL and RPL at the same time. With regard to the Co-RPL DIO, a corona identifier (C_ID) is identified in the messages. It makes it possible to obtain information on the distance from the sender node DIO to its corona well. This distance is calculated using a single metric namely the number of nodes. The LQI value is saved by the DIO message receiver. This value makes it possible to know the link between the sender and the recipient in order to choose the right route during the transfer of the data packets. A node's preferred parent will be chosen based on the candidate's corona ID and uses the link LQI value as the tiebreaker. When a Co-RPL node receives a DIO message, it sends the latter immediately. The Trickle timer is neglected at this time. When a new neighbor is discovered, a DIO message is instantly sent. This operation makes it possible to say that the RPL protocol is reactive, but it suffers from the overload which increases during the exchanges. Because of this overhead, the authors recommend using a fixed periodic timer to transfer DIO messages. To avoid blocking during the transfer of a data packet on a route, if a mobile node cannot transmit it in the upstream direction, it is better to forward it to another node with a higher corona level and advises its children that it is unavailable to forward packets. The packet will maintain to be transmitted until a route to a receiver is found. So Co-RPL avoids packet loss when certain routes are temporarily inaccessible.

Co-RPL presents an improvement of RPL which uses the corona mechanism. It supports RPL router mobility. It needs minimal add-ons because it reuses the same control messages as RPL. The actors approved that the RPL standard is limited on QoS guarantees for MWSNs. Additionally, Co-RPL has minimum packet loss rates that are less than 45%. As for the end-to-end delay, it is less than 2.5 s and the average power consumption is less than 50%. Then Co-RPL is effective for low-power and lossy mobile wireless sensor networks.

In [22], actors designed an RPL-based multipath routing protocol named DMR. The latter is intended for mobile sensor networks. In the construction of DMR, researchers adopted the design of RPL without any modification. The DMR constructs DAGs and provides path redundancy when DIO messages that have an indication of link rank and link quality are broadcast. This allows mobile nodes come up with several alternative paths on the different routes. If there is a local problem, to ensure its repair, the actors proposed adding sibling nodes to each node's routing table. If there are no nearby sister nodes, the node detects a broken link and can perform a global repair by requesting that its sink rebuild a new DODAG.

4 Securing the Protocol

Protocols in LLNs face several threats. Several attacks come from inward the LLN [23]. Not all traditional communications encryption techniques are effective in ensuring data security and protection. It is important to define threats and identify countermeasures to establish that the routing protocol works properly. These measures must comply with safety standards. The implementation of security measures for LLN responds to specific issues. The services and protocols used make it possible to adapt with many nodes.

RPL's standard version includes security and self-healing capabilities to ensure that the network runs smoothly. Data confidentiality and integrity are ensured by the security mechanism. Indeed, there are three security options available on an RPL network: unprotected, pre-installed, and authenticated. To secure data exchanges in the non-secure mode, link/transport layer security materials are used. They use pre-shared keys for pre-installed and authenticated modes.

The RPL self-healing techniques guarantee that the RPL protocol is secure. Loop detection and repair are mechanisms used when transporting data. The detection is performed using a route verification and validation tool whose data will be transferred. Each data packet, without a doubt, includes an IPv6 extension header containing flags indicating the way that the packet should be manipulated by RPL routers [24]. These flags are used to determine if any inconsistencies have been detected. Thus, they make it possible to define the appropriate measures to be taken such as local or global repair according to the instances. These mechanisms improve RPL resilience and help to deal with attacks.

The IETF ROLL working group outlines exigencies and security aim for the RPL protocol [25]. In the literature, no specific security model has been suggested. The only thing currently offered [26] is in-app key management for pre-configured devices. The key allows authentication of devices putting on the RPL network. The feebleness of the RPL's security design is a cause of the presence of external attacks. Threat analysis is discussed by several actors [27], they presented ways to meet the challenges of the threat. Furthermore, studies [27–29] have shown that the RPL protocol can suffer from many routing attacks namely, Route Falsification, Routing Information Replay, Sybil, Sinkhole, Blackhole, Greyhole and Version Number.

The [30] presents a security model that identifies several features to ensure good security communication like integrity, confidentiality, access control, availability and authentication. The [30] illustrates areas to cover and identifies sensitive areas that require defense to adjust network security. The model predicts the classification of attacks and threats that influence the availability, confidentiality and integrity of routing protocols. The suggested prototype is adapted to the 6LOWPAN environment. It discusses the properties of security to improve routing security in RPL.

To enhance the security level of the RPL protocol, several techniques are performed. The [31] cited several internal attacks that threaten the RPL, so he carried out a very detailed analysis of these attacks. They studied also the RPL Rank property. The latter is used to eschew route control *overhead* to the root and forbid routing loops during packet forwarding. Among the threats we find an inside attacker who compromises a node within an RPL network, he can inject data packets with spiteful intentions. The RPL protocol is limited because control messages from a parent node may not know a child node. At this time, a child node cannot decide which parent to select from the list of potential parents. This implies that it is possible to select a malicious parent node, then the data packet cannot reach the receiving node. Stakeholders suggested the design of integrating a surveillance system into RPL that helps a child node observe, evaluate, and select trustworthy parents.

Studies performed in [32] show that an insider attacker can conceive a Sybil like replica of the DODAG RPL root node with similar reachability as the DODAG root node. This attack can weaken the security of the RPL network. The attacker forwards a higher RPL instance version number to get to the DODAG root node, the goal on the one hand is to create a route topology around it. And on the other hand, spreading a superior RPL instance version by the attacker, which induces nodes of the original RPL network to join the new DODAG root node. So, the attacker can spy on and manipulate the nodes underneath easily. To deal with this type of attack, in [32], the authors suggested a system that adds protection against RPL version number attacks. The authentication system transmits superior RPL instance version numbers or sends falsely low DIO rank values.

5 Conclusion

The sensor network has a wide potential with various practical and useful applications. However, there are still many challenges that need to be addressed to ensure that these networks work efficiently in real applications. For this, several protocols are suggested to perform data routing in wireless sensor networks such as the RPL protocol. In this article, we have cited several objective functions that help to choose the right route to the destination by taking into account node and link metrics like:

ETX and energy. This article reviews some contributions using in RPL protocol, they are focusing on those related to the optimization of topology, security.

References

1. G. Montenegro, N. Kushalnagar, J. Hui, D. Culler, Transmission of IPv6 packets over IEEE 802.15.4 networks, in *Internet Suggested Standard RFC 4944* (September 2007)
2. T. Watteyne, K. Pister, D. Barthel, M. Dohler, I. Auge-Blum, Implementation of gradient outing in wireless sensor networks, in *IEEE GLOBECOM* (2009)
3. F. Osterlind, A. Dunkels, J. Eriksson, N. Finne, T. Voigt, Crosslevel sensor network simulation with COOJA, in *Proc. of LCN* (2006), pp. 641–648
4. Contiki COOJA site. http://www.sics.se/contiki
5. N. Accettura, Performance analysis of the RPL routing protocol, in *Proceedings of the 2011 IEEE International Conference on Mechatronics* (Istanbul, Turkey, April 13–15 2011)
6. T.P. Thuber, Objective function zero for the routing protocol for low-power and lossy networks, in *IETF RFC 6552* (2012)
7. O. Gnawali, P. Levis, The minimum rank with hysteresis objective function, in *IETF RFC 6719* (2012)
8. O.K. Patrick, N. Emmanuel, D. Thomas, et al., Energy-based routing metric for RPL. RR-8208, in *Paris, France: Institut National de Recherche en Informatique et en Automatique (INRIA)* (2013), pp. 1−14
9. B. Pavkovic, A. Duda, W.J. Hwang, et al., Efficient topology construction for RPL over IEEE 802.15.4 in wireless sensor networks. Ad Hoc Netw. **15**, 25−38 (2014)
10. M.A. Lodhi, A. Rehman, K.M. Khan, et al., Multiple path RPL for low power lossy networks, in *Proceedings of the 2015 IEEE Asia Pacific Conference on Wireless and Mobile (APWiMob'15), Aug 27−29, 2015, Bandung, Indonesia* (IEEE, Piscataway, NJ, USA, 2015), pp. 279−284
11. W.S. Tang, X.Y. Ma, J. Huang, et al., Toward improved RPL: a congestion avoidance multipath routing protocol with time factor for wireless sensor networks. J. Sens. 8128651/1−11 (2016)
12. C. Ma, J.P. Sheu, C.X. Hsu, A game theory based congestion control protocol for wireless personal area networks. J. Sens. 6168535/1−13 (2016)
13. M.A. Alsheikh, S. Lin, D. Niyato et al., Rate-distortion balanced data compression for wireless sensor networks. IEEE Sens. J. **16**(12), 5072–5083 (2016)
14. C. Sergiou, V. Vassilios, HRTC: a hybrid algorithm for efficient congestion control in wireless sensor networks, in *Proceedings of the 6th International Conference on New Technologies, Mobility and Security (NTMS'04), Mar 30–Apr 2, 2014, Dubai, United Arab Emirates* (IEEE, Piscataway, NJ, USA, 2014), p. 5
15. J. Jin, M. Palaniswami, B. Krishnamachari, Rate control for heterogeneous wireless sensor networks: characterization, algorithms and performance. Comput. Netw. **56**(17), 3783–3794 (2012)
16. P.O. Kamgueu, E. Nataf, T. Djotio, Energy-based routing metric for RPL, in *Research Report Inria, RR-8208* (2013)
17. O. Gaddour, A. Koubâa, M. Abid, Quality-of-service aware routing for static and mobile IPv6-based low-power and lossy sensor networks using RPL. Ad Hoc Netw. Sci. Direct J. (2015)
18. B. Sharkawy, A. Khattab, K.M.F. Elsayed, Fault-tolerant RPL through context awareness. in *2014 IEEE World Forum on Internet of Things (WF-IoT)* (2014), pp. 1–5
19. H. Lamaazi, N. Benamar, J. Antonio J, Study of the impact of designed objective function on the RPL based routing protocol, in *Adv. Ubiquitous Netw. (Lect. Notes Electr. Eng.)*, vol. 2 (2016), p. 397
20. Y. Chen, J. Chanet, K. Hou, H. Shi, G. De Sousa, A scalable context-aware objective function (SCAOF) of routing protocol for agricultural low-power and lossy networks (RPAL). Sensors 19507–19540 (2015)

21. O. Gaddour, A. Koubaa, R. Rangarajan, O. Cheikhrouhou, E. Tovar, M. Abid, CoRPL: RPL routing for mobile low power wireless sensor networks using corona mechanism, in *Proceedings of the 9th IEEE International Symposium on Industrial Embedded Systems (SIES)* (2014), pp. 200–209. https://doi.org/10.1109/SIES.2014.6871205. O. Gaddour, A. Koubâa, {RPL} in a nutshell: a survey, Comput. Netw. 56

22. K.-S. Hong, L. Choi, Dag-based multipath routing for mobile sensor networks, in *2011 International Conference on ICT Convergence (ICTC)* (2011), pp. 261–266

23. P.O. Kamgueu, Survey on RPL enhancements: a focus on topology, security and mobility. Comput. Commun. (submitted to Preprint) (February 19, 2018)

24. J. Hui, J. Vasseur, The routing protocol for low-power and lossy networks (RPL) option for carrying RPL information in data-plane datagrams, RFC 6553 (Suggested Standard) (Mar. 2012)

25. D. Airehrour, SecTrust-RPL: a secure trust-aware RPL routing protocol for internet of things. Futur. Gener. Comput. Syst. (2018). https://doi.org/10.1016/j.future.2018.03.021

26. T. Winter, P. Thubert, A. Brandt, J. Hui, R. Kelsey, P. Levis, K. Pister, R. Struik, J. Vasseur, R. Alexander, *RPL: IPv6 Routing Protocol for Low-Power and Lossy Networks* (Internet Engineering Task Force (IETF), 2012)

27. T. Tsao, R. Alexander, M. Dohler, V. Daza, A. Lozano, M. Richardson, *A Security Threat Analysis for the Routing Protocol for Low-Power and Lossy Networks (RPLs)* (Internet Engineering Task Force (IETF), 2015)

28. L. Wallgren, S. Raza, T. Voigt, Routing attacks and countermeasures in the RPL-based internet of things. Int. J. Distrib. Sens. Netw. **2013**, 11 (2013)

29. A. Mayzaud, A. Sehgal, R. Badonnel, I. Chrisment, J. Schönwälder, A study of RPL DODAG version attacks, in *Monitoring and Securing Virtualized Networks and Services*, ed. A. Sperotto, G. Doyen, S. Latré, M. Charalambides, B. Stiller, vol. 8508 (Springer, Berlin, Heidelberg, 2014), pp. 92–104

30. D.B. Parker, Restating the foundation of information security. Comput. Audit. Updat. **1991**, 2–15 (1991)

31. L. Anhtuan, J. Loo, A. Lasebae, A. Vinel, C. Yue, M. Chai, The impact of rank attack on network topology of routing protocol for low-power and lossy networks. Sens J (IEEE) **13**, 3685–3692 (2013)

32. A. Dvir, T. Holczer, L. Buttyan, VeRA-version number and rank authentication in RPL, in *2011 IEEE 8th International Conference on Mobile Adhoc and Sensor Systems (MASS)* (2011), pp. 709–714

Analysis of Cybersecurity Risks and Their Mitigation for Work-From-Home Tools and Techniques

Obaidullah and Muhammad Yousaf

Abstract This chapter is about analyzing cybersecurity risks and their mitigation for work-from-home considering the COVID-19 situation, concerning the tools and techniques being used to run the organization's operations. This chapter will help you understand the utilizations of online stages as a home office, and it will clear all the issues that emerge during this pandemic circumstance. Everything is discussed in insights concerning the dangers of tools and applications, which has picked up the business's goal shockingly. This is the worst pandemic that ever happened to humanity because every company and department suffers from this tragedy. Moreover, cybercriminals are constantly looking for new attack vectors. Already they have attempted to exploit the servers of many corona research centers. Also, they have strived to take over video conferencing platforms like zoom. Alleviation of those all-outsider applications and their answers are covered in this study work, which will assist you to make your home environment and your home office safe.

Keywords Covid-19 · Cyber threats · Information warfare · Risk and mitigation · Work from home · Zero trust network

1 Introduction

According to journals and other publications, COVID-19 suddenly put IT personnel and users in a challenging environment. Cybersecurity is facing a different type of attacks with a new shape. COVID-19 is the main factor in redefining the shapes of these attacks. The advantages and disadvantages of work from home is defined by [1], which indicates the benefits of shifting to the home environment.

Obaidullah (✉) · M. Yousaf
Riphah Institute of Systems Engineering, (RISE) Riphah International University, Islamabad, Pakistan
e-mail: obaid.ullah@riphah.edu.pk

M. Yousaf
e-mail: Muhammad.yousaf@riphah.edu.pk

© The Author(s), under exclusive license to Springer Nature Switzerland AG 2022   41
M. Ouaissa et al. (eds.), *Big Data Analytics and Computational Intelligence for Cybersecurity*, Studies in Big Data 111,
https://doi.org/10.1007/978-3-031-05752-6_4

Covid-19 has reshaped the perception that individuals and organizations are covered [2]. During our analysis, we have discovered that massive data of different regions in the US are analyzed for switching work from the office to work from home scenario [3].

It's essential to know about technology takeover during this pandemic that robots replace human employees. Now different type of operations are being performed by drones, artificial intelligence, cloud computing and chatbots usages are increased. Still, it brought different type of vulnerabilities and exploits to the business community [4]. Criminals are doing their best to enhance their techniques and tricks to breach someone's home or office network security with pretty good tools and scripts that may have devastating types of attacks [5].

This chapter will help you how to secure your home office and organization office against cyber-related threats. Criminals are doing their best to enhance their techniques and tricks to breach someone's home or office network security with pretty good tools and scripts that may have devastating types of attacks. However, state actors are always ready for such type of attack to react on time and make security more enhanced. Moreover, invariably there will be vulnerability which could be used as a death threat to an organization. These tragedies and events could be done only for their profits, and advantages are taken from the COVID-19 pandemic. The critical threat to the organization during the response to COVID-19 is those platforms that have already been used for several reasons and multiple times since the past. The highest ratio of these attacks is generated through Phishing, Vishing, Social Engineering with business Emails, Remote Access Threats, Securing password policies, Remote Conferencing, and personal devices for work purposes. These third-party Software's/Applications are concern with the highest number of users, such as conferencing tools, software development tools, cloud infrastructure, office management, and online tools [6].

There are not only a few departments or associations that are contaminated by this coronavirus. Indeed, it reached each corner of the world, and every organization is hit it by. The health department is not secure, and each day, counterfeit news is coming that they have designed a vaccine accessible on the dark web, which was absolutely phony news [7]. The education department is an important market for hackers since they have moved from ground studies to online platforms and acquired third-party applications for directing their classes. Zoom was the notable application for the conference call, and their end-to-end encryption was only a puzzle. That is why hackers found another technique to hit those associated with the zoom application, and their personal meetings become not any more encrypted.

By moving from an office environment to a home environment, different difficulties have been raised, which could be found in any type of organizational operations such as, Account operations, human resource operations, performing administration obligations from a home office. Aside from these all-greatest challenges are looked at by the information technology workers. Each association needs every minute of every day accessible, assistance with secure organization, and no interference during correspondence and tasks. Every employee does not have a technology-related background, and they can be deceived easily by the miscreants, because consistently hackers are attempting to discover a loophole or weak point in organization assets or network. Only machines do not commit blunders or errors human does too due to lust, laziness, greed, and lack of knowledge. For that reason, they are focusing on such types of employees who can be redirected without any hurdles. This work will help any foundation worker, regardless of whether they have an IT background or no. Following these steps and procedures discussed in this chapter will make your home network and office network both protected and secure.

2 Related Work

According to the recent publication, (have you been a victim of COVID-19-related Cyber incidents, survey, taxonomy, and mitigation strategies) [8]. Their chapter mentioned that COVID-19 themed cyberattacks and categorized them into four categories: the first category is disrupting the services, the second type is financial goals, another one is information theft, and finally fear ware.

Due to covid-19, everything is control and managed by remote configuration, while their documents do not cover important aspects to discuss how to verify users? How to validate their devices which does not contain any malicious applications? Do they follow those limitations which were assigned to them? These are the major key points that have been raised due to the pandemic situation, and their documents do not cover it. This document has solved these problems, and the solution is provided too. No one is trustworthy and no device on the network should be considered legitimate until its authenticated and verified [9].

Hackers have sharpened their arsenal and reshaped the tools and techniques for the attack. If we analyze this situation from a highly close point of view, every organization focuses on the hackers in view of the pandemic circumstance. Nobody is protected and secure because everybody is associated with the web so that nobody can go outside for any purpose and strict orders are given by high authorities to remain at home, and because of this request, everybody is attempting to keep their self-occupied with various kind of actions over the web. Due to this pandemic, there is an enormous increment of web clients around the globe, and an exceptionally high measure of traffic is produced due to remaining at home. Travel is fundamentally prohibited everywhere on the globe. It is another disadvantage for every department. Social distance is strictly requested from government specialists, and everyone is attempting to keep them-self disengaged too [9].

Everybody is attempting to sit in a corner and keep themselves self-occupied with devices on the internet. The ratio of Reliance on online platforms is high due to Essential government services, business, education, banking sector, health care, and finally, entertainment. If we have a look from another perspective which is a risk of cyberattacks and it could be found in multiple types of attacks and mainly focused was data breaching of any organization, denial of service happened to many platforms such as US health and human services department servers were under attack during the covid-19 period [10]. The financial loss happened to many well-known associations, and the famous story came to the media during this pandemic which has happened to Hammersmith medicines research. This is a London-based coronavirus testing facility that was reportedly affected by the ransomware. Hackers request a ransom amount that the mentioned organization did not pay. Consequently, their confidential data of thousands of patients were published online [11].

3 Zero Trust Model

The zero trust models main concept is that no one is trustworthy on the network, and every device should be verified, validated, and authenticated. There is no default trust for internal and external networks, and IT professionals consider outsider networks untrusty. The attack can be easily generated from the local network as well, where malicious insiders can inject something specious easily [12]. Traditionally, we have installed many devices for filtering of the frames and packets like Firewall, Anti-Virus, Intrusion detection system (IDS) and intrusion prevention system (IPS), also unified threat management (UTM) [13].

Packets are transmitted from source to destination just by following the traditional way, checking the packets header, and each layer is binding its information which is taking a new step and going ahead to the next layer. Packets cannot trust because they check the headers, and the zero-trust model does not trust the packets. Security professionals considering the internal network trusted while external is considered untrusted. For that purpose, they implement every security type to make them more secure from the outside network, which is an external network, and we consider them as an Internet. However, for zero network model, all type of traffic is considered untrusted [14]. This is because of the security mechanism, which the mentioned model follows. To implement the zero-trust model, one should consider the following steps. First of all, security professionals must verify and secure all resources that may contain hardware and software before installing any application. It is compulsory to check it because third-party software may contain some malicious functionalities that will share your data directly with the developer. Also, the hardware version must be kept updated because who knows your sophisticated hardware contains some deficiency where all organization network traffic relies on your security device. Secondly, limit and strictly enforce access control.

Strick rules are essential because every employee is not loyal to the organization, and some of them are not caring about every notice. Insertion of the USB should be strictly banned in any system because rubber dicky scripts are open source and everyone can build their rubber ducky USB and systems are easily getting deceived by rubber ducky mechanism because it considers as actual hardware like keyboard or mouse while it could be human interface device (HID) which may increase risk of your network [15]. Furthermore, inspect logs of all networks that can be easily maintained by the tools built for the network analysis, and Splunk is the famous one acquired by well-known organizations [16].

The best thing about the zero-trust model is that no matter who created the traffic and where it was generated if someone is connected from anywhere in the world, their generated traffic would be considered a threat until the traffic is verified and authorized. Also, it has another feature where traffic will be inspected and make sure it is secure and does not contain any malicious data inside the packet. Zero trust model demand is from security professionals to treat the local traffic and make them protected.

Figure 1 focus is on the externally connected devices for security measures if your local network contains some malicious insider whose intention is to damage your organization assets, network, or reputation. How will you deal with that scenario? That is why the zero-trust model does not trust any traffic, whether generated locally or globally [17]. To examine more around the zero-trust model, the following idea is zero trusts in access control. Technical groups are making policies and sending them to specialized groups for usage, and they are implementing. However, there is still a chance of enforcement from inside employees that they can change anything

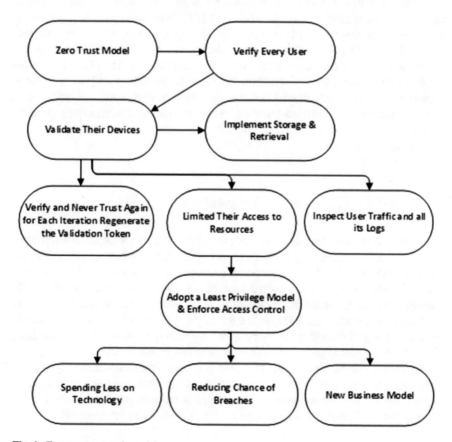

Fig. 1 Zero trust network model

they would like to do with the data and enhance the confidential data. The best part of zero trust model usage does not expressly characterize jobs based on access control. It follows the appropriate identity of appropriate access and administration methodology to confirm representative access rights. In traditional security models, once roles are made, tasks are assigned, then administrators are limiting the access to resources, and through those rules and regulations, employees are working in their limited domain. Nevertheless, the zero-trust model does not stop here.

It goes one step ahead, which is the requirement of storage and retrieval of the data, and it will be checked continuously to inspect the user traffic for a sign of suspicious activities [16]. Because sometimes we observe the users and follow their activities and based on those activities, we are judging and starting belief on them and gradually allowing them for further activities. But in the zero-trust model, they do not trust users. Instead of trusting users to do the right things, the zero-trust model verifies that they are doing the right things? Trust and verify are converted to verify and never trust. For that purpose, users can be verified through its anomalous behavior, and they will never be trusted. For each iteration, a new token will be generated after authentication and verification [17]. Such as downloading/uploading a vast file type that is not performed daily, such type of activity can be observed and analyzed. Also, frequently accessing the system in those timing which is not a routine can be suspicious activity, and it will be analyzed. The current situation based on COVID-19, where work from the home scenario is at its peak, and every organization has changed its environment, and employees are working from home. it is highly recommended to follow zero trust model where they should not be dependent on the dedicated network, and they can work from anywhere anytime without any hurdles and problem.

3.1 Traditional Networks Shortcomings

According to the zero-trust architecture journal covered by Dibanjali Gosh, it indicates that traditional security methods classify everything, including users, devices, and applications inside the local corporate network, as untrustworthy. This statement shows that local networks may contain multiple types of threats and flaws, which will lead to breaches of the organization's security. To identify the users and devices in traditional networks, they have installed virtual private networks and network access controls through which they are trying to ask for their credentials. If provided, then they are letting access of the network to them which is full of risk.

Many incidents have occurred because of these traditional network configurations because attackers are trying to find a loophole in their hardware and software and bring changes accordingly into them and exploit those organizations which have already installed the mentioned devices on their organization premises [9].

3.2 Principle of Zero Trust

To elaborate on the security functionality, one should never trust someone or something simply because it is behind the enterprise wall, and nobody knows which type of breaches may occur. Everything on the network should be considered a threat, like users, devices, and the network itself should be considered a threat. Consistently this formula should be implemented that everything on the network should be verified, validate and permission should be given very limitedly as much they are requested that much should be allowed to them. This phenomenon is essential to know that zero-trust network architecture follows the "never trust, always verify" statement, which is the reason for success for this newly introduced platform, rising surprisingly [9]. One more critical point is necessary to be discussed: analytics, which needs to be implemented with the help of solid algorithms. Additionally, filtering mechanisms need to be acquired to implement some roles and regulations. Finally, logging events should be generated to capture the logs of every event, and for the next session credential, a newly generated token of the event and previous logs should be checked to allow or deny their access to the network.

3.3 Logical Components of Zero Trust Architecture

Three main components need to be discovered for the zero-trust network. The first thing we need to know is Policy Engine, the Second point is Policy administration, and the final critical point is the policy enforcement point. When the request is generated, this will be checked accordingly in the policy engine to check whether grant access or deny it. Then policy administrator will be check because it can create a path for the communication between the subject and resources of the organization. In the end, the policy enforcement point can terminate, enable, establish and monitor the connection between both entities. It will terminate the connection immediately if some abnormal activity is observed on the network. This can be done with the help of machine learning and user behavior analytics. Some data will be feed to the database, and numbering will be assigned to them, and accordingly, the decision will be taken whether to grant the access or terminate the request [9].

4 Secure Access Service Edge "SASE"

SASE is a network architecture that combines WAN capabilities with cloud security functions, in collaboration with secure web access, cloud access security brokers, firewall, DNS, remote browser isolation, web application and API protection as a service, and finally, zero trust model network access. This term was generated by Gartner in the 2019 networking hype cycle report [18]. Before going ahead to

SASE, we need to discuss a bit more about traditional network security, which is currently acquired by organizations and industry, and currently, such traditional network concepts and implementations are not completing the need for security. Every organization's demand is now a day a connection without any interruption for their users 24/7, which is not necessary from where they are online and through which device, they would like to access their data [19].

With heavy incrimination of the remote users and software as a service (SaaS) application, it confirms that data is transferring from physical data centers to the cloud computers available somewhere in the globe remotely and can be accessed from any device in the meantime. This heavy traffic generated by the worldwide public users' needs a new security implementation and concept that can provide non-stop service to the clients that do not have any interruption during accessing time. SASE convergence is made of the two methodologies and combined their functionalities to make one new security bond [18]. Figure 2 indicates the convergence of the SASE, both functionalities are used. Likewise, network as a service and network security as a service Functionalities of the network as a service is collected and determined Like SD-WAN, Carriers, Content Delivery Network (CDN), WAN Optimization, Bandwidth Aggregators, and finally, the Networking Vendors [20].

In network security as a service, some steps are to be performed and observed because each of them may contain some critical data that will harm your network. However, these factors are analyzed in SASE convergence because the network receives once data. first of all, it will scan the data, an alarm will be generated, or it will be terminated. Suppose it contains sensitive data, which is the initial step of the sensitive data awareness [20]. Furthermore, threats will be deducted due to the convergence of many vendor's functionalities, such as, network security which will determine every user's behavior through their activities.

A cloud access security broker (CASB) is a service between two entities: an organization and cloud provider infrastructure [21]. The basic concept of the CASB is like a security guard who is hired by the organization and assigned a task. If the person is fulfilling the security requirement, allow him to enter the organization premises. Else do not let him even cross the limited area. The same scenario is defined for the CASB, which is supposed to check the security measures between sender and receiver [21].

One more functionality is defined: Secure Web Gateway (SWG), and this is used for protection against those threats, which are internet-born threats like phishing, malware, and botnet attacks [23].

It's clearly understandable by Fig. 2 that any user who wants to access the data from anywhere around the world with secure web access can easily maintain it by enabling a secure web access feature, which is the requirement of every association nowadays due to social engineering techniques. No one can understand the devil-minded concept behind social engineering attacks, where this sophisticated type of attack can deceive every background user. Nevertheless, due to secure web gateway (SWG) now accessing web security is more enhanced. Additionally, the zero-trust model network concept is implemented where no user is authorized without authentication. A simple formula is generated for the ZTM network. Once a request is

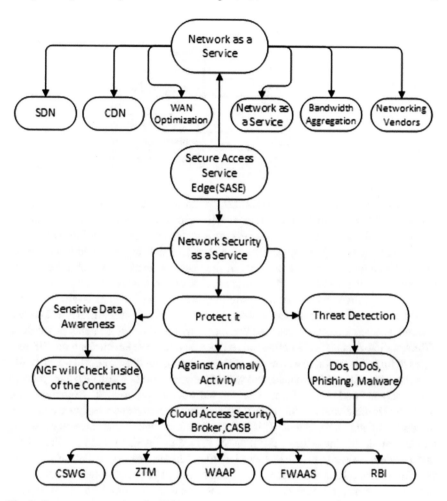

Fig. 2 Secure access service edge [22]

generated, first of all, your authentication is mandatory. If you have passed this phase, then you are supposed to be checked for authorization to see whether you have permission to access such type of data which you have requested or no. when both phases are done, you are identified and got verified you will have access to the resources, but still, there are some rules which need to be followed and those are limitation of the access to the resources, you cannot use all resources. The least privilege model is set, which will implement the enforced access control. For that purpose, the zero-trust model will reduce the chances of breaches [13].

As long as the web application uses WAAPaaS, it will provide protection against WAF, API security, botnet management, and DDoS protection, which will help decrease the chances of web-facing applications [24]. Furthermore, FwaaS, which

Table 1 Abbreviations

Abbreviations	
CSWG	Content secure web gateway
ZTM	Zero trust model
WAAP	Web application API protection
FWaaS	Firewall as a service
RBI	Remote browser isolation
SDN	Software define network
CDN	Content delivery network

stands for Firewall as a service, is utilized for advanced protection, including next-generation firewall capabilities. Also, it will help to maintain the URL filtrations and advance threat protection. Those firewalls which are installed traditionally in the organization does not have any ability to protect your data against of advance attacks because it cannot read inside contents of the packet to examine which type of data is available in the packet, whether its actual data or it is a malware [25].

Traditional firewalls are reading only the surface, which means it will read only the extensions; if a file contains some.exe extension, it will discard the packet. But firewall as a service is the advanced version; it will read inside contents of the data. Which will indicate it is a genuine file or it is a payload. We are clicking on different types of links and trying to download one thing during our browsing, but accidentally a package is getting downloaded, and without any intentions, those packages contain some malware and worms through which our devices get affected. Sometimes we become a zombie, and the third party utilizes our system resources, but we are still not aware of such an incident. This can happen to us due to our careless browsing and download of things from the internet [25]. But SASE brought a solution for this accident as well, and they have included remote browser isolation (RBI) in their framework. RBI is a specific implementation of browser isolation that occurs remotely by moving the execution of all browsing activity from the user's computer to a remote server [26]. Once this implementation is done to the users, user activity can be hosted in the cloud or on-premises organization network, not harming the cloud environment. This is because the remote user is isolated from the physical desktop and network. They may not become a threat from web-based attacks [26].

Table 1 is giving full names of the abbreviations which are utilized in entire chapter.

5 Cybersecurity Framework for Work-From-Home

NIST is identifying and developing cyber security framework which is widely used by the sophisticated organizations and it has defined some critical key points which were adopted by the industry and organizations were following those all-rules regulations.

This Cybersecurity framework will address how the organization will work from home securely. Which depends on the technology tools and techniques. Therefore, the organization must address a comprehensive cybersecurity framework in Telework, Remote Access, and Bring Your Own Devices (BYOD) solutions. This framework is aligned with the global best practices used to address the home environment's work [27].

Figure 3 indicates the following recommendations for all the users working from their home environment. The organization needs to understand the concept that the cyber world is getting more complex, and it is becoming difficult for the organizations to address all the threats associated with any environment. The telework threats can be mitigated using the encryption techniques at different levels (data at rest, data in transit, and data in process) [29]. To secure the information, it is important to classify the information as sensitive and non-sensitive; as the data is more sensitive, more protection needs to be implemented.

So, securing the data at rest can be done by encrypting the hard drives locally, the sensitive data should not be stored at the local systems, and there should be multi-factor authentication at all levels, which may need some protection in the form of virtual private networks [30]. In an organization, it's more important to know about the following points and need to be considered very strictly before any incident happened. One should identify the risk for an organization which could be in different formats like asset management and its property of the company which needs to be secure on a priority basis [28].

These assets may include hardware, software, data, information, contracts, projects, deals, and many more critical things as well which need high security. Another form of security organization need is in form of organization building, fixed

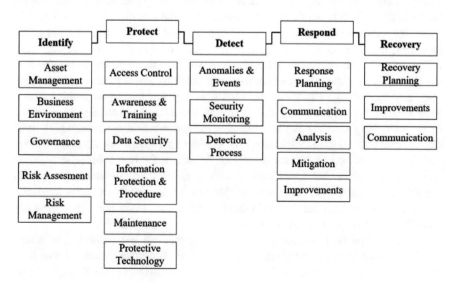

Fig. 3 NIST framework [28]

assets like furniture devices and machines. From a business point of view, this point should be very clear that the business environment should have some strict rules and regulations where governance needs to be implemented at any cost. Still, there are some cases where security may breach or maybe some natural disaster can damage the asset of the organization which needs to be re-asset and make more strong management for such incidents [28]. Once the identification phase is covered and the plan is set accordingly then higher management needs to have a focus on the protection of the assets. The initial point to start this stage is access controls. There must be access controls for each and everything whether it's part of hardware, software, or employees. Employees are the biggest assets of any company and they need the most protection as compared to any other sort of asset. But should be clear that awareness is the key to success in every platform because every employee has a different type of background and position and they may not be able to know everything that's why awareness sessions are very important.

To discuss more, there should be some terms and policies for each department of the organization which needs to be implemented like the IT department's initial focus should be on those policies which cover data protection terms. Information protection and the procedure is the thing that will lead the technical team into a lead position. Because as much your organization employee's data is secure and not available publicly that much security will be enhanced of the department and assets of the company. Every bad guy is looking for a small piece of information that can be penetrated accordingly against the organization and protection of user's information and their data need to be protected and strictly policy needs to be implemented. The first two stages are basic but important which need more focus and the third stage needs more consideration which is the detection phase. This is very critical and needs strong technical knowledge regards network security and cyber security because no buddy knows which type of attack is planned and what bad guys are targeting.

In the organization network administrators and system administrators are following different techniques but recommended step is to check the logs and traffic on a routine basis. Likewise, the traffic during the morning is normal and the percentage can be seen in the normal range but during peak hours the percentage may vary accordingly. But if the same traffic is captured during not office hours it means there is something wrong with the network and traffic which can be penetrated by the third party and they might be utilizing our organization resources which need strong protection. There are different software and hardware available in the market which can be utilized like a new generation firewall is the best solution nowadays. NGF has the best quality which can scan the data inside of the document as well. Traditional firewalls and antiviruses can check the extension and its public key only. But next-generation firewall beauty is that it can scan the data inside of the document as well which can decrease the threat level.

The critical point arises during the response phase, this is the crucial time where the IT department can be under stress and feel fear. The exactly should know how to respond and react during a critical situation which can be seen nowadays everywhere during a pandemic. The best step is to have a strong response plan and secure communication channel during such type of incident. The start point should be analysis that how the attack was conducted, and which loopholes were targeted. How the assets were exploited, and which type of loss has been done to the company. Logs should be traced, and network traffic must be observed to know exactly from where the attack was generated and how much loss can be covered. From this point of view, it's important to know the mitigation for next time. Improvement of the network and its security is more mandatory now.

The final part is to know about the recovery from the attack which type of recovery can be done and which steps need to be performed to recover the data, information, assets, and resources.

According to "The Evolution of Fileless Malware" [31] fileless malware is a more impacted vector of attack. It's very crucial to capture such types of attacks because fileless malware can attach itself to a genuine process and execute itself. Even most antiviruses cannot identify the file-less malware during attack. Now, this is very challenging for the Information technology department to tackle down such a type of attack. One more thing which is hard to fight is a ransomware attack that will encrypt your all data and the only solution for the decryption is to pay them in form of digital currency else it's almost impossible to break the key and decrypt your data. Proper documentation needs to be created and it has to be shared with the higher management. How the attack was conducted? Which steps are taken for the protection of the mentioned attack? How the attack was detected and how the technical team respond accordingly need to be clear for the higher management. Finally, how the data and information were recovered from this incident and how much the organization has paid in form amount and data lost [31].

Figure 4 illustrates the following architecture is design for a work from home scenario, and there are some significant points that need to be considered. And the

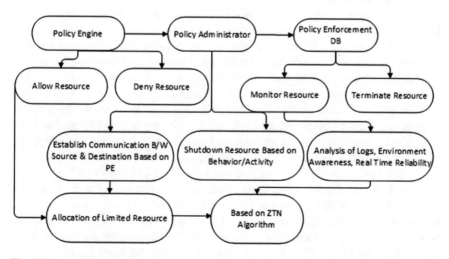

Fig. 4 Proposed framework

first step is starting from the policy engine where some rules regulations are pre-defined whether to allow the user or resource or not. If requirements meet, further steps will be taken to enable them and give access to them; else, the request will be terminated. Once the first step is cleared, then another stage is the policy administrator that is supposed to make decisions based on PE permission. The policy administrator provides a connection for resources between sources and destinations found on the pre-defined policies of the policy engine. If some behavior or activity is, consider which is against the policy on the spot abrupt decision to shut down the connection for connected resources. Both policy engine and policy administrator will be compared, and limited resources will be allocated to them, which are pre-defined [32].

The policy enforcement database is the main thing of this architecture, where each action and log will be monitored. If somehow some resource is going against the defined rules and performing those activities which are not supposed to perform it suddenly their connection will be terminated. For the next session, both logs will be analyzed from the previous records where the connection was terminated, and the newly established connection. If still some abnormal activity is performed, then account, user, the resource will be suspended for a short period of time, and this can be followed for upcoming two more sessions. If abnormal activity is still performed by the end-user, device, or network, that connection will be banned permanently.

But if the connection is going smoothly still every log, Environment awareness, environment security, real-time reliability will be considered a threat for the network. These decisions can be taken based on the zero-trust network algorithm.

In work from home scenario, secure communication with an external network is a security risk that can be minimized through encryption technologies to protect information conditionality and integrity of communication and secure end-point authentication by identifying endpoints. This can be done by using a virtual private network to access specific endpoints securely.

If we assume that any telework could be breached with a malicious type of code and that will lead to failure because that malware may infect your systems in such a systemic way that either firewall, intrusion detection system, and intrusion prevention system will not detect. Similarly, any system file could be infected through malware, such as Sys32, or it will hit some windows. API's will lose their control and may give complete control to the attacker [33]. Similarly, we do not know about the performance or activity of that malware that what it is supposed to do with our machine that may be controlled by C&C server, or it may be infected our machines with some adware that will be showing us irrelevant ads. It is also possible that a piece of code that is malicious will take overall control of the attacker's system.

To prevent such attacks, one should learn that which type of link should be clicked and which type of file should be open. There is a different type of tools which can be used to detect malicious file and links before opening the link put it in www.vir ustotal.com, also understand the context of any received emails, you can identify the sender by checking the details of the email by checking the header and also check the grammatical mistakes inside the email which shows that the sender is not genuine. Also, make sure that do not share any personal information with anyone online without confirming who will receive this information [34]. As the home network is not under any firewall and security controls, so the threats from the external networks increased. Therefore, it is necessary to restrict access to the organizational assets for remote users. That should be limited for the users who need limited access rights to do certain jobs online. Giving more access rights to a user increases the risks to the organizational assets, and the processes become more vulnerable because the access rights are being used excessively. End-point devices need to be configured securely at all levels, e.g., Hardware-level, Operating system level, Windows Applications and services level, and Users files and applications [34].

As the system becomes, more functional the more vulnerable system would be. To address these concerns, the end-user needs to secure the system from unknown physical access, windows unused services need to be turned off. Unnecessary ports need to be turned off, unnecessary tools should not be available inside the working system, the system must be up to date, and should have enabled the firewalls, which should be configured securely. The tools should be limited that need to be used during working hours, email clients' needs to configured indeed using the security protocols and Virtual Private Networks. The browser should be appropriately configured not to reveal personally identifiable information of the user. This can be done using the browser "Firefox" or "Brave". The Firefox has multiple extensions that can help end-user to maximize privacy and security, like "duckduckgo Privacy Essentials", "AdBlocker", "Cookie AutoDelete", "HTTPS Everywhere", "Privacy Badger", and "NoScrip" etc. [35].

It is recommended that servers are to be hosted by trusted and authorized administrators. If servers are not fully patched, then there is an excellent chance of the server being compromised. To protect such a thing, all servers should be patched and up to date and should be placed in organizational network premises. Inbound traffic should be restricted single point of entry through which traffic is to be monitored and permitted into the internal organizational network. By doing all that, servers will be protected from outside threats, which can cause server exploitation [36]—considering the network placement of remote access servers. It is recommended to place a server in organizational network premises to restrict a single-entry point through which traffic is to be monitored and through which inbound traffic can be monitored and permitted into the internal organizational network [37].

6 SASE and ZTN Combination

SASE and ZTNA will lead your organization to a very secure platform, likewise, no one has the intention to fully rely on physical appearance to office 100% because in this pandemic scheduling was changed and business strategies were completely changed. Not only does secure access service edge improve connectivity and secure communication it also develops productivity.

Before organizations were focusing on physical security walls and barriers to stop cyber threats and other types of malicious activities but nowadays due to improving technology and strong communication channels through ZTNA and SASE firewalls are somewhere in the cloud. The same strategy can be followed for the zero-trust network architecture.

Figure 5 defined the two major steps for ZTNA one is user authentication another one is user authorization. If a user would like to utilize the resources of the organization, he/she needs to prove they are genuine users. For that purpose, there is a process through which can be reached till authorization level. First of all, the identification phase will begin which will check the username and password of the existing user. In most cases, employees leave the organization after their resignation still their accounts exist and many of them use those resources which are not supposed to touch them due to their resignation. But this is a failure of the department and they're not properly implemented policies. There should be a policy and regulation if a user is supposed to resign or they are fired by the department on the spot their all accounts and resources should be frozen and they cannot use any of those platforms to which he/she had access to it. In ZTN the system is checking the active directory and looking for the provided credentials checking the authentication as well, whether it's correct, valid, and genuine then it will proceed to another stage which is mapping of the data. If there is no problem and credentials were accurate the system will map the data and proceed to access control which is a very important stage. In the access control step, the system will check the available resources and controls for the mentioned user.

Every user has a specific domain where he/she is supposed to work and utilize the provided resources but some policies will be implemented on them. If we elaborate this stage with an example, it could be like the following example. A user is allowed to consume the internet 4 GB which cannot exceed the limit but somehow that user has utilized the limit with genuine traffic it can be considered as an exceptional case. Because their call logs will be captured and analyzed based on their activity more utilization will be granted. This can be done because of the trust evaluation engine. If a user is having the same issue frequently then accordingly their logs history will be stored in the trust evaluation engine. Based on the current scenario we can define

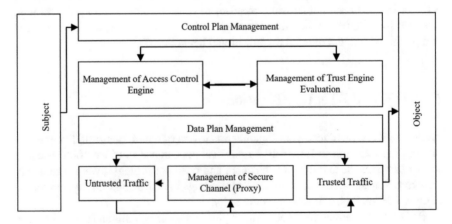

Fig. 5 Proper communication channel

our resource access into two sections which can be the control plane and data plan [38].

As we can see it's clearly drawn on the Fig. 5 control plan management is further divided into sections one is the management of the access control which can have a focus on all access control which may include data access, information access, hardware, and software access. Furthermore, this control engine is dependent on the evaluation engine which contains all types of logs and history. Every user that entering to their domain and proves their authentication phase will be stored in the trust evaluation engine. This storage may vary on users' activity over the network and utilization of the available resources [38]. It's the state will not be static because on daily basis this storage level can be changed and accordingly one token will be generated and issued. If a system is a shutdown or logged off for some reason again a user is supposed to prove their identity provide their credentials and the system should identify and authenticate it. One token can be used for one time not several times this is the beauty of this marvelous platform and its most secure. For the very first time, every user and every type of traffic will be untrusted for ZTN. If a user would like to send their request from source to destination they need to go through several phases. A subject could be any sort of device or machine which is connected to the internet and using the organization domain resources to reach their destination which can be any type of application or server that need to be filtered by trust access control, trust evaluation engine, a secure channel which could be any type of proxy and finally the traffic will become a trusted.

7 Trust Management and Challenges in ZTN

Management of the cloud itself is a challenge due to several reasons, up to now we have to define strong management and best policies for the organization's security and provide many solutions. But, now it's very important to describe some trust management scenarios in a cloud. Initially, it is important to know sending a request from source to destination needs some sort of connectivity which may never have been connected to such a device before and your traffic needs to transfer through different platforms and itself is a challenge. Man in the middle attack can have multiple vectors in form of attack [39].

The most important thing is secure data access. It's very important to provide trust for the end-user that how much your data is secure and the platform is trustworthy. One more point needs to be clear that there will be several types of ISP's, which are connected in the network but how many more cloud nodes are connected and every node may have its policy. This is now the biggest challenge to tackle in such scenarios. Every organization has its policy, rules, and regulations but the challenge is connectivity and compatibility with another node available on the cloud. Some issues may face by vendors and end-users during contact with the cloud services. Companies need to detect every anomaly and malicious activity over the network.

Data increment is the biggest challenge which is increasing day by day and management needs more strong planes and protection techniques to make sure the end-user that your data is fully covered and secure. Once the data is managed now it needs protection as well because no one knows for how long this data will be stored on the storage devices and it needs availability as well. The best solution for data protection is cryptographic techniques where data format can be changed and it can be encrypted. This solution can be followed in both scenarios' local storage at the organization's end and remote storage which can be available somewhere in the world in form of cloud storage [39, 40].

Multifactor authentication is used in every step of ZTN to authorize and authenticate the users. In traditional network design, every user is basically connected with an active directory and their authentication can be done from the local system which is no more the best solution. In a zero-trust network, you can use your files and services from all over the globe just you need to have access to a system that is connected to the internet and provide your identification to go ahead and start utilization of the applications and services. If a person is assigned the least privileges in the department and very limited resources are assigned to them [39]. There is check and balance available in the department, how this can be ensured in the cloud where systems are available somewhere in the world. The actual answer is bots. Some best algorithms are designed for this purpose to have check and balance and strictly follow the policies and this task is being performed by the machines all the time with the best result. One more scenario where bots are performing actively and making prevention of the blunders. In organizations, there is a help desk team available from the IT department who is responsible for every technical problem and they are solving those issues.

If we assume a user has forgotten a password and they would like to change their system password or ID password, they call the helpdesk for help. For sure someone from the mentioned department will ask some questions to verify whether this user is genuine or not and this account belongs to him/her or not. A technician may ask for username, employee ID, department, and designation. Through these technical questions, the IT department will accept that yes this ID belongs to him/her and they will go-ahead to the next stage. But how this process can be done in the cloud where no physical appearance is available and departmental discussion will arise during such type of circumstances.

But this is a challenging scenario for the cloud users because if someone has an idea about those critical questions and they know the exact answer for them for sure a person can impersonate and trick a bot by providing the exact information but still there is hope because ZTN is providing another solution as well. It checks the logs based on the user history and verifies whether the request is coming from a new device or an old device and the behavior of the incoming traffic can be checked as well, if any sort of anomaly is caught the connection will be suspended due to security ineligibility.

At the end data ZTN has defined some categories for the data and information like which type of data is how much sensitive and after how many checks and balances and multiple authentications that data will be available in the best part of the zero-trust network [39].

8 Challenges for SASE

It's understood that SASE is the best emerging service solution and ZTN is the main and most important part of the secure access service edge which is the best solution of traditional security and replacement of virtual private networks (VPNs). If someone would like to connect with another platform, they need to use an API that will connect you easily. Any platform can be connected with any other third-party solution provider platforms like cloud services and application services through API's. We need some API's to use for connecting with another application or server through the mentioned platform. But bad guys are always looking for an opportunity that has a weak point and API is the biggest weak point that can be exploited easily. Developers are trying to develop their API's with best practices but red team members are always looking for that vulnerable software, application, service which has a small loophole and they are trying to manipulate and exploit the entire system.

It's clear that API is considered as glue for any framework to be connected and this connectivity may lead your full secure system into a vulnerable system. Apart from the developer team who are involved in API development for several projects, there should be another team that should look for API security breaches. They should hire a team to break their API before bad guys get access to that weak and vulnerable application programming interface [40]. A very strong platform needs a small weak point to shut down the entire network so API is also one of those small weak points which need most consideration at the moment.

Due to API misconfiguration, the following risks may increase. Increment in attack surface, threat movement over the network, dis-connectivity, unwanted traffic, and data loss. To elaborate the increment in attack surface which will lead your entire organization into trouble due to one small misconfiguration of application programming interface that was used for a better approach to another platform or connectivity to another cloud service will make you fall into trouble. If this suspicious action is captured on your organization's premises, then there are chances of attack and due to these chances always bad guys will be looking for interrupting your organization's network and resources which will burden your entire network with unwanted traffic.

If somehow one of the API's is compromised which is a glue for another framework will lead your company's confidential data into loss and the reputation of the company will be damaged with this small incident. Before going ahead with utilizing any sort of API's this is recommended to check its vulnerabilities and security level by your security team.

One more thing which needs to be considered is data protection techniques. Data can be found in different formats like data at rest, in transit, and use. First of all, let's assume that someone has saved their confidential documents through these secure channels and it has reached the final destination which is a storage device. Such type of data which is at rest must be available on the hard disk and it needs some sort of security the best solution is encryption of the data. If someone has physically accessed the hard disk and tried to manipulate the data from the hard disk it should be protected and the form of the data must be changed from readable to unreadable format. Such a type of algorithm should be implemented on the data which can give a tough time to the one who has gained physical access to the data and it cannot be decrypted [39, 40].

Another scenario where data can be leaked is in transit. This stage has different formats to transfer the data from one location to another location within the data center different types of attacks are supposed to happen like MITM, eavesdropping session hijacking, and many more. Now this stage needs more consideration and best practices to transfer the data in the best manner to don't lose or damage the data. Finally, in the transit stage, both layers are supposed to cover the data with the encryption method. The application layer needs to follow the TLS mechanism for data protection and Data Link Layer which is supposed to encrypt each frame for the best security solution.

The critical point is how to secure your data that is already in use and is not saved yet in any storage whether it's in local system storage or it's another cloud platform. There are some services available on the internet which can protect your such data which is currently in use. The following platforms and services can be utilized for best practices like software-as-a-service, platform-as-a-service, infrastructure-as-a-service. And there are two more solutions available which can be avail and best results are expected from them and they are azure confidential computing and centralized storage of application secrets [40].

9   Security Analysis of Tools and Techniques to Support IT Services

The expanding fame of the Internet has made more individuals affected by inadequate software security. Because of the poor security in today's software, large groups of people and companies have been disturbed by malicious software. The expanded cost brought by malware has caught the advertisement companies' focus due to such type of greed resulting in more exposure to software insecurities.

The Internet has made data on programming hacking freely accessible. Several websites are available where payloads and exploits are openly available against any framework and infrastructure. Famous of them are "www.exploit-db.com" and "www.cve-mitre.com". Any recent incident related to cybersecurity exploits could be found there without any hurdle. When a security exploit is found in a platform by any security researcher, there are security companies, through their platforms, such security flaws can be reported. Like bug crowd, hacker one is top listed. Against those reports, heavy bounties are paid to them. That is the main reason where security researchers are trying to find the loophole in their programs.

Other category programmers are intentionally creating malicious scripts for several reasons, like gaining access, remote code execution, spyware, and adware. The following table contains information about those tools and techniques which every organization uses for their daily operations, and due to some reasons, they are not following the standards and making updates on time which create a loophole for the hackers to inject the payload into their systems, websites, and devices and quickly access can be gained by the bad guys for different intentions.

Table 2 explains the example, pay the board, charge the executives, or spending the executives. Also, a few apparatuses and procedures could be utilized to run their normal activities efficiently with no issue for every one of them.

Table 2 Tools and techniques to support IT services

IT services and functionalities	Sub-domains	Tools and techniques
HR operations	Email marketing	MailChimp
	Attendance management	Online attendance system
	Hiring	Glassdoor
Accounts operations	Salary management	Salary management system
	Tax management	Office 365
	Budget management	Centage/Prophix
IT operations	Software development	Microsoft Visual Studio
		SQL Server 2017
		Android Studio
		Microsoft Azure
		MS Teams
		Outlook
	Software testing	Microsoft Visual Studio
		Test Complete
		SQL Server 2017
		Android Studio
		Microsoft Azure
		Microsoft Azure
	Software deployment	Microsoft Visual Studio
	Software maintenance	SQL Server 2017
	Software solution	Android Studio
		Microsoft Azure
		MS teams
		Outlook
		MS excel
		MS Visio
		Dropbox
Administration	File sharing	Google drive
		Mega
		OneDrive

Table 3 Cybersecurity risks associated with these tools

Rank	Tools	Associated risks
1	Microsoft Azure	CVE-2019–1372 which can lead an attacker to run remote code execution [41]
2	MS Office 365	CVE-2019–1200 found in version 2019 [42]
3	Microsoft Visual Studio 2019	Critical vulnerability in MS Visual Studio 2019 version 16.0 [42]
4	SQL Server 2017	Found critical vulnerability in MS SQL Server 2017 version 01 [43]
5	MS Outlook	Insecure protocols being used, IMAP, POP3
6	Redmine 3.4.3	CVE-2017-18026 which can lead an attacker to execute XSS [44]
7	Accounts management system	Multiple privacy concerns regarding the access rights of the organizational staff
8	Android Studio	Organization need to update it into latest version
9	Online attendance system	Having some critical bugs in the login panel

In Table 3 each classification of the system contains some danger and weaknesses which may drive you to fall into inconvenience. Like utilizing the online resources may be influenced by some exploitations, and still, you are running such uses will damage your association assets. The above table clarifies some tools that are needed to be updated due to their associated risk, which is covered in their updates. Still, prominent associations are accessible in the business community that is burning heavy amounts of dollars, financial plans on the framework, and physical security. However, they are not focusing on the current digital war, which is a cold war. However, if you become a victim of the attack for sure, all your endeavors will disappear forever within minutes.

Table 4 cover this section is essentially covering mitigation of the risks which are related to your tools. Every single day attacks are generated on some advanced applications, and the red team is consistently prepared to react to them. In some cases, we do not update that application due to apathy, insatiability, desire, or possibly different reasons. On account of this little episode that is not refreshing on time may lead your organization into trouble.

Table 5 is defining association that has its standards and guidelines, and they are making their approach and plans. They are tenderizing changes like a half year, short plans, and long-term plans from 1 to 3 years. Due to Coronavirus, those strategies are presently practically pointless in light of the fact that their arrangements were for the whole organization, not for singular substance. This record will help you correlate the current alleviation methods and proposed solutions, too [53].

Table 6 defines the large acknowledged that moving from on-site to home office got some effectiveness and limitations by working from home. Like a company

Table 4 Potential mitigation of each identified risk

Tools	Associated risks	Mitigation of risks
Microsoft Azure	CVE-2019–1372 which can lead an attacker to run remote code execution	Update to the latest available version to mitigate the risk [41]
MS Office 365	CVE-2019–1200 found in version 2019	Vulnerability can exploit the user with remote code execution [45]
Microsoft Visual Studio 2019	Found critical vulnerability in MS Visual Studio 2019 version 16.0	Need to update to MS Visual Studio 2019 version 16.4 [42]
SQL Server 2017	Found critical vulnerability in MS SQL Server 2017 version 01	Need to update MS SQL Server 2017 to the latest version [43]
MS Teams	CVE-2019-5922 attacker can gain privileges via a Trojan horse DLL	Update to the latest platform of the MS Teams to mitigate the risk [46]
MS Visio	CVE-2016-3364—Microsoft Office Memory Corruption Vulnerability	Update the MS Visio to the latest version and implement all configuration [47]
MS Outlook	Insecure protocols being used, IMAP, POP3	Use the latest version of MS Office 365 along with the VPN and PGP keys to maximize the privacy and security [48]
Redmine 3.4.3	CVE-2017-18026 which can lead an attacker to execute XSS	Update to the latest version 4.1.1 to mitigate the risk [44]
MailChimp	CVE-2014-7152 which could lead the attacker to execute XSS attack	Update the MailChimp to latest available version to mitigate the risk [49]
Android Studio	Organization have been using an older version of the Android Studio 3.6 and below	It is recommended to update to the latest version available 4.1 to avoid the risks [50]
Online attendance system	Having some critical bugs in the login panel	Need to upgrade the version [51]

needs to get the event logs, they must have employee browsing history if an incident occurred. Another risk is implementing information technology policies that cannot be easily implemented at their home office. These all variables are discussed in the above table.

10 Conclusion

This chapter covers a pandemic situation, and due to this situation, many challenges and threats have also occurred for organizations where they have been shifted from

Table 5 Comparison of proposed solution with the existing mitigation techniques [52]

Controls	Existing mitigation techniques	Proposed solution
Information security policy	Organization needs polish their policy for the IT operation, tools and techniques to avoid exploitation and vulnerabilities	These vulnerabilities can be mitigated through the implementation of the information security
Access control policy	Suddenly conversion from on-site to online platform may experience different type of risk problems	Due to the work from home environment the policy needs to implement at all levels to address the access controls of the employees and the executives
Information backup	Information backups are being organized and properly maintained but the data retention and destruction policy is not available	Data retention and destruction policy should be implemented so that the vulnerabilities in the operational prospective could be mitigated
Security of network services	For remote access organization has not implemented secure network policy which needs to be implemented timely	Communication between the users and the system needs to be secure through the secure network protocols and the tunneling mechanisms
Electronic messaging	Organization has strict policy against the online social media platforms for the communication and coordination	WhatsApp and Signal has capability of encrypted communication which will improve the communication and coordination
Outsourced development	Organization's policy against the outsourcing is not implemented, which has many operational and administration flaws	The organization needs to make a proper policy against the outsourcing of the project to address the confidentiality of the projects e.g., using the secure drive

office environment towards the home environment with several types of security threats and risk. We have Identified all the tools and techniques used by the organization's employees and associated risks to those tools and techniques that can lead to significant incidents for the organization. We have reached this point by identifying risks that which types of flaws/weaknesses are available in the current tools and techniques. This study will address best practices and solutions for the organization, which can help address the current identified flaws or vulnerabilities at all levels. The available vulnerabilities can be at the policy level. It can be at the management level, operations level, outdated technology level. This will also address that we need to update our tools if there is another alternative solution instead of one currently used by employees. The proposed solution addresses the overall methodology of identifying, protecting, detecting, responding, and recovering from cyber incidents and, if there are somehow residual risks are left. The chapter will also help you remove

Table 6 Effectiveness and limitations of the existing mitigation controls [52]

Existing controls	Effectiveness and limitations
Information technology policy	Organization could be affected due to not implementing the information security policy therefore the organization has a lot of technical and operational vulnerabilities
Access control policy	Due to the work from home environment, there are a lot of risks has been identified which can lead to the overall corruption of the organization's operations
Information backup	The data retention and destruction policy should be improved so that the important data could be retained for longer period
Event logging	An effective logging policy needs to be implemented so that the incident response can be done effectively
Electronic messaging	Organization has strict policy against the online social media platforms for the communication and coordination
Outsourced development	Organization's policy against the outsourcing is not implemented, which has many operational and administration flaws

the known risks, how you will respond to the incidents, and how to mitigate those risks. This chapter also recommends shifting to online platforms and acquiring those resources and services that can be available 24/7, like the zero trust model (ZTM) and secure access service edge (SASE), which is highly demanded during this pandemic.

References

1. A. Purwanto et al., Impact of work from home (WFH) on Indonesian teachers performance during the covid-19 pandemic: an exploratory study. Int. J. Adv. Sci. Technol. **29**(5), 6235–6244 (2020)
2. A. Kramer, K.Z. Kramer, The potential impact of the Covid-19 pandemic on occupational status, work from home, and occupational mobility. J. Vocat. Behav. **119**, 103442 (2020). https://doi.org/10.1016/j.jvb.2020.103442
3. E. Brynjolfsson, J.J. Horton, A. Ozimek, D. Rock, G. Sharma, H. TuYe, Covid-19 and remote work: an early look at us data, in *Clim. Chang. 2013—Phys. Sci. Basis*, vol. 2220 (2020), pp. 1–30. https://www.nber.org/system/files/working_papers/w27344/w27344.pdf
4. T. Weil, S. Murugesan, IT risk and resilience-cybersecurity response to COVID-19. IT Prof. **22**(3), 4–10 (2020). https://doi.org/10.1109/MITP.2020.2988330
5. H. Wijayanto, I.A. Prabowo, Cybersecurity vulnerability behavior scale in college during the Covid-19 pandemic. J. Sisfokom (Sistem Inf. dan Komputer) **9**(3), 395–399 (2020). https://doi.org/10.32736/sisfokom.v9i3.1021
6. U.K National Cyber Security Centre, 10 steps to cyber security, in *U.K Natl. Cyber Secur. Cent.* (2016), pp. 1–16. https://www.ncsc.gov.uk/collection/10-steps-to-cyber-security/introduction-to-cyber-security/executive-summary%0Ahttp://www.smartprotect.eu/resources/report1.pdf
7. M. Kartikay, Dark web has become a marketplace for "Vaccine" and other pandemic scams, https://www.bloomberg.com/news/articles/2020-11-11/dark-web-has-become-a-marketplace-for-vaccines-and-other-pandemic-scams. Accessed 11 Nov 2020

8. S. Hakak, W.Z. Khan, M. Imran, K.K.R. Choo, M. Shoaib, Have you been a victim of COVID-19-related cyber incidents? Survey, taxonomy, and mitigation strategies. IEEE Access **8**, 124134–124144 (2020). https://doi.org/10.1109/ACCESS.2020.3006172
9. R. For, D. An, E. Security, Roadmap for deploying an enterprise security model
10. D.E. Sanger, N. Perlroth, M. Rosenberg (ed.), *Hackers Attack Health and Human Services Computer System*, https://www.nytimes.com/2020/03/16/us/politics/coronavirus-cyber.html. Accessed 16 Mar 2020
11. Ryan Gallagher and Bloomberg, Hackers 'without conscience' demand ransom from dozens of hospitals and labs working on coronavirus. GMT+5, https://fortune.com/2020/04/01/hackers-ransomware-hospitals-labs-coronavirus/. Accessed 1 April 2020
12. M.K. Pratt, *What is Zero Trust? A Model for More Effective Security*, https://www.csoonline.com/article/3247848/what-is-zero-trust-a-model-for-more-effective-security.html
13. G. Leonard, No title, in *IDS vs IPS vs UTM—What's the Difference?*https://cybersecurity.att.com/blogs/security-essentials/ids-ips-and-utm-whats-the-difference
14. C. Decusatis, P. Liengtiraphan, A. Sager, M. Pinelli, Implementing zero trust cloud networks with transport access control and first packet authentication, in *Proc.—2016 IEEE Int. Conf. Smart Cloud, SmartCloud 2016* (2016), pp. 5–10. https://doi.org/10.1109/SmartCloud.2016.22
15. F.A. Barbhuiya, T. Saikia, S. Nandi, An anomaly based approach for HID attack detection using keystroke dynamics, in *Lect. Notes Comput. Sci. (including Subser. Lect. Notes Artif. Intell. Lect. Notes Bioinformatics)*, vol. 7672 LNCS (2012), pp. 139–152. https://doi.org/10.1007/978-3-642-35362-8_12
16. A. Tabona, *The top 20 free Network Monitoring and Analysis Tools for sysadmins* (2018), https://techtalk.gfi.com/the-top-20-free-network-monitoring-and-analysis-tools-for-sys-admins/
17. N. Surantha, F. Ivan, *Secure Kubernetes Networking Design Based on Zero Trust Model: A Case Study of Financial Service Enterprise in Indonesia*, vol. 994 (Springer International Publishing, Berlin, 2020)
18. W.-J. Herckenrath, *Secure Access Service Edge (SASE)* (2019), https://www.catonetworks.com/sase/
19. A. Kaur, V.P. Singh, S.S. Gill, The future of cloud computing: Opportunities, challenges and research trends, in *Proc. Int. Conf. I-SMAC (IoT Soc. Mobile, Anal. Cloud), I-SMAC 2018* (2019), pp. 213–219. https://doi.org/10.1109/I-SMAC.2018.8653731
20. *What does CDN stand for? CDN Definition*, https://www.akamai.com/us/en/cdn/what-is-a-cdn.jsp
21. C. Lawson, V.R. Analyst, S.D. Analyst, No title, in *Magic Quadrant for Cloud Access Security Brokers*, https://www.gartner.com/doc/reprints?id=1-24H6BNY4&ct=201028&st=sb
22. R. Srinivasan, *SASE Will Redefine Network and Cloud Security: So What Does it Mean?*https://www.forcepoint.com/zh-hant/blog/insights/forcepoint-gartner-sase-converging-network-cloud-security. Accessed 02 Oct 2020
23. A. Masood, J. Java, Static analysis for web service security—Tools & techniques for a secure development life cycle, in *2015 IEEE Int. Symp. Technol. Homel. Secur. HST 2015* (2015), pp. 1–6. https://doi.org/10.1109/THS.2015.7225337
24. T. Bush, *What Is The Difference Between Web Services and APIs?*https://nordicapis.com/what-is-the-difference-between-web-services-and-apis/#:~:text=There-you-have-it%3A-an,all-APIs-are-web-services
25. Z. Zhou, T. Song, Y. Jia, A high-performance URL lookup engine for URL filtering systems, in *IEEE Int. Conf. Commun., no. 60803002* (2010). https://doi.org/10.1109/ICC.2010.5501982
26. J. Mitchell, Reported web vulnerabilities "In the Wild" data from aggregator and validator of NVD-reported vulnerabilities (2010)
27. A.P.P. de Oliveira Cardoso, SP 800-114 Revision 1, User's Guide to Telework and Bring Your Own Device (BYOD) Security, in *Inovar com a investigação-ação, Inovar com a Investig.* (2014). https://doi.org/10.14195/978-989-26-0666-8.NIST
28. P. Paganini, *Introduction to the NIST Cyber Security Framework for a Landscape of Cyber Menaces*, https://securityaffairs.co/wordpress/58163/laws-and-regulations/nist-cybersecurity-framework-2.html. Accessed 20 April 2017

29. M.P. Souppaya, K.A. Scarfone, NIST: 800-114: User's guide to telework and bring your own device (BYOD) security, in *NIST Spec. Publ.*, vol. 800 (2016), p. 114, https://nvlpubs.nist.gov/nistpubs/SpecialPublications/NIST.SP.800-114r1.pdf

30. E. Barker, Q. Dang, S. Frankel, K. Scarfone, P. Wouters, Guide to IPsec VPNs, in *Draft NIST Spec. Publ. 800-77 r1* (2019), p. 161, https://csrc.nist.gov/publications/detail/sp/800-77/rev-1/draft

31. D. Patten, The Evolution to Fileless Malware (2017), p. 13, https://infosecwriters.com/Papers/DPatten_Fileless.pdf

32. J. Seefeldt, What's New in NIST Zero Trust Architecture, 217–232 (2012). https://doi.org/10.1007/978-3-642-35755-8_16

33. M. Souppaya, K. Scarfone, Guidelines for managing the security of mobile devices in the enterprise, in *NIST Spec. Publ. 800-124, Revis. 1* (2013), pp. 1–30, http://nvlpubs.nist.gov/nistpubs/SpecialPublications/NIST.SP.800-124r1.pdf%5Cnpapers3://publication/doi/10.6028/NIST.SP.800-124r1

34. E. Gabber, P.B. Gibbons, Y. Matias, A. Mayer, How to make personalized web browsing simple, secure, and anonymous, in *Lect. Notes Comput. Sci. (including Subser. Lect. Notes Artif. Intell. Lect. Notes Bioinformatics)*, vol. 1318 (2015), pp. 17–31. https://doi.org/10.1007/3-540-63594-7_64

35. K. Scarfone, P. Hoffman, Guidelines on firewalls and firewall policy recommendations of the national institute of standards and technology, in *Nist Spec. Publ*

36. J. Franklin, et al., NIST Special Publication 1800-4—Mobile device security, in *NIST Spec. Publ.* (2019), https://nvlpubs.nist.gov/nistpubs/SpecialPublications/NIST.SP.1800-4.pdf

37. R. Shustin, Azure remote code execution vulnerability, https://portal.msrc.microsoft.com/en-US/security-guidance/advisory/CVE-2019-1372. Accessed 10 Aug 2019

38. Qi An Xin and Gartner, Zero trust architecture and solutions. (1), 1–21 (2020), https://nvlpubs.nist.gov/nistpubs/SpecialPublications/NIST.SP.800-207-draft.pdf, https://www.gartner.com/teamsiteanalytics/servePDF?g=/imagesrv/media-products/pdf/Qi-An-Xin/Qi-An-Xin-1-1OKONUN2.pdf

39. B. Chen et al., A security awareness and protection system for 5g smart healthcare based on zero-trust architecture. IEEE Internet Things J. **8**(13), 10248–10263 (2021). https://doi.org/10.1109/JIOT.2020.3041042

40. E.R. Onainor, SASE challenges include network security roles, product choice. J. Surg. CI Res. **5**(1), 47–55 (2014), https://www.techtarget.com/searchnetworking/tip/SASE-challenges-include-network-security-roles-product-choice

41. Aka, Microsoft outlook remote code execution vulnerabilty, https://www.cvedetails.com/cve/CVE-2019-1200/. Accessed 14 Aug 2019

42. A buffer overflow vulnerability exists in the Microsoft SQL Server that could allow remote code execution on an affected system. https://msrc.microsoft.com/update-guide/vulnerability/CVE-2018-8273

43. Microsoft Teams allows an attacker to gain privileges via a Trojan horse DLL in an unspecified directory, https://cve.mitre.org/cgi-bin/cvename.cgi?name=CVE-2019-5922. Accessed 10 Jan 2019

44. Allow remote attackers to inject arbitrary web script or HTML via the update_options action to wp-admin/admin-ajax.php, https://www.cvedetails.com/cve/CVE-2014-7152/. Accessed 30 Sept 2014

45. Diagnostic Hub Standard Collector, Visual studio standard collector elevation of privilege vulnerabilty, https://www.cvedetails.com/cve/CVE-2019-0727/. Accessed 16 May 2019

46. Microsoft Office Memory Corruption Vulnerabilty, https://www.cvedetails.com/cve/CVE-2016-3364/. Accessed 14 Sept 2016

47. S. Ruoti, K. Seamons, Johnny's journey toward usable secure email. IEEE Secur. Priv. **17**(6), 72–76 (2019). https://doi.org/10.1109/MSEC.2019.2933683

48. Allow remote attacker to execute arbitrary commands through Mercurial adapter, https://www.cvedetails.com/cve/CVE-2017-18026/. Accessed 10 Jan 2018

49. This critical android security threat could affect more than 1 billion devices: What you need to know, Davey Winder Senior Contributor, https://www.forbes.com/sites/daveywinder/2020/05/26/critical-android-data-stealing-security-threat-confirmed-for-almost-all-android-versions-strandhogg-google-update-warning/?sh=6663328a2c82. Accessed 26 May 2020

50. L. Kamelia, E.A.D. Hamidi, W. Darmalaksana, A. Nugraha, Real-time online attendance system based on fingerprint and GPS in the smartphone, in *Proceeding 2018 4th Int. Conf. Wirel. Telemat. ICWT 2018* (2018), pp. 1–4. https://doi.org/10.1109/ICWT.2018.8527837

51. M. Mylrea, S.N.G. Gourisetti, A. Nicholls, An introduction to buildings cybersecurity framework, in *2017 IEEE Symp. Ser. Comput. Intell. SSCI 2017—Proc.*, vol. 2018 (2018), pp. 1–7. https://doi.org/10.1109/SSCI.2017.8285228

52. K. Zhang, Y. Mao, S. Leng, S. Maharjan, Y. Zhang, Optimal delay constrained offloading for vehicular edge computing networks, in *IEEE Int. Conf. Commun.* (2017), pp. 1–6. https://doi.org/10.1109/ICC.2017.7997360

53. S. Writer, *Pulse Connect ushers PERSOL towards Zero Trust architecture*, https://www.frontier-enterprise.com/pulse-connect-ushers-persol-towards-zero-trust-architecture/. Accessed 02 Oct 2020

54. A. Calder, NIST cybersecurity framework—Aligning to the NIST CSF in the AWS cloud, in *NIST Cybersecurity Framew* (2019), https://d1.awsstatic.com/whitepapers/compliance/NIST_Cybersecurity_Framework_CSF.pdf

A Systemic Security and Privacy Review: Attacks and Prevention Mechanisms Over IoT Layers

Muhammad Ayub Sabir, Ahthasham Sajid⬤, and Fatima Ashraf

Abstract In this contemporary era internet of things are used in every realm of life. Recent software's (e.g., vehicle networking, smart grid, and wearable) are established in result of its use: furthermore, as development, consolidation, and revolution of varied ancient areas (e.g., medical and automotive). The number of devices connected in conjunction with the ad-hoc nature of the system any exacerbates the case. Therefore, security and privacy has emerged as a big challenge for the IoT. This paper provides an outline of IoT security attacks on Three-Layer Architecture: Three-layer such as application layer, network layer, perception layer/physical layer and attacks that are associated with these layers will be discussed. Moreover, this paper will provide some possible solution mechanisms for such attacks. The aim is to produce a radical survey associated with the privacy and security challenges of the IoT. The objective of this paper is to rendering possible solution for various attacks on different layers of IoT architecture. It also presents comparison based on reviewing multiple solutions and defines the best one solution for a specific attack on particular layer.

Keywords Internet of things · Security and privacy · IoT layers · Attacks with solution mechanism

1 Introduction

Internet of things has a combination of various devices; the things which are connected through the internet are sensors and actuators. Our world today contains billions of sensors and electronic devices that constantly track, store, compile and analyze vast amounts of personal information. This information may include our

M. A. Sabir · F. Ashraf
University College of Management and Sciences, Khanewal, Pakistan

A. Sajid (✉)
Department of Computer Science, Faculty of ICT, BUITMS, Baluchistan, QUETTA, Pakistan
e-mail: ahthasham.Sajid@buitms.edu.pk

© The Author(s), under exclusive license to Springer Nature Switzerland AG 2022
M. Ouaissa et al. (eds.), *Big Data Analytics and Computational Intelligence for Cybersecurity*, Studies in Big Data 111,
https://doi.org/10.1007/978-3-031-05752-6_5

position, Browsing habits on the contact list, and details on health and fitness. The Internet of Things plays a major part in the day-to-day life of every individual. It's a service that allows transmissions from person to object or object to object. As IoT is the incorporation of frequent heterogeneous networks so, to establish reliable connection among nodes is challenging. The IoT comprises of sensors, smart devices, networks, cloud computing, all connected by common standards. Every band put susceptibilities and threats to security. Discouragingly, the accessibility move towards the outlay of security and privacy risks that private, personify data can pose extensive harm to our property, reliability and personal security if it is uncovered by unauthorized agent. Such tools also include assets which are provided by their suppliers at different stages with their supply chain of development, in addition to our personalized data. Such modes include fuses, firmware, and debugging. Million dollars of stolen intellectual property will be lost through unauthorized access also it can cause misuse of these resources. Such security vulnerabilities can be disastrous with the widespread implementation of these apps. IoT applications are used in many fields as illustrated in Fig. 1 such as environmental monitoring, home automation, transportation, health care and medical systems, and so on. In this regard, data ensuring and safety of privacy is the largest challenge. There are some main technologies of IoT which includes RFID technology, Sensor technology, embedded system technology, and Nanotechnology. Therefore, construction technology is posing main threat.

Fig. 1 Importance of IoT [1]

1.1 IoT Security Challenges

Due to the extensive consequences on daily life, each and every IoT appliances are susceptible for challenges of security and privacy such as authenticity, confidentiality and integrity tec. Security problems are split up into various classifications, like confidentiality of data, observing and tracing down the activities, sidestep the malicious insiders, hijacking of services or procedures, phishing, fraudulent activities and exploitation etc. [2, 3]. Information Technology's (IT) security reflects [4–6] reflects three features which includes availability, integrity and confidentiality. These are called as CIA triad, and these are also the key purposes [7]. Data security refers to the confidentiality, although integrity is not changing the data while it is transmitting [8, 9]. Additionally, whenever it is essential to transmit data smoothly, the availability will be helpful. Among multiple security difficulties, the necessary challenges are:

- Data Privacy and Security: Data have to be secured, and hidden so that the data will be safe from stealing and from the hackers, while transmitting data seamlessly.
- Insurance Concerns: Many companies are setting up IoT devices which are helping to gather data to make decision about insurance.
- Technical Concern: Now a day's excessive use of IoT devices generates lot of traffic, so it requires large network and storage capacity.
- Need of common Standard: Although, there are numerous standards for IoT devices but the most backbreaking factor is to connect the authorized and un-allowed devices.

In IoT, whole security is dependent on Three-layer architecture. If physical layer is in control then it will be easy to get access and attack to other nearby devices with the help of original node. On the other side of spectrum, devices that have online ability can be easily hacked, if device has no virus protection then these devices are used as "bots" to deliver malicious code [6]. Resent survey of HP shows that 70% of IoT devices are not secure. Therefore, one of the major concerns regarding IoT devices are security and privacy because these two things show that communication between devices and internet is reliable. So this paper presents some threats, attacks and vulnerabilities that are associated with three-layer architecture and provide possible solution regarding these issues. The main contribution of this paper as follows:

1. This paper presents security survey of IoT layered architecture.
2. Review of many IoT security approaches, to tackle security issues and challenges of individual layer.
3. Complete analysis of the structure and working flow of these techniques and solutions along with their pros and cons.
4. Comparison is made through table that will give proper direction, which technique is best solution of every layer's security and how it is different from others as a result.

1.2 Key Role of Layers in IoT

The following section presents the introduction of Three-Layer architecture. Application layer consists multiple application such as smart energy saving, smart home and smart cities. Network layer is integrated with communication and transferring data accurately which is collected through sensor nodes. Conversely, last tier is physical which contain various hardware technologies like RFID devices, sensor etc.

1.3 Physical Layer/Perception Layer

In three-layer architecture of IoT, the Physical layer comes in the top bottom layers. Identifying devices provide connectivity and provide service discovery are main responsibilities of physical layer [3]. In IoT manners, Wi-Fi, Bluetooth, ZigBee are physical devices that are used to connected two devices through internet either directly or indirectly. These are used in IoT because of low energy consumption and low power for connectivity furthermore, every device has a tag which has unique identity to connect with other devices. Perception layer is a part of physical layer but there is a bit different between them that perception layer sense data through sensors and collect information to environment. It has many sensors like temperature sensor, vibration sensor, RFID that allow devices to sense other objects.

1.4 Network Layer

As like physical layer, network layer is responsible for communication and connectivity through different communicational protocols: there is no doubt that IoT are having not such protocols and standards but some like MQTT and CoAP are being used in IoT. The main goal of network layer is to transfer data in between devices [8]. With the help of wireless sensor physical layer send information or data to network layer then it transmit to any particular processing system.

1.5 Application Layer

Application layer consist the services that are provide through IoT. Application layer is known as service-oriented. It store information or data in his database and retrieve information when user need it for example Smart home, healthcare and smart cities as illustrated in Fig. 2 below.

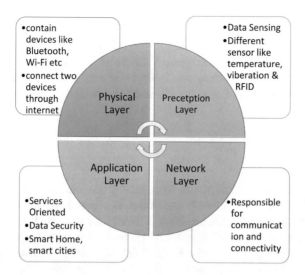

Fig. 2 Four-layer architecture for IoT

2 Attacks and Their Countermeasures on Layer Architecture

2.1 Network Layer

Network layer is liable for the property of the internet of things infrastructure [4, 5, 10]. This layer conjointly transfers information to the next upper layer by gathering information from perception layer. For communication the wired or wireless medium is used. Moreover, the foremost technologies which are used are ZigBee, Wi-Fi, Bluetooth, 3G, and so and so forth [5, 6]. Network layer has several attacks, usually moving organization of labor and data sharing between devices [6]. Just like the other Network Layer model this one includes network boundaries, communication channels, and network. Controlling, data maintenance, and intellectual process, and are particularly to blame for the communication and property of all the devices in IoT system through the support of multiple communication protocols. The foremost common protocols that are presently being employed are MQTT three [5] and also the affected Application Protocol (CoAP) [6]. It's at recesses this layer that the congregated data from the Physical Layer are transmitted to any specific information science system at intervals the network misuse Wireless Sensors [6] or to an out of doors network over existing communication structures just like the net or a Mobile Network.

2.2   Network Layer Attacks

Figure 3 below illustrate different types of attacks which could be done over network layer in IOT Applications. Moreover, Table 1 will render the summary of different studies regarding Network Layer Attacks and provide the best prevention technique as well.

2.2.1   Sybil Attack

An attack in which an isolated attacker can literally take our network, but the attacker faking it to be bunches of other nodes, it happens in peer to peer network. It is very pragmatic attack because attacker sends a lot of fake requests to network and the network does not really know either request is original or fake. Single user pretends many fake or Sybil (create various account from different IP addresses) identities. In this regards attacker can completely take over the peer network Sybil attack could affect the performance, resource utilization and data integrity [21].

2.2.2   Countermeasure

Solution for this problem is a trending research area in this decade. Many researcher are doing working on it such as, improving the security nodes in smart grids Najafabadi et al. [5] has proposed a technique. Similarly, Demirbas et al. [11] have a technique for detection, through RSSI (Received Signal Strength Indicator) value that can get from neighboring nodes. Douceur's approach is also very popular that used as a solution of Sybil attack. The above mentioned mechanisms are most expensive for resource constrained devices in respect of storage and processing. As compare to these approaches this paper suggest the best solution for Sybil attack is "behavioral profiling" [21] as mentioned in Table 1. With the help of network parameters they develop a behavioral profile for every node. As a result of these profiling nodes that have less trusted values are rejects, through reliable nodes, packets are routed again.

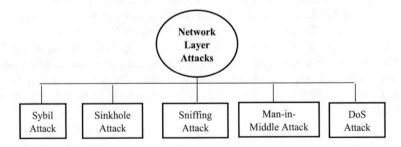

Fig. 3 Types of network layer attacks

Table 1 Summary of network layer attacks and its prevention

Attacks name	Challenges	Research comparison	Best prevention technique
Sybil attack	Creating and sending lot of fake requests	Demirbas et al. [11]: Sybil detection technique by using RSSI Sharmila and Umamaeshwari [12]: energy and hop count based detection Najafabadi et al. [5]: detection for smart grid by traffic	Behavior profiling "Develop a behavioral profile for every node"
Sinkhole attack	Select specific node for attack and reroute data toward it	Krontiries et al. 1271: rule based approach "creating two rules and implemented in IDS" Roy et al. [13]: dynamic trust management system	For sinkhole detection using EM (Extra monitoring node) and RSSI (Received single strength indicator) technique
Sniffing attack	Intercepting traffic between two hosts	Dubey et al. [14]: propose an efficient partition technique for web based files Op, html, php), text (word, text files) and PDF files Qadri and Pandey [15]: tag based client side detection	Prevention through log file: log have content of data separately by frequency, space and position as well
Man in the middle	Information modification between two parities without their knowledge	Aziz and Hamilton [16] MitM detection system that uses arrival time of packets to infer the possibility of MitM Liu et al. [17] use computing resources known as Registration authority (RA), who is responsible authenticating	Encrypt communication with either symmetric or asymmetric algorithms e.g. AES, DES, RSA, Diffie-Hellman, ECC, etc.
Denial of service (DOS)	Sending huge non-relevant request for making server unavailable for particular user	Agah et al. [18] game theoretic approach Mohi et al. [19] A Bayesian game approach Zhang et al. [20] Message observation mechanism (MoM)	Ingress/egress technique: ingress filtering use to filter the malicious traffic and egress filtering use to discard these malicious traffic to local network

2.3 Sinkhole Attack

In sinkhole attack the attacker compromise with one network node and treat it as an attacker furthermore, it is totally upon routing matrix (that protocol mostly uses) this attacker node send fake information and do its best to fascinate whole traffic of other neighboring meeting points (called nodes), this happens in wireless sensor network [22]. After managing traffic, it launches a particular attack then all the data packets will pass it before going to base station.

2.3.1 Countermeasure

There are many traditional approaches to overcome this problem but these approaches are categorized such as key management based, rule-based and anomaly based. In Papadimitriou et al. [23] where RESIST protocol has been invented the key management approach is used there. Sharmila and Umamaheswari [24] proposed a message digest algorithm for anomaly based sinkhole detection furthermore, Krontiris et al. [25] proposed detection rules for rule based detection. The above mentioned techniques are not suitable for IoT low end devices. In contrast of these above EM and RSSI technique is best one solution. To detect sinkhole attack Extra Monitor node (EM) and Received signal strength indicator (RSSI) are being used. The basic functions of EM are: provide high communication range along with calculating the RSSI of node it send to base station with next hop and source ID. Visual graphical map (VGM) is also used in this process; RSSI values are used by base station to measure VGM. The position of node can be indicated through this VGM. After this identification when base station receiver another changed value from EM, because EM send updated RSSI value then base station identify that packet flow are not matched to previous one so it indicate there is a sinkhole attack [22].

2.4 Sniffing Attack

It a generic attack, sniffing is basically intercepting traffic between two hosts which has got verity based on mode of attacks. It is usually used to control the traffic and snatch confidential data. LAN network attack is harmful and easy to do. Malicious user can easily steel confidential information because of network traffic sniffing. Two types of sniffing are passive sniffing and active sniffing. Passive sniffing is possible in 'hub' cases while we use hub devices (hub use to broadcast data or message), attack or hacker no need to put extra effort to get traffic from network. If you get access to hub then sniffing performance is so easy. Conversely, active sniffing are used in switch cases, in switches if one computer intend to communicate to other then it will not broadcast the data it will directly connected to 2nd computer. If attacker wants to listen that traffic, then attack will make his land card in promiscuous mode.

2.4.1 Counter Steps

One of the most appropriate techniques is proposed in Namrata Shukla et al. [2] for revealing this attack. According to this method a log file has been created which have content of data separately by frequency, space and position as well. On the other hand, by using the substitution method they encrypt the data then pass to receiver side along with, they send the log file that is organized in a way of cryptic message as well. It is summarize in Table 1. Conversely, recipient matches data conforming to frequency, position or message, for a sense if it does not match. The error can be observed by them, and they will never approve that file. Several other mechanisms have been proposed like Tag based client side detection [15], server side content sniffing attack prevention technique [26].

2.5 Man-in-Middle Attack

In environment of server client, Man-in-middle attack is generally perceived. A game called cup telephonic is used to understand man-in-middle attack, in which two friends can talk to each other through cup in a string. While conversation a third unknown person come and cut the string, put two cups in between string and start listing conversation. This third person intercept the network and possible to alter the communication message. Same happens in this attack, an attacker insert himself in a network or start interrupting the conversation or steel personal information. This attack can be possible where sending and receiving process occurs.

2.5.1 Countermeasures

According to above statements, one major factor is involved in this attack which is authentication. If we have a command in this manner we can resolve this problem. Multiple techniques such as DNS spoofing, Gateway spoofing, DHCP spoofing, IP address spoofing, stealing of port are used in man-in-middle attack. Majority of the mechanisms that are used against this attack are authentication mechanisms like secret question, 2-step authentication, voice recognition, biometric, public key infrastructure etc. Many other modern mechanisms are involved such as: one time password authentication (time synchronized), MitM detection system [16], multifactor authentication to protect login and software toolbar etc. Furthermore one of the best countermeasures is communication with symmetric or asymmetric algorithms such as AES, DES.

2.6 DOS Attack

Denial of service attack is one of the perilous attacks for user in which attacker make device or computer unavailable for it user by sending multiple request, sending wrong password furthermore, by disconnecting host, it degrades the network performance. DOS is used to get unauthorized access along with it help to control the system. Many attacks are including in DOS such as: TCP SYN attack, HTTP flood attack, SIP flood attack, UDP flood attack, slow request/response attack and bandwidth depletion attack.

2.6.1 Countermeasures

Ingress/egress is one of the best approaches which can overcome DOS attack. With the help of spoofed IPs, these techniques prevent traffic. Ingress filtering use to filter the malicious traffic and egress filtering use to discard this malicious traffic to local network. It work like if a user enter network, ingress allow traffic to enter in network (which match with predefined range of domain prefix) furthermore, if user use IP spoofing (which does not match with defined prefix) then egress discard this in the routers [27]. Various other mechanisms are used to prevent this attack like Route based packet filtering, Game Theore tic Approach [18], An Ant Based Framework, Protection using KDS [28], Hop count filtering, path identifier, history based filtering, honey pots and load balancing etc. The above techniques depends on nod's malicious behavior, routing algorithms, parameter analysis and what parameter has considered for detection of attacks so, due to that factor Ingress/egress is a best one technique. Table 1 will more justify these above studies.

2.7 Perception Layer

Perception layer is one of the bottom layers which are used for the purpose of information or data collection from WSN, heterogeneous devices; sensors type real world objects, humidity, and temperature etc. Single purpose of perception layer is to identify address. Figure 4 below illustrates various types of attacks possible on this layer. Table 2 will explain comparative studies and provide the best prevention technique against these following attacks.

2.7.1 RF Jamming

One of the basic attacks in perception layer in which attacker send Radio signals to create interfere between readers and legitimate tags. RFID tags are used to create

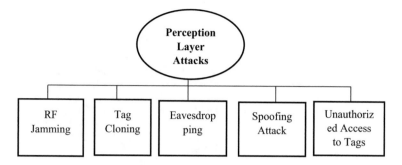

Fig. 4 Perception layer attacks

interruption between communications and prevent radar to communicate all with tags within its range [7].

Countermeasures

Solution of RF jamming is dependent on the type of jammer either reactive or proactive. Many researcher are still working on wireless anti-jamming techniques [36] such as Signal strength technique, Regulated Transmitted Power, using of LDPC (Low density Parity Check) codes [24], Ultra-Wide Band Technology and Using Reed-Solomon codes with respect to 802.11b techniques has been proposed for jamming detection. The above mentioned are working in various direction like some are focusing on the designing of physical layer, some are on sensor network and some are on MAC layer. On the other hand Frequency-Hopping Spread Spectrum is one of the best techniques for RF jamming in which radio signal spread through different switching among many frequency channels.

2.7.2 Tag Cloning

Cloned tag is a duplicate copy of original one. The middle man mean the attacker build a tag same like original when the reader read tag it is impossible to him to differentiate between them. With the help of this tag the attacker can sense data, captured information and modify it as he intend.

Countermeasures

There are two basic approaches prevention and detection is used to avoid tag cloning. Having encryption and cryptograph technology, prevention method provides security against tag cloning but this technique is not effected for low cost tags [10]. Conversely, detection is used in low cost tags to handle clone tag issues. Several other methods

Table 2 Summary of perception layer attacks and its prevention

Attacks name	Challenges	Research comparison	Best prevention technique
RF-jamming	Hurdles to exchange data through frequency jamming	Wood et al. [29]: this study posed the issue of jamming detection in the loose context of the utility of the communication channel Proakis [30] for jamming resistant focuses on the design of physical layer technologies, such as spread spec turn	Frequency-Hopping Spread Spectrum technique: "radio signal spread through different switching among many frequency channels"
Tag cloning	Head off data flow between tags	Bu et al. [31] GREAT using Aloha-based anti-collision protocol to find irreconcilable collisions Bu et al. [32] BASE: ID cardinality and tag cardinality Bu et al. [32] DeClone using a hybrid design of slooted aloha and tree traversal to determine collisions	EPC (Electronic product code) technique: in which a specific identity has given to a particular physical object
Eavesdropping	Interrupted data during; transmission over HTTP	Nguyen et al. [4]: a hybrid prevention method for eavesdropping attack Li et al. [33]: a framework that taking account into various channel conditions and antenna models	Anonymous forward-secure mutual authentication protocols (AFMAP)
Spoofing attack	Fake IP packet which behave like an original one and broadcast it	Herzberg et al. [34] proposed the concept of trusted credential area (TCA) of the browser window Felten et al. [35] an attacker stays between the client and the target site such that all web pages destined to the user's machine are routed toward the attacker's server	Access control list (ACLs): system or resources are granted to operate Secure Scout Layer (SSL): for secure transmission

such as: GREAT, BASE and DeCloneare used to overcome tag cloning. But as we have discussed in Table 2, clone detection through EPC is a best approach to prevent tag cloning [31]. In EPC (electronic product code) a specific identity has given to a particular physical object, which may be used to track all king of object. So the attacker cannot build that specific identity and will not be able to steal your data.

2.7.3 Eavesdropping

Eavesdropping is one of the perilous attack in perception layer in which the attacker are sitting somewhere in network and can steal information or traffic, through peace of software. The attacker insert a piece of software in the form of malware or other method to compromised device later on he can easily retrieve data through this piece of software. Due to wireless era, it become very easy to get personal information like password through Eavesdropping.

Countermeasures

Several RFID private authentication protocols are used to overcome eavesdropping problems. Network division is one of the best approaches for revels eavesdropping. Before establishing connection, every device should be verified so network access control mechanisms are used in this manner. Network segmentation, network monitoring, security technologies like firewall, VPNs and anti-malwares are most important. On the other hand Anonymous Forward-Secure Mutual Authentication Protocols (AFMAP) and RWP authentication purpose are also effective for this problem [4].

2.7.4 Spoofing Attack

In spoofing, attackers create a fake IP packet which behave like an original one and broadcast in RFID system. It shows like authentic source and create security hole in system furthermore, attacker can get access to network and send malicious data, wrong information to system through this hole.

Countermeasures

Several researchers are working on spoofing attack. With the aid of filtering mechanisms (which can filter the outgoing and incoming packets) the spoofing attacks could be defended moreover, Trusted Credential Area (TCA) technique [34], web-spoofing attack, patch method [37] are used. But these approaches have ambiguities: TCA is much costly and web spoofing is not well for hand-held devices as well. So, one of the best approaches is access control list (ACLs) and Secure Scout Layer

(SSL) authentication mechanisms that are used to decrease the risk of spoofing attack [9].

2.7.5 Unauthorized Access to Tags Attack

As we can observe it by name that an unauthorized person can access your data. With the increasing rate of technology the usage of RFID tags are increasing too in commercial and industrial application for quickly, powerful and flexible response. Different types of tags are used to shear information but fake RFID reader fail this authentication mechanisms moreover, he can keep record of personal information from tags and can modify, delete or access this confidential information.

Countermeasures

With the help of hash and encryption algorithms YapingZaing [38] proposed protected data exchange protocol which is used to provide privacy and avoid information leakage. Transmission of information, authentication identity and disconnect communication are main three parts of this algorithm. Many other mechanisms such as: sequence encryption mechanisms are used to resolve this issue as mentioned in Table 2.

2.8 Application Layer

Application layer consist the services that are provide through IoT. Application layer is known as service-oriented. It store information or data in his database and retrieve information when user need it for example Smart home, healthcare and smart cities. Various types of attacks over this layer are presented in Fig. 5 below.

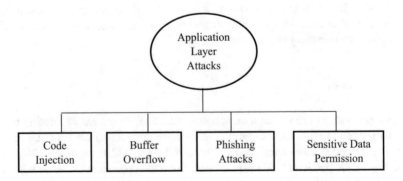

Fig. 5 Types of application layer attacks

2.8.1 Code Injection

Code injection is used for several purpose, get personal information or data, to control system and spread worm. HTML script injection and shall injection are common code injections. In this attack Malicious code will be injected in application furthermore, the injected code is able to compromising privacy property, correctness and security and database integrity. It could be used to steal data and provide authentication control [2]. Code injection attacks are used to losing the control on system or even used to completing shutdown.

Countermeasures

The best prevention technique is WAVES which is known as black box testing are proposed by Hunag and his companions for testing the web application Vulnerabilities regarding code injection or SQL. Web crawler are used in this techniques that identify the point at where SQL can performed, after that it create specified points, points that based on attack techniques. When attack happened, WAVES then monitor the application's response moreover, it used machine learning techniques to improve and prevent code injection attacks. Several other techniques like, Combined Static and Dynamic Analysis [4], Taint Based Approaches, URL scan, HP SCRAWLR are used to prevent code injection [39].

2.8.2 Buffer Overflow

Memory storage which holds data temporary is called buffer furthermore, it holds data in its transfer time from one location to other. When the data volume exceeds the storage capacity of memory buffer then buffer overflow occurred. The result of buffer overflow may cause the application crashes, memory access error and generate incorrect results. Buffer overflow give permission to attacker to overwrite the memory of application. This change can distract the path, release private information and be responsible for damage files.

Countermeasures

Many solutions such as static symbolic execution [40], evolutionary calculation method [41], Data execution prevention, Structured Exception Handler Overwrite Protection (SEHOP) are also used to overcome this predicament. In contract these, Table 3 present the one main solution of buffer overflow is Address Space Location Randomization. In this scenario, it arbitrarily hither and thither, according to the data region of specific addresses space locations. Classically, to get the information about the zone of executable code mostly a buffer overflow attack happens, but it can be purposeless by randomizing address spaces.

Table 3 Literature survey of application layer attacks

Attacks name	Challenges	Research comparison	Best prevention technique
Code injection	Inject malicious code to application or HTML source	Gould et al. [40]: JDBC-Checker is a technique for statically checking the type correctness of dynamically-generated SQL Queries McClure et al. and Cook et al. [41, 42]: new query development paradigms: SQL DOM and safe query objects, use encapsulation of database queries to provide a safe and reliable way to access databases	Black box testing: web crawler are used that identify the point at where SQL can performed, WAVES then monitor the application's response
Buffer overflow	Overwrite the memory of application	Duraes et al. [43]: proposed an automatic identification method for buffer overflow vulnerability of executable software without source code Dudina et al. [44]: used static symbolic execution to detect buffer overflow vulnerabilities Rawat et al. [45]: proposed an evolutionary calculation method for searching for buffer overflow vulnerability	Address space location randomization (ASLR): randomly moves around the address space locations of data regions. Typically, buffer overflow attacks need to know the locality of executable code, and randomizing address spaces makes this virtually impossible
Phishing attack	Using of disguised sources to steal credential information	Chen and Guo [46]: End-host based anti-phishing algorithm Atighetchi and Pal [47]: a framework based on attribute based checks for defending against phishing attacks Iliev et al. [48]: phishing prevention approach based on mutual authentication is provided	Anti-phishing authentication (APA) technique: it uses 2-way authentication and zero-knowledge password proof

2.8.3 Phishing Attack

It is very parlous attack for application layer. E-mail and other communication applications are hit list in this manner. In punishing attack, attacker pretends his self like
original user forgetting personal information of user like password of credit card
details. It is a common type of cyber-attack. Fake E-mails that have send through
attacker and designed to lure victim. The message looks like original one and received
from trusted sender. If receiver fills the attacker's requirements then attacker can
easily get access to your password or other information.

Countermeasures

The best prevention technique is Anti-Phishing Authentication (APA) technique
to detect and prevent real-time phishing attacks. It uses 2-way authentication and
zero-knowledge password proof. Users are recommended to customize their user
interfaces and thus defend themselves against spoofing [49]. There are many other
techniques for preventing phishing attacks such as end-host based anti-phishing algorithm [48], phishing prevention approach based on mutual authentication is provided
[48], but these methods are depend on type of data just like in attribute based Anti-
phishing provides different kind of checks like URL check, image attribute check.
In URL check 'it check the URL of any page either it is new or not furthermore,
for checking the similarity of legitimate page to suspected page image attribute
check are used. The only ability of attribute checking technique is to detect unknown
phishing attack and new one. Same like this, many other are used such as Identity
Based Anti-Phishing Approach, Content Based Anti-Phishing Approach and genetic
Algorithm-based Anti-phishing Techniques.

2.8.4 Sensitive Data Permission

This attack refers to manipulate the sensitive data, illegal access and violation user
privacy as well. Attacker will get access and will be entered in permission model
then they have given permission to attacker to exploiting vulnerabilities in permission
model. Now attack have right to control application. In this manner, the privacy of
user will be violated because smart devices sends data to smart application so, due
to the lake of sufficient protection it will not complete securely. Many unauthorized
used can get this sensitive data.

Countermeasures

As we have mentioned In Table 3 there are several techniques could be used such as:
train your user to avoid clicking unknown and malicious links, many software tools
and spam filter are available in market to prevent E-mail from suspicious person,

extensions and ads-on must be enable to prevent users to clinking malicious links and more pivotal is two factor authentication must be applied.

3   Open Issues

This review indicates some attacks and vulnerabilities regarding three tire layered architecture and provide some possible solutions as well. We recommend that other security and privacy concerns should be considered as like middleware layer security in IoT, Mobility first Architecture in IoT, security and privacy in ICN based architecture for IoT and security and privacy regarding mobile IoT. These attacks index is calculated from IoT environments. Conversely, the above recommendation will find more ways in IoT security manners, if these will perform with respect to time and training.

4   Conclusion

IoT technology and things are surging day by day conversely, with this development the only hurdle in Internet of Things progress is security. Privacy of data is another challenge with the development of IoT. Internet of things (IoT) systems specifically concentrate on various security challenges, to design appropriate perception for network's security, and also different security frameworks i.e. system security. The basic concept of this paper is to emphasize the security issues and also their challenges of IoT in different layers, and reflecting the security concern in various layers and their possible corrective measures. It presents every attack at IoT layered architecture, with its working and provides possible solution separately; furthermore, we also discussed other possible solution those can also use to resolve the challenge according to situation and help to boost IoT performance too. It shows the paramount importance of security in developing viable IoT solutions. It also presents clear comparison based on working and characteristics of every solution technique of single layer individually. I hope this article will help you in selecting secure IoT technique for layered architecture in your organization.

References

1. https://justcreative.com/internet-of-things-explained/
2. B. Gautam, J. Tripathi, S. Singh, A secure coding approach for prevention of SQL injection attacks. Int. J. Appl. Eng. Res. **13**(11), 9874–9880 (2018)
3. M. Conti, N. Dragoni, V. Lesyk, A survey of man in the middle attacks. IEEE Commun. Surv. Tutor. **18**(3), 2027–2051 (2016)

4. T.H. Nguyen, M. Yoo, A hybrid prevention method for eavesdropping attack by link spoofing in software-defined Internet of Things controllers. Int. J. Distrib. Sens. Netw. **13**(11), 1550147717739157 (2017)
5. S.G. Najafabadi, H.R. Naji, A. Mahani, Sybil attack Detection: improving security of WSNs for smart power grid application, in *Smart Grid Conference (SGC)* (IEEE, 2013), pp. 273–278
6. H. Çeker, J. Zhuang, S. Upadhyaya, Q.D. La, B.H. Soong, Deception-based game theoretical approach to mitigate DoS attacks, in *International Conference on Decision and Game Theory for Security* (Springer, Cham, 2016), pp. 18–38
7. K. Grover, A. Lim, Q. Yang, Jamming and anti–jamming techniques in wireless networks: a survey. Int. J. Ad Hoc Ubiquitous Comput. **17**(4), 197–215 (2014)
8. L. Gao, Y. Li, L. Zhang, F. Lin, M. Ma, Research on detection and defense mechanisms of DoS attacks based on BP neural network and game theory. IEEE Access **7**, 43018–43030 (2019)
9. F. Qi, F. Bao, T. Li, W. Jia, Y. Wu, Preventing web-spoofing with automatic detecting security indicator, in *International Conference on Information Security Practice and Experience* (Springer, Berlin, Heidelberg, 2006), pp. 112–122
10. H. Kamaludin, H. Mahdin, J.H. Abawajy, Clone tag detection in distributed RFID systems. PLoS ONE **13**(3), e0193951 (2018)
11. M. Demirbas, Y. Song, An RSSI-based scheme for sybil attack detection in wireless sensor networks, in *Proceedings of the 2006 International Symposium on World of Wireless, Mobile and Multimedia Networks* (IEEE Computer Society, 2006), pp. 564–570
12. S. Sharmila, G. Umamaeshwari, Energy and hop based detection of sybil attack for mobile wireless sensor networks. Int. J. Emerg. Technol. Adv. Eng. **4**(4), (2014)
13. D.S. Roy, A.S. Singh, S. Choudhury, Countering sinkhole and blackhole attacks on sensor networks using dynamic trust management, in IEEE Symposium on Computers and Communications, (ISCC) (IEEE, 2008), (pp. 537–542)
14. A. Dubey, R. Gupta, G.S. Chandel, An efficient partition technique to reduce the attack detection time with web based text and PDF files. Int. J. Adv. Comput. Res. (IJACR) **3**(9), 80–86 (2013)
15. S.I.A. Qadri, K. Pandey, Tag based client side detection of content sniffing attacks with file encryption and file splitter technique. Int. J. Adv. Comput. Res. **2**(3), 215 (2012)
16. B. Aziz, G. Hamilton, Detecting man-in-the-middle attacks by precise timing, in *Third International Conference on Emerging Security Information, Systems and Technologies* (2009), pp. 81–86
17. J. Liu, Y. Xiao, C.P. Chen, Authentication and access control in the internet of things, in *32nd International Conference on Distributed Computing Systems Workshops (ICDCSW)*, (IEEE, 2012), pp. 588–592
18. A. Agah, K. Basu, S.K. Das, Preventing dos attack in sensor networks: a game theoretic approach, in *IEEE International Conference on Communications (ICC)*, vol. 5 (IEEE, 2005), pp. 3218–3222
19. M. Mohi, A. Movaghar, P.M. Zadeh, A Bayesian game approach for preventing dos attacks in wireless sensor networks, in *WRI International Conference on Communications and Mobile Computing (CMC'09)*, vol. 3 (IEEE, 2009), pp. 507–511
20. Y.-Y. Zhang, X.-Z. Li, Y.-a Liu, The detection and defence of dos attack for wireless sensor network. J China Univ. Posts Telecommun. **19**, 52–56 (2012)
21. A. Rajan, J. Jithish, S. Sankaran, Sybil attack in IOT: modelling and defenses, in *International Conference on Advances in Computing, Communications and Informatics (ICACCI)* (IEEE, 2017), pp. 2323–2327
22. G.W. Kibirige, C. Sanga, A survey on detection of sinkhole attack in wireless sensor network. (2015). arXiv:1505.01941
23. A. Papadimitriou, L.F. Fessant, C. Sengul, Cryptographic protocols to fight sinkhole attacks on tree based routing in WSN, in *5th IEEE Workshop on Secure Network Protocols (NPSec)* (IEEE, 2009), pp.43–48
24. G. Noubir, G. Lin, Low-power DoS attacks in data wireless LANs and countermeasures. SIGMOBILE Mob. Comput. Commun. Rev. **7**(3), 29–30 (2003)

25. I. Krontiris, T. Dimitriou, T. Giannetsos, M. Mpasoukos, (2008). Intrusion detection sinkhole attacks in wireless sensor network, in *IEEE International Conference on Wireless and Mobil e Computing, Networking and Communications, WIMOB'08* (IEEE, 2008), (pp. 526–531)
26. A. Barua, H. Shahriar, M. Zulkernine, Server side detection of content sniffing attacks, in *IEEE 22nd International Symposium on Software Reliability Engineering* (IEEE, 2011), pp. 20–29
27. T. Mahjabin, Y. Xiao, G. Sun, W. Jiang, A survey of distributed denial-of-service attack, prevention, and mitigation techniques. Int. J. Distrib. Sens. Netw. **13**(12), 1550147717741463 (2017)
28. V.S. Dines Kumar, C. Navaneethan, Protection against denial of service (dos) attacks in wireless sensor networks. Int. J. Adv. Res. Comput. Sci. Technol. **2**, 439–443 (2014)
29. A. Wood, J. Stankovic, S. Son, JAM: a jammed-area mapping service for sensor networks, in *24th IEEE Real-Time Systems Symposium* (2003), pp. 286–297
30. J. G. Proakis, *Digital Communications*, 4th edn. (McGraw-Hill, 2000)
31. K. Bu, X. Liu, J. Luo, B. Xiao, G. Wei, Unreconciled collisions uncover cloning attacks in anonymous RFID systems. Inf. Forensics Secur. IEEE Trans. **8**(3), 429–439 (2013)
32. K. Bu, M. Xu, X. Liu, J. Luo, S. Zhang, M. Weng, Deterministic detection of cloning attacks for anonymous RFID systems. IEEE Trans. Ind. Inform. **11**(6), 1–1 (2015)
33. X. Li, H.N. Dai, Q. Zhao, An analytical model on eavesdropping attacks in wireless networks, in *IEEE International Conference on Communication Systems* (IEEE, 2014), pp. 538–542
34. A. Herzberg, A. Gbara, TrustBar: Protecting (evenNaive) Web Users from Spoofing and Phishing Attacks. CryptologyePrint Archive (2004). Report 2004/155
35. E.W. Felten, D. Balfanz, D. Dean, D.S. Wallach, Web spoofing: an Internet Con Game, in *20th National Information Systems Security Conference* (1997)
36. W. Xu, W. Trappe, Y. Zhang, T. Wood, The feasibility of launching and detecting jamming attacks in wireless networks, in *Proceedings of the 6th ACM International Symposium on Mobile Ad Hoc Networking and Computing* (2005), pp. 46–57.
37. A. Adelsbach, S. Gajek, J. Schwenk, Visual spoofing of SSL protected web sites and effective countermeasures, in *Proceedings of Information Security Practice and Experience*, LNCS 3469 (2005), pp. 204–216
38. W. Zhang, B. Qu, Security architecture of the Internet of Things oriented to perceptual layer. Int. J. Comput. Consum. Control (IJ3C) **2**(2), 37–45 (2013)
39. P. SaiKiran, E. SureshBabu, D. Padmini, V. SriLalitha, V. Krishnanand, Security issues and countermeasures of three tier architecture of IoT-a survey. Int. J. Pure Appl. Math. **115**(6), 49–57 (2017)
40. C. Gould, Z. Su, P. Devanbu, JDBC checker: a static analysis tool for SQL/JDBC applications, in *Proceedings of the 26th International Conference on Software Engineering*. (IEEE 2004), pp. 697–698
41. R.A. McClure, I.H. Kruger, SQL DOM: compile time checking of dynamic SQL statements, in *Proceedings of the 27th International Conference on Software Engineering (ICSE)*. (IEEE, 2005), pp. 88–96
42. W.R. Cook, S. Rai, Safe query objects: statically typed objects as remotely executable queries, in *Proceedings of the 27th International Conference on Software engineering* (2005), pp. 97–106
43. J. Durães, H. Madeira, A methodology for the automated identification of buffer overflow vulnerabilities in executable software without source-code, in *Latin-American Symposium on Dependable Computing*. (Springer, Berlin, Heidelberg, 2005), pp. 20–34
44. I.A. Dudina, A.A. Belevantsev, Using static symbolic execution to detect buffer overflows. Program. Comput. Softw. **43**(5), 277–288 (2017)
45. S. Rawat, L. Mounier, An evolutionary computing approach for hunting buffer overflow vulnerabilities: a case of aiming in dim light, in *European Conference on Computer Network Defense* (IEEE, 2010), pp. 37–45
46. J. Chen, C. Guo, Online detection and prevention of phishing attacks, in *First International Conference on Communications and Networking in China* (IEEE, 2006), pp. 1–7
47. M. Atighetchi, P. Pal, Attribute-based prevention of phishing attacks, in *Eighth IEEE International Symposium on Network Computing and Applications* (IEEE, 2009), pp. 266–269

48. D. Iliyev, Y.B. Sun, Website forgery prevention, in *International Conference on Information Science and Applications* (IEEE, 2010), pp. 1–8
49. M. Sharifi, A. Saberi, M. Vahidi, M. Zorufi, A zero knowledge password proof mutual authentication technique against real-time phishing attacks, in *International Conference on Information Systems Security* (Springer, Berlin, Heidelberg, 2007), pp. 254–258

Software-Defined Networking Security: A Comprehensive Review

Meryem Chouikik, Mariyam Ouaissa, Mariya Ouaissa, Zakaria Boulouard, and Mohamed Kissi

Abstract With the growth in the use of information technology, there is a huge increase in traffic flowing through networks due to the large number of connected devices and modern internet applications, such as social networking and sharing of documents. Network administrators must manage a wide range of data formats, service types and devices, which is difficult with traditional network management tools that were not designed to cope with scalable topologies at very high speeds large scale. The concept of Software Defined Networking (SDN) is the solution to meet the needs of users of these network services and applications. This approach centralizes and simplifies network management, allowing administrators to orchestrate and automate it through a central software control interface without physically accessing hardware components. As SDN technology gains traction and more internet providers and data center administrators gradually adopt it, there is growing interest in the security issues that may arise with regard to its deployment in production. In this chapter, we present a comprehensive review of the SDN technology includes the architecture, applications, benefits and the programmable networks. In addition, we discuss the security issues of this technology.

Keywords SDN · OpenFlow · Security · DoS · IDS

M. Chouikik · Z. Boulouard · M. Kissi
LIM, Hassan II University, Casablanca, Morocco
e-mail: zakaria.boulouard@fstm.ac.ma

M. Kissi
e-mail: mohamed.kissi@fstm.ac.ma

M. Ouaissa (✉) · M. Ouaissa
Moulay Ismail University Meknes, Meknes, Morocco
e-mail: mariyam.ouaissa@edu.umi.ac.ma

M. Ouaissa
e-mail: mariya.ouaissa@edu.umi.ac.ma

© The Author(s), under exclusive license to Springer Nature Switzerland AG 2022
M. Ouaissa et al. (eds.), *Big Data Analytics and Computational Intelligence for Cybersecurity*, Studies in Big Data 111,
https://doi.org/10.1007/978-3-031-05752-6_6

1 Introduction

By introducing programming to the network, Software-Defined Networking (SDN), often referred to as a revolutionary new idea in computer networking, promises to dramatically simplify network control and management while also enabling innovation. SDN technology provides speed and agility when deploying new enterprise applications and services. And, thanks to a platform capable of meeting the needs of the most demanding networks, both current and future, SDN solutions comply with the key characteristics of these solutions, namely flexibility, policy execution, and programmability [1].

Computer networks are usually built from a large number of network devices such as switches, routers, firewalls, etc. with many complex protocols embedded with them. Policies are set up by network administrators to respond to a variety of network events and application scenarios. These high-level policies are manually translated into low-level configuration commands. Because these extremely complex tasks are frequently completed with very limited tools, controlling network management and performance is a difficult and error-prone task [2].

Programmable networks have been proposed as a means of facilitating network evolution. SDN is a new networking paradigm in which forwarding hardware is separated from control decisions. This separation results in a network architecture that is more flexible, programmable, cost-effective, and innovative [3].

There are numerous research themes based on the use of traditional security techniques to protect networks from various types of attacks, but the techniques using the SDN paradigm to improve security are new, promising, and widely discussed.

In this chapter, we offer an overview of the SDN technology. First, we begin by presenting the concept, the architecture, the main types of applications and the main components defining SDN. After, we present the programmable networks. Then, we discuss the impact and issues of security in SDN networks and possible solution to avoid intrusions.

2 Overview of SDN

With the help of SDN, network design and management have become more innovative in recent years. This technology appears to have appeared out of nowhere, but it is actually part of a long history of making computer networks more programmable. For the evolution of selected programmable networks over the last 20 years, as well as their chronological relationship to network virtualization advances. The plot is divided into three parts. Each has made their own contribution [4]:

(1) Active networks (from the mid-1990s to the early 2000s) introduced programmable functions into the network, resulting in greater innovation.

(2)   From 2001 to 2007, there was a separation between the control plane and the data plane, which resulted in open interfaces between the control plane and the data plane.

(3)   The OpenFlow API and network operating systems (from around 2007 to 2010) were the first to see widespread adoption of an open interface, and they were designed to make this separation scalable and practical. It is also worth noting that network virtualization has played an important role in the evolution of SDN.

2.1   Definition and Concept

SDN allows the physical separation of control and transport in the network and introduces the notion of scheduling. Software-Defined Networking (SDN) is an emerging architecture that is dynamic, manageable, cost-effective, and adaptable. This architecture decouples the intelligent part of the network, namely the control plane and the data transfer plane, thus allowing the network to become directly programmable and the architecture to be abstract for network applications and services [5].

In SDN, network intelligence is logically centralized in software controllers or control plane, and the transmission network equipment becomes a simple packet transmission device or the data plane that can be programmed through an interface opened. One of the earliest implementations of this open interface is called OpenFlow. The OpenFlow protocol is the fundamental element of SDN, thanks to it the SDN architecture is:

Directly programmable: the network control is directly programmable because it is separate from the transfer functions.

Agile: The abstraction between the control plane and the data transfer plane allows administrators to dynamically adjust traffic flow across the network to meet changing needs.

Centralized management: Network intelligence is centralized in SDN controllers, which maintain a unified global view of the network.

Programmatically configured: SDN enables network administrators to quickly configure, manage, secure, and optimize network resources via applications developed and deployed on top of the SDN architecture [6].

2.2   SDN Architecture

SDN allows applications to be aware of the network by taking a new approach to network architecture. In a traditional network, a network device such as a router or switch contains both the control plane and the data plane. The control plan determines the route that will take traffic through the network, while the data plan is the part of the network that actually carries the traffic [7]. By separating these two planes, the network equipment can be configured from the outside by management (not specific to the software vendor) that transforms the network from a closed system to an open system as shown in Fig. 1. The following list defines and explains the elements of the SDN architecture:

- SDN Application: SDN applications are programs that allow networks to communicate in an explicit direct and programmable manner. Network behavior is controlled via an SDN controller through the NorthBound Interface (NBI). An SDN application consists of a logical application and one or more NBI Drivers.
- SDN Controller: The SDN controller is an application that manages flow control to enable smart network functionality. SDN controllers are generally based on

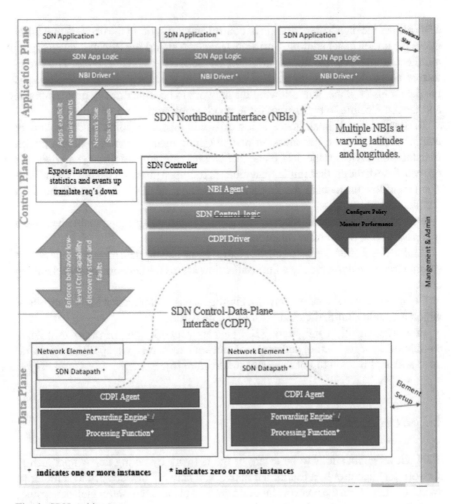

Fig. 1 SDN architecture

the Openflow protocol which allows the controller to communicate with other elements of the architecture through an Application Programming Interface (API).

- The controller is the kernel of an SDN network, all communications between applications and devices must go through the controller.
- SDN Data plane: This is the part of a network that carries the user's traffic. The data plan allows the transfer of data to customers, the management of multiple conversations through multiple protocols, and manages conversations remotely. The data plan traffic moves through the routers.
- SDN Control to Data-Plane Interface (CDPI) or Southbound Interfaces: This is the defined interface between a controller and an SDN Datapath and provides at least one programmatic control of all shipping operations, statistical reports, and event notification.
- SDN Northbound Interfaces (NBI): SDN NBI are interfaces between SDN applications and SDN controllers and provide a view of the abstract network and give direct expression to network behavior.

2.3   Benefits of SDN

SDN is getting a lot of attention, and we can quite easily find very optimistic adoption prospects among the various analysts. A founding principle of SDN is to decouple a switch's control plane from its data plane. In view of the promises of the SDN, it seems obvious that he has a bright future ahead of him. However, for the adoption to materialize, the subject must be clarified. Decoupling a switch's control plane from its data plane is a founding principle of SDN [8]. Thanks to this principle, operation is carried out centrally and via an application. More openness as a result, the network becomes both more open and simpler:

- A switch with very limited intelligence, and simply responsible for switching packets
- Centralized management of all network equipment (regardless of the choice of manufacturers) via an application: an SDN controller.
- A communication protocol between the switches and the SDN controller. (Openflow is a standardized protocol that emerged as the SDN approach surfaced).
- Allows the manual control interface in network equipment to be replaced by a programming interface. Allowing automation of tasks such as configuration and management of flow control policies and also allows the network to dynamically respond to application requirements.
- Unified view with centralized visibility including network information.
- Makes the network programmable and flexible which allows us to manipulate the infrastructure elements as a single entity and not individually.

3 Towards a New Approach

3.1 State of Traditional Networks

Current networks have undergone a major transformation in recent years due to the great demand from Datacenters, the virtualization of new application architectures and the transition to higher speeds.

Network performance challenges are increasing. They are linked to the size, complexity and mobility of modern network environments, so networks are very complex to manage.

However, this demand increases the complexity of managing and monitoring networks, network switches have always operated on the basis of hardware processors to route data, which has relatively limited the IT team's control over data traffic through the network.

The performance and scalability of current-generation networks are significantly limited, preventing the flexibility needed by network managers. So, it's time to take a better approach and move past these limits [9].

3.2 Limits of the Current Architecture

Today's networks are undergoing major change, for many reasons, which forces network infrastructures to become more flexible and programmable so that the network can dynamically respond to changing user needs and manage applications and services on demand [10]. Network architects are restricted by the limitations of networks, to cite:

- Complexity: Changing the network configuration requires time and manual work, as it relies on making changes from the Command Line Interface (CLI). However, networks are more complex, requiring data to be transmitted regularly.
- Inconsistent policy: The implementation of a network that respects company policy such as security and data access require the configuration of thousands of equipment and machines. For example, adding a virtual machine can take hours and in some cases days, the network administrator must reconfigure all of the ACLs throughout the network. The complexity of today's networks makes it very difficult to apply a consistent approach in terms of security, QoS,
- Scalability: Today, with the proliferation of devices and communication channels, the growth of data, business and customer expectations are increasing, so networks must become more scalable, agile and programmable.
- Adaptation of new services: Service providers are always looking to deploy new services and offer more options to their customers, however they are hampered and blocked by product life cycles which are defined by OEMs who can reach three years or more.

The lack of standards and the non-compatibility of interfaces also limit the ability to adapt services to customer needs. This mismatch between market needs and network capabilities is bringing the industry to a tipping point.

3.3   New Needs and Challenges

The limitations cited above lead to thinking about a new solution that facilitates the management of hardware and software resources, in order to optimize costs, resources, and promote the evolution of the IS, by simplifying its daily management. The need for this change is due to the following new IT trends:

- Security: IT will be able to make an impact by implementing stronger policies, requiring continuous monitoring and tighter access controls to limit potential harm, better detect and stop unwanted activity, but also recover more quickly and efficiently after incidents.
- Mobility: Phones and wearable devices are now part of an extended computing environment that includes things like consumer electronics and connected displays in the public and workspace.
- It is increasingly the general environment that will have to adapt to the demands of the mobile user, and this will continue to create significant management challenges for IT organizations as they lose control of the end devices of users.
- The Evolution of Cloud Services: The Cloud has undergone a major evolution and many companies have switched the operation of their IT services to Cloud services hosted in Datacenters in order to lower production and operating costs, in addition to complexity, the architecture of cloud services must be developed in an environment that meets business requirements.
- Change in traffic pattern: The traffic pattern has changed significantly in storage centers unlike client–server applications where most of the traffic passes between a single client and a single server, newly developed applications have access to different databases thus creating a machine-to-machine data flow before sending the result to the end user based on the classic "client–server" model.
- At the same time customers are continually disrupting the traffic pattern by using different types of devices and from different places to connect.

All of these features and new uses are disrupting traffic behavior in companies especially in the DSI.

4   Programmable Networks

In the current operation of IP networks, each network device such as IP router, Ethernet switch performs its control plane and data plane functions. SDN proposes to create a central point that manages the control plane, while the physical switches/routers only have to support the data plane. To do this, the Open Networking Foundation (ONF) implemented OpenFlow. This is a standard protocol used by the controller to send instructions to the switch that program their data plane and get information from those switches so that the controller can have a logical global view (abstraction) of the physical network. This view is used for all the decisions that the control plane must make (routing, traffic filtering, load sharing, address translation, etc.). The SDN Controller provides an API to "SDN applications" that includes this global view. These applications implement, via the controller, services such as flow routing, implementation of QoS policies for flows, end-to-end flow security, flow filtering, etc. [11].

This approach allows flexible and rapid implementation of new network applications that emulate network appliances.

4.1 OpenFlow Concept

OpenFlow is the first standard communication protocol that is defined between the control and data layers of an SDN architecture. It is a new network approach based on Ethernet switches, which sets up a wide field of experimentation on the real network without having to disrupt its operation or spend additional costs on other types of infrastructure.

The basic idea is as follows: The technology takes advantage of the fact that most modern Ethernet switches and routers have high-speed flow tables for implementing firewalls, NAT, QoS, and collecting statistics. These flow tables are specific to each manufacturer, but we can identify a set of features that are implemented on most routers and switches in order to be exploited by OpenFlow.

4.2 Why OpenFlow?

OpenFlow guarantees the separation between the control plane (responsible for associating a decision with the packet) and the data plane (responsible for the transfer of the packets in the network) as well as the treatment of the packets like data flows, where a flow of packets is defined as a combination of Layer 2, 3, and 4 headers. The data plane must be implemented in OpenFlow switches, which will alleviate the burden on switches that will only be responsible for forwarding packets. The control plane is ensured by a separate machine (the controller), which performs the processing necessary to determine the decisions to be applied to a data flow.

This separation between the control plane and the data plane will make it possible to achieve very high throughput. Another advantage of this separation is that the components of the equipment responsible for the data plane will be optimized for the transfer of the packets, which avoids weighing down the processing path with functionalities that come under the control plane.

4.3   OpenFlow Protocol

This is an open protocol; this standard is used by the controller to transmit and add instructions to the switch. The latter (called flow entries) are rules with a "pattern" (source or destination IP, mac address, TCP port, etc.) and a corresponding action (reject the packet, transmit on port x, add a VLAN header, etc.).

These rules are used to program the data plane and get information from these switches so that the controller can have a logical global view (abstraction) of the physical network. This view is used for all the decisions that the control plane must make (routing, traffic filtering, load sharing, address translation, etc.).

Since OpenFlow is an open standard, it can therefore be used by any Open-Flow compatible controller to communicate with any OpenFlow compatible network equipment, independent of the manufacturer.

OpenFlow protocol messages: Three types of messages can appear between the switch and the controller. They are:

- Controller->switch messages: These represent the most important category of OpenFlow messages.
- Asymmetric messages: Transmitted indifferently by the controller or the switch through the secure channel without having been requested by the other entity.
- Asynchronous messages: Initiated by the switch and used to update the controller:

4.4   Component of an OpenFlow Network

OpenFlow offers a new architecture well suited to virtual network environments, the main idea is to communicate the two planes that have been separated, the controller can serve the entire network through OpenFlow using the network connections. The network is essentially formed of one or more OpenFlow switches, one or more controllers, and finally the protocol.

OpenFlow defines all the messages exchanged between controllers and switches and allows communication to be standardized.

4.5   SDN Controllers

A controller is an application in SDN whose mission is to provide an abstraction layer of the network and present the latter as a system. The SDN controller enables intelligent networking, it quickly implements a change on the network by translating a global request into a follow-up operation on the equipment, it relies on the OpenFlow protocol which allows the controller to communicate to the switches or send the packets via APIs, and also configure network devices, choose the optimal network path and better throughput for application traffic [12].

Indeed, the controller plays the role of the brain of the network, it acts as an operating system for the latter, facilitates management in an automated way by offering a centralized end-to-end view of the network and it makes it easier to integrate and administer enterprise applications.

A number of controllers have been created, the vast majority of which are open source and support the OpenFlow protocol. These controllers differ in terms of programming languages, OpenFlow versions supported, techniques used such as multitasking, and performance metrics such as throughput. We present below a non-exhaustive list of SDN controllers:

- NOX and POX: Are the first two OpenFlow controllers. NOX is programmable in C++ while POX uses Python language
- Maestro: Uses multitasking technology to perform low-level parallelism, keeping a simple programming model for application developers. He reaches his performance by distributing core tasks to available threads and minimizing memory consumption.
- Beacon: Is a controller developed by Davide ERICSSON it is modular and based on java.
- SNAC: To manage network rules, a web application is used. To configure network devices and control their events, a flexible rule definition language and simple interfaces have been integrated.
- RISE: Designed for large-scale network experiments, RISE is a Trema-based controller. The latter is a Framework programmed in Ruby and C. Trema that provides an integrated testing and debugging environment, including a set of development tools.
- Ryu: Is an SDN controller capable of configuring network equipment using OpenFlow, NetConf, and OF-config.
- Floodlight: Is This software-defined network controller, backed by Big Switch Corporation, provides a central management point for OpenFlow networks and can manage devices such as openvswitch seamlessly. It also supports a wide range of OpenFlow physical switches, so it greatly simplifies network management. It is under Apache license and written in java.
- ONOS (Open Networking Operating System): It is a recently released network operating system or SDN controller that focuses on service provider use cases, it is implemented on different layers operating as a cluster, it ensures the scalability, high availability, performance, and abstractions to facilitate the creation of applications and services.
- OpenDaylight: This is an open-source project supported by several network providers. Opendaylight is an SDN control platform developed by java, it was created to address the issue of the controller which is a key element of an SDN architecture, the objective of this consortium is to combine industrial and academic efforts to develop a truly usable controller that agrees to deploy it on any hardware platform and operating system.

5   SDN Security Based on IDS

For a long time, data security involved only the encryption of data for researchers, but with the arrival of the Internet at the end of the 1980s, the need to have more efficient tools to secure data has been felt. Therefore, to solve these problems, firewalls and IDS/IPS have been introduced for the protection of data, services and applications in today's networks, whose threats and attacks have become too important. Despite the widespread use of firewalls and IDS/IPS, these security mechanisms networks also have their limits, even if the recent versions allow to have much more evolved functionalities. Network security requires continuous efforts to anticipate in order to adapt to modifications of threat trends to prevent, minimize or avoid future attacks. In any type of network, security must be consistent, dynamic, scalable, and end-to-end. Then making security dynamic and intelligent enables better identify threats through predictive analyzes and respond in real-time if necessary [13].

5.1   Definition of IDS

An Intrusion Detection System (IDS) it is a tool among the important tools used for network security, capable of to monitor in real time the packets passing through the network in order to detect those which seem suspicious.

The main objective of the IDS is to protect the network against intrusions and to ensure its Security IDS in the networks, as well to collect information on the intrusions, to ensure a management centralized alerts and perform an initial diagnosis of the nature of the attack, furthermore allowing and reacting actively and quickly to the attack to slow it down or stop it.

There is a possibility of IDS detection error which varies from system to system and exactly it is about false positive alarms and false negative alarms. An alert is said to be false positive if the IDS detects legitimate activity as malicious activity while an alert is said to be false negative if the IDS accepts malicious activity as an activity legitimate.

Otherwise, IDS is a mechanism intended to identify abnormal or suspicious activities on the analyzed target (a network or a host). It is an action of prevention and intervention on the risks of intrusion in order to detect attacks that a system (computer network) may suffer, therefore it would be necessary to have specialized software whose role is to monitor the data passing through this system, and who is able to react if data seems suspicious.

The IDS is called Network IDS if it is executed on a firewall, on a special device placed at the input of a serial or probe network. The principle is to control the network activity (content, volume, etc.) by analyzing and interpreting the packets circulating on the network.

The equipment must have a large packet processing capacity because it necessarily inspects all packets in real-time if placed in series. We call Host IDS if it is installed on a server to monitor system activity (logs, files, processes, etc.). Network IDS and Host IDS are complementary. A network IDS is much broader in terms of scope than the Host IDS. An IDS Application is a particular application on a machine.

The IDS makes it possible to highlight a possible intrusion in a network. After detection of a threat the IDS must collect as much information as possible (target, source, process, etc.), then archive and trace the threat. He must also prepare the response to an attack by putting it in quarantine for example.

It should also be noted that the multiple intrusion detection and prevention technologies are complementary and work for different types of environments, with some giving better results [14].

5.2   Types of IDS

Intrusion detection systems can be classified into two categories [15]: either the network traffic; we speak of IDS network or NIDS (Network based IDS)), or the activity of the machines; we speak of IDS System or HIDS (Host based IDS)).

- Network-type Intrusion Detection Systems (NIDS): a NIDS listens to all network traffic, analyzes it and generates alerts if it detects packets that seem malicious. Most NIDS are also called inline IDS because they analyze the flow in real-time.
- Host-type Intrusion Detection Systems (HIDS): a HIDS analyzes in real-time the flows relating to a machine on which it is installed as well as the event logs and generates alerts. It detects activities that seem malicious.

5.3   Methods of Detections of IDS

There are major intrusion detection methods: the detection method anomalies compared to a usual traffic profile, based on a comparison of traffic and signature-based method of known attacks by checking headers and payload of packets.

- Anomaly-based IDS: this method is based on the detection of anomalies compared to a usual traffic profile; the IDS will discover the normal operation of the system to be monitored with a learning phase. From these behaviors, the system can trigger an alert when out-of-profile events occur.
- Signature-based IDS: it is based on a comparison between the traffic and a base of signature, looking at packet headers and payload. The advantage of this method is that it generates far fewer false alarms. The problem with this method is that it does not detect new threats without regularly updating its signature database.

Signature-based IDS observers and monitors packets inside the Network plus detection of attacks by looking for patterns evidently as byte sequences in network traffic, otherwise known as malicious instruction sequences used by malware as well compared packets with pre-configured and pre-determined attacks patterns known as signatures, this terminology come from anti-virus software, and this refers to these detected patterns as signatures. Notably, signature-based IDS could detect known attacks, as well as it is difficult to detect new attacks, for which no pattern is disposable.

- Dynamic protocol analysis: it is based on comparing predefined profiles of generally accepted definitions of benign protocol activity, for each protocol state, with observed events, to identify deviations.

5.4 Examples of IDS

Intrusion detection systems are essential for network security certainly it is possible to detect intrusion attempts based on the basis of a signature or the learning of behavior for the flows present in the network. There are several IDS open-source that can be used for detection and prevention against any type of attack among them:

- Snort: is an intrusion detection and prevention system (NIDS/NIPS), created by Martin Roesch in 1998. This system has established itself as the most popular intrusion detection system more used. Its commercial version has given it a good reputation among companies.

Snort can analyze network traffic in real time and is equipped with intrusion detection technologies such as protocol analysis and "pattern matching." It can detect a wide range of attacks, including denial of service, array overflow, stealth port scanning, and attempts to fingerprint operating systems. Snort has three modes of operation:

- Packet sniffer: in this mode, Snort observes the packets circulating in the network and displays them continuously on the screen as "tcpdump";
- Packet logger: in this mode, Snort records the packets circulating on the network in directories on disk. You can use this mode to limit the logs with certain criteria such as IP address range or protocol;
- NIDS: in this mode, Snort analyzes the network traffic, compares this traffic to the rules already defined by the network administrator and defined actions to perform.

- Bro: is an open-source UNIX-based NIDS which was founded by Vern Paxson in 1998. Bro passively monitors network traffic in order to find malicious feeds. It detected intrusions by first analyzing network traffic, then it runs event-driven parsers that compare the activity with patterns considered malicious. Suspicious traffic analysis includes the detection of specific attacks (signature and events) and unusual (abnormal) activities.
- Suricata: is an open-source intrusion detection and prevention system that has was developed by the Open Information Security Foundation (OISF). The beta version was released in December 2009 while the first stable release took place in July 2010. Suricata was created to bring new ideas and technologies in the field of intrusion detection. OISF provides Suricata with a set of intrusion detection

and prevention rules. Suricata is able to use rules to from other systems such as Snort to provide the best possible set of rules.

6 Literature Review

In this section, we focus primarily on research efforts that have addressed Denial of Service (DoS) attacks [16] in SDN networks to mitigate their impact on performance of the network. First, we expose the different mitigation techniques, which are developed in various studies. Then, we have recourse to solutions which exploit the IDSs in their mitigation strategies against DoS attacks as used in ours. In addition, we present proposed traffic sampling techniques for further minimize the analyzed traffic in the networks, especially in the case of attacks of DoS. We end this section with a comparative study that compares our proposed solution, SDN-Guard, to those that already exist based on their targeted objectives to minimize during DoS attacks.

6.1 Mitigation Techniques Without the Use of IDS

- FlowRanger: Lei Wei and Carol Fung [17] propose FlowRanger, a system of flow storage that helps detect and mitigate the impact of DoS attacks. FlowRanger is implemented in the SDN controller.
- IP address filtering technique: Another solution has been proposed in [18] to protect SDN networks against DDdS attacks and which is based on the IP filtering technique. The mechanism consists of analyzing the behavior and activities of the user in the network in order to calculate the number of connection attempts of his address IP.
- FloodGuard: FloodGuard is a new platform for defending against DoS attacks in networks SDN proposed by Wang et al. [19]. The objective of authors is to avoid saturation of the switching tables of the switches of the transmission plan as well as controller overload. To achieve these goals, they implement two supervision and flow management modules in the controller.
- Autonomous management technique: In this solution [20] the authors take advantage of the SDN technology's centralization and programmability to provide an autonomous management system in which the Internet Service Provider (ISP) and its customers collaborate to mitigate the impact of DoS attacks. The ISP collects threat information reported by customers for use in enforcing security policies and updating switch flow tables in the network with this solution.
- AVANT-GUARD: AVANT-GUARD [21] is a new platform that allows the control plane to be more resilient and scalable against attacks that cause the saturation of the control plane such as the attack of floods TCP-SYN (DoS). In this work, the aim of the authors is to avoid the saturation of the bandwidth from switch to

controller and increase controller response rate requests arriving from switches in the data plane.

- SLICOTS: Mohammadi et al. [22] proposed SLICOTS as another solution to mitigate the impact of TCP-SYN flood attacks in SDN networks. It is presented as a security module embedded in the control scheme. This module monitors all incoming TCP requests on the network in order to detect and prevent TCP-SYN flood attacks, as well as to block malicious hosts.
- Combination OpenFlow and sFlow: Giotis et al. propose in [23] a method that combines the OpenFlow protocol specifications and standard sFlow traffic monitoring technology. This method provides an efficient mechanism for detecting anomalies and mitigating their effects in real-time SDN environments. The solution is comprised of three major modules.

6.2 Mitigation Techniques with the Use of IDS

In addition, our solution leverages an Intrusion Detection System (IDS) to analyze the traffic and detect DoS attacks in the SDN network. In this section, we expose some research works that propose mitigation strategies that use IDSs for traffic analysis and detection of DoS attacks. This work is based on the alerts provided by an IDS in order to properly manage traffic and mitigate the impact of DoS attacks on the SDN network performance.

- BroFlow: The BroFlow mechanism presented in [24], is a detection and prevention system for Distributed Intrusion for SDN environments. This system uses the IDS Bro (The Bro Network Security Monitor, 2017) to monitor and analyze network packets by sensors implemented in network switches. The aim of the authors is to optimize the placement of Bro sensors by minimizing their number and maximizing coverage network for each sensor. Two types of Bro sensor are deployed in the architecture, the virtual network BroFlow sensors deployed in virtual network and the infrastructure sensor BroFlow installed in a physical machine. The role of these sensors is to monitor the network, attack detection and controller alert.
- Collaborative approach: The purpose of this approach is to provide early warning of potential threats with high accuracy, Tommy Chin et al. [25]. The authors provide a new attack detection technique. This technique consists of coordination between the monitors (IDS) distributed over the networks, a correlator, which is the centralized controller, and the switches. With this solution, when the IDS detects anomalies in the traffic, it sends an alert to the correlator (SDN checker) to check for suspects by looking at packets in more detail network to find additional attack signatures, make the appropriate decision and update the switch flow table to mitigate the impact of an attack.
- SnortFlow: SnortFlow is a system offered in [26] and which is dedicated to the prevention of virtual environments such as cloud computing against detected

intrusions. This system aims to detect intrusions, select appropriate countermeasures and reconfigure security networks. Cloud to mitigate or prevent attacks. To achieve this goal, Xing et al. combine how Snort works as a network IDS (NIDS) with the power of programmability offered by SDN technology to deploy Snort-Flow. This solution is based on the SnortFlow server component that collects alerts from Snort agents.

- Traffic sampling techniques: The DoS attack as well as the DDdS, aim to disrupt, or totally paralyze, the functioning of the SDN network by bombarding it with erroneous requests. This type of attack can affect the performance of the network by saturating its resources such as bandwidth, space storage in the switch TCAM tables and controller processing capacity. In the context of our work, we exploit an IDS to analyze the traffic in the network and detect whether it is malicious or not, based on its behavior.

6.3   Discussion

This subsection presents a comparative study between our SDN-Guard and the different works already covered in the literature review. Table 1 compares our proposed solution, SDN-Guard, to existing ones in relation to their objectives of

Table 1 SDN-guard versus existing solutions

Approach	Management of traffic		Attack mitigation minimizing		
	Sample traffic	BW of network	Controller Load	BW	Storage
SDN-guard	✔	✔	✔	✔	✔
FlowRanger	✘	✘	✔	✘	✘
IP address filtering	✘	✘	✔	✔	✔
Autonomous management technique	✘	✘	✘	✘	✘
FloodGuard	✘	✔	✘	✔	✔
AVANT-GUARD	✘	✘	✔	✔	✔
SLICOTS	✘	✘	✔	✔	✔
Combination of and sFlow	✔	✔	✔	✔	✘
BroFlow	✘	✔	✘	✘	✔
Collaborative approach	✘	✘	✘	✘	✔
SnortFlow	✘	✘	✘	✔	✔
Suspicious traffic sampling	✔	✘	✘	✘	✘

performance targets during DoS attacks. None of the existing solutions can simultaneously reduce controller load, bandwidth between the controller and the switch noted by Bandwidth (BW) and avoid flooding the switching tables of the switches. In this comparison, we aim to achieve simultaneously these objectives in order to maintain an acceptable network performance during DoS attacks.

7 Conclusion

Software-Defined Networks or SDN is a paradigm that separates the control plane from the data plane, transforming the concept of a software-driven network and the challenge to keep it secure while keeping them open to their customers evidently IDS is one of the best technology used to provide the security needs ids continuously monitor the network to identify potential incidents and record related information in logs, resolve incidents and report them to security administrators. Additionally, some networks use IDS to identify problems with security policies and deter individuals from violating those policies. IDS have become essential to most companies security infrastructure precisely because they can stop attackers while they gather information from your network. Today's attacks are so sophisticated that they can defeat the best security systems, especially those protected only by encryption or firewalls. Unfortunately, these technologies alone cannot thwart today's attack.

References

1. A. Shaghaghi, M.A. Kaafar, R. Buyya, S. Jha, Software-defined network (SDN) data plane security: issues, solutions, and future directions. Handb. Comput. Netw. Cyber Secur. 341–387 (2020)
2. S.K. Tayyaba, M.A. Shah, O.A. Khan, A.W. Ahmed, Software defined network (sdn) based internet of things (iot) a road ahead, in *The International Conference on Future Networks and Distributed Systems* (ACM, 2017), pp. 1–8
3. A. Prajapati, A. Sakadasariya, J. Patel, (2018, January). Software defined network: Future of networking, in *2018 2nd International Conference on Inventive Systems and Control (ICISC)*, IEEE, (2018), pp. 1351–1354
4. A. Voellmy, J. Wang, Scalable software defined network controllers, in *ACM SIGCOMM 2012 conference on Applications, technologies, architectures, and protocols for computer communication* (ACM, 2012), pp. 289–290
5. Y. Li, M. Chen, Software-defined network function virtualization: a survey. IEEE Access **3**, 2542–2553 (2015)
6. M. Yang, Y. Li, D. Jin, L. Zeng, X. Wu, A.V. Vasilakos, Software-defined and virtualized future mobile and wireless networks: a survey. Mob. Netw. Appl. **20**(1), 4–18 (2015)
7. J. Matias, J. Garay, N. Toledo, J. Unzilla, E. Jacob, Toward an SDN-enabled NFV architecture. IEEE Commun. Mag. **53**(4), 187–193 (2015)
8. M. Ojo, D. Adami, S. Giordano, A SDN-IoT architecture with NFV implementation, in *2016 IEEE Globecom Workshops (GC Wkshps)* (IEEE, 2016), pp. 1–6
9. V.G. Nguyen, A. Brunstrom, K.J. Grinnemo, J. Taheri, SDN/NFV-based mobile packet core network architectures: a survey. IEEE Commun. Surv. Tutor. **19**(3), 1567–1602 (2017)

10. M.S. Bonfim, K.L. Dias, S.F. Fernandes, Integrated NFV/SDN architectures: a systematic literature review. ACM Comput. Surv. (CSUR) **51**(6), 1–39 (2019)
11. O. Flauzac, C. González, A. Hachani, F. Nolot, SDN based architecture for IoT and improvement of the security, in *2015 IEEE 29th international conference on advanced information networking and applications workshops* (IEEE, 2015), pp. 688–693
12. Q. Waseem, S.S. Alshamrani, K. Nisar, W.I.S. Wan Din, A.S. Alghamdi, Future technology: software-defined network (SDN) forensic. Symmetry **13**(5), 767 (2021)
13. S. Seeber, L. Stiemert, G.D. Rodosek, Towards an SDN-enabled IDS environment, in *2015 IEEE Conference on Communications and Network Security (CNS)* (IEEE, 2015), pp. 751–752
14. S. Scott-Hayward, G.O'Callaghan, S. Sezer, SDN security: a survey, in *2013 IEEE SDN For Future Networks and Services (SDN4FNS)* (IEEE, 2013), pp. 1–7
15. J.C.C. Chica, J.C. Imbachi, J.F.B. Vega, Security in SDN: a comprehensive survey. J. Netw. Comput. Appl. **159**, 102595 (2020)
16. M.H. Khairi, S.H. Ariffin, N.A. Latiff, A.S. Abdullah, M.K. Hassan, A review of anomaly detection techniques and distributed denial of service (DDoS) on software defined network (SDN). Eng. Technol. Appl. Sci. Res. **8**(2), 2724–2730 (2018)
17. L. Wei, C. Fung, FlowRanger: a request prioritizing algorithm for controller DoS attacks in Software Defined Networks, in *2015 IEEE International Conference on Communications (ICC)* (IEEE 2015), pp. 5254–5259
18. N.N. Dao, J. Park, M. Park, S. Cho, A feasible method to combat against DDoS attack in SDN network, in *International Conference on Information Networking (ICOIN)* (IEEE, 2015), pp. 309–311
19. H. Wang, L. Xu, G. Gu, Floodguard: a dos attack prevention extension in software-defined networks, in *45th Annual IEEE/IFIP International Conference on Dependable Systems and Networks* (IEEE, 2015), pp. 239–250
20. R. Sahay, G. Blanc, Z. Zhang, H. Debar, Towards autonomic DDoS mitigation using software defined networking, in *SENT 2015: NDSS Workshop on Security of Emerging Networking Technologies* (Internet society, 2015)
21. S. Shin, V. Yegneswaran, P. Porras, G. Gu, Avant-guard: Scalable and vigilant switch flow management in software-defined networks, in *The 2013 ACM SIGSAC Conference on Computer and Communications Security* (ACM, 2013), pp. 413–424
22. R. Mohammadi, R. Javidan, M. Conti, SLICOTS: an SDN-based lightweight countermeasure for TCP SYN flooding attacks. IEEE Trans. Netw. Serv. Manage. **14**(2), 487–497 (2017)
23. K. Giotis, C. Argyropoulos, G. Androulidakis, D. Kalogeras, V. Maglaris, Combining OpenFlow and sFlow for an effective and scalable anomaly detection and mitigation mechanism on SDN environments. Comput. Netw. **62**, 122–136 (2014)
24. M. Lopez, U. Figueiredo, A. Lobato, O.C. Duarte, Broflow: Um sistema eficiente de detecção e prevenção de intrusão em redes definidas por software, in *Anais do XIII Workshop em Desempenho de Sistemas Computacionaise de Comunicação* (SBC, 2014), pp. 108–121
25. T. Chin, X. Mountrouidou, X. Li, K. Xiong, Selective packet inspection to detect DoS flooding using software defined networking (SDN), in *2015 IEEE 35th International Conference on Distributed Computing Systems Workshops* (IEEE, 2015), pp. 95–99
26. T. Xing, D. Huang, L. Xu, C.J. Chung, P. Khatkar, Snortflow: a openflow-based intrusion prevention system in cloud environment, in *Second GENI Research and Educational Experiment Workshop* (IEEE, 2013), pp. 89–92

Detection of Security Attacks Using Intrusion Detection System for UAV Networks: A Survey

Khaista Rahman, Muhammad Adnan Aziz, Ahsan Ullah Kashif, and Tanweer Ahmad Cheema

Abstract UAV-network is basically referred to flying ad hoc networks. Mobile ad hoc network is considered the main concept behind UAV-network. Unmanned aerial vehicles form a high speed topological flying network. As due to innovations in various fields of study, security is very much important. Therefore, UAV-networks need cyber secure flying air space for communication. Due to high mobility in FANETs DoS/DDoS security attacks can be deployed by intruders to take access of data. Denial of service is also called third party attack. While, distributed denial of service is a dangerous threats to UAV-network which attempt to unbalance the behavior of network. UAV's are having many applications in civil and military fields. FANETs are quite adaptable with diverse background than other traditional networks. However, DoS/DDoS attacks make vulnerable UAV-networks by broadcasting false information to hijack the entire system. Though, FANETs structure consists of base station, UAVs and satellite. This paper presents the comprehensive survey regarding different IDS for UAV-networks. Also explains architectural overview of FANETs which include wireless communication technologies and protocols for optimal communication. Detection of normal and abnormal data packets can be easily identified using intrusion detection system. Apart from that security attacks, high accuracy and monitoring of data packets will be the prime focus especially in the area of flying ad hoc networks.

Keywords FANETs · UAVs · IoT · IDS

K. Rahman (✉) · M. A. Aziz · T. A. Cheema
Department of Electronic Engineering, School of Engineering and Applied Sciences (SEAS), Isra University, Islamabad Campus, Islamabad, Pakistan
e-mail: malakand73@gmail.com

A. U. Kashif
Department of Electrical Engineering, International Islamic University Islamabad, Islamabad, Pakistan

© The Author(s), under exclusive license to Springer Nature Switzerland AG 2022
M. Ouaissa et al. (eds.), *Big Data Analytics and Computational Intelligence for Cybersecurity*, Studies in Big Data 111,
https://doi.org/10.1007/978-3-031-05752-6_7

1 Introduction

UAV-networks using 5G communication technologies will play significant role to improve channels. FANETs are derived from mobile ad hoc networks which move in three different directions. However, there can be either single or multi-UAV system [1]. UAVs are now commercialized which use to have different applications like aerial photography, surveying various areas, agriculture crops monitoring and rescue operations. Also, in few decades' agriculture crop management using drones has improved decision making process. This approach reduces cost and helps to spray the crops properly. Instead of satellite, UAV images have high resolution which is quite beneficial for mapping analysis [2]. Though, product delivery from one place to another is introduced by Amazon using mini-drones. Progressive economies must have better transportation mechanism while delivering goods from one place to another. COVID-19 has influenced every field, but transportation industry has invested a lot on UAVs to boost their businesses [3].

Improving the standards in between UAV-to-UAV and UAV-to-base station routing protocols use to enhance communication channels. Wireless based technological protocols which include Bluetooth, 2G, 3G, 4G, 5G and beyond can be used to secure communication in UAV-networks. In addition routing techniques like DSR, AODV, MDART, ZRP and natured inspired ACO find best path from source to destination in FANETs [4]. Aerial ad hoc networks receive data packet information from base station and satellite. Therefore, IoT based connected UAV-network need novel approach to eliminate congestion. Improved sequencing heuristic-DSDV is proposed which must have incremental updates, loop free and updated routing table information mechanism. Nomadic mobility model is utilized in UAV-network to cover different zones which optimize communication [5].

Wireless communication is the backbone technology for UAV-networks. First generation was earlier the most low cost but having limited area to cover. Wireless local area networks lies in second generation networks for shaping the optimize communication links between nodes. 3G and 4G have subsequently high data rates in comparison with traditional techniques. OFDM is the basic model for 5G communication networks which is having high capabilities with better connectivity standards. While 6G is quite adaptable and flexible which improve connected things with smart intelligence approach [6].

Intruder launch different cyber-attacks like sink hole, DNS, DoS, DDoS which directly affect to the unavailable node. However, ping of death is just like DoS attack to hijack the network through malicious data packets. Intrusion detection system is the most widely used approach to identify security attacks. There exist three types of IDS which include signature, anomaly and hybrid for giving alert messages to base station. While, IoT based network is designed to evaluate the performance utilizing DoS/DDoS attacks. Moreover, analytical approach using anomaly-IDS is implemented to address miss detection over queue length [7]. As discussed earlier aerial networks are quite vulnerable due to that intelligent detection system is proposed which relies on Markov chain distribution. The network topology consists of UAVs, wireless technology like ZigBee and novel IDS is designed to secure communication within UAVs. Queue length approach is maintained using binomial Markov chain distribution and detection of DoS/DDoS are easily identified [8].

This paper mainly contributes towards attack identification where major points are as under:

- Topological structure and architectural design of UAV-networks
- Routing protocols for FANETS
- Different types of intrusion detection system for UAV-network
- Security attacks like DoS/DDoS detection in aerial ad hoc networks
- Mobility models for UAV-networks.

2   Literature Survey

This section is focused on literature study regarding UAV-networks which are as below.

Cognitive IoT-networks utilize IEEE 802.15.4 to provide better channel communication. Therefore, unlicensed wireless networks use various technologies for heterogeneous networks. End-to-end channel path failure, throughput and source routing is improved [9].

Multiple UAV application covers more geographical area than single UAV network. In multiple UAV application more the one sub network has been created among UAV. All data has been passed through backbone UAV to reach to its desired destination. In multi-UAV strong communication link is needed between UAV and base station for better utilization of network resources and special type of hardware is needed.

UAV based IoT networks have revolutionize the field of sports for monitoring players. FANETs are having UAV nodes which collect information from ground IoT sensor nodes and send to base station. As aerial vehicles move in three different directions (x, y, z) to observe sport players performance. Therefore, Ant-hocnet is proposed for efficient communication among UAV-network [10]. The vision of 6G implementation will change the dynamics of wireless networks. 6G will reconfigure artificial intelligence, machine learning and will improve connectivity in between different wireless nodes. 6G technologies will enhance block-chain, quantum communication, UAVs, big data, and higher energy in wireless autonomous system [11]. Inam Ullah Khan et al., designed a topology for FANETs which consist of nine UAVs. As aerial vehicles are moving away from base station signal strength becomes weak. Therefore, machine learning technique called decision tree is utilized to improve receive signal strength indicator from base station to UAVs. Also, proposed solution is compared with wireless sensor networks localization technique using 2-D and 3-D environment [12]. Fisheye state routing protocol is deployed on the environment of UAV-network which has improved throughput, packet drop rate, congestion and average end-to-end delay in comparison with TORA, OLSR, AODV, DSR and DSDV. FSR routing mechanism is based on neighbor discovery, information dissemination and route optimization [13]. Muhammad Abul Hassan et al., deployed FSR routing protocol to improve energy efficiency in flying ad hoc networks. Packet delivery, bandwidth and throughput are utilized to evaluate FANET performance [14]. UAV-network is having unique features but quite open to security attacks. FANETs are having many applications where security is unsatisfactory. Therefore, hybrid intrusion detection system is formulated for traffic analysis to monitor data packets. Hybrid is basically the integration of anomaly and signature to detect real traffic based flooding. Paparazzi system is specially used for real time experimentation which is decomposed into three segments which include ground segment, airborne and communication segment [15]. Centralized system have limited bandwidth and power. Ad hoc networks are vulnerable to security attacks which violate passively and actively attacking directly. Malicious workstations send false information where worm hole tunneling, DoS and black hole attack are launched [16].

The mentioned research papers describes related issues regarding flying ad hoc networks which include security, privacy, energy, reliability, mobility patterns and network traffic management. Apart from that cyber security attacks like DoS and DDoS must be detected using intrusion detection system. Due to these problems flying ad hoc networks nodes might be hijacked or either has unbalanced energy level which affect overall network.

3　Architectural Design of UAV-Networks

UAV-network is combination of aerial vehicles, base station and satellite. However, communication among nodes can be made possible in ad hoc behavior. UAV-network effectively covers specified area which is under range of wireless technologies like IEEE 802.11 or zigbee. In centralized approach UAVs can only be connected with more than one base station. Ground base station is utilized to divert the flow of data traffic and reduce failures. Centralized approach is having many problems like dedicated bandwidth, latency and attack vulnerability. Although, multi-UAVs are selected which connect multi-group aerial vehicles to ground base stations? Intergroup and intra communication standards provide optimal performance [17].

4　Routing Protocols for FANETs

For better and effective communication in FANETs routing protocols are very important. Traditional routing protocols are modified with the passage of time. However, FANET routing protocols are divided in three basic categories [18, 19]. Figure 1, describes the categorization of routing protocols.

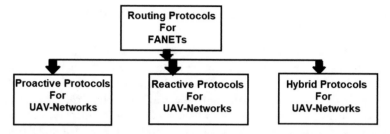

Fig. 1 Categorization of routing protocols

4.1 Proactive Protocols for UAV-Network

Proactive routing protocols are considered table driven which use to maintain one or more routing tables. DSDV, OLSR and POLSR use to establish one hop communication which handles searching, updating, adding and removing [20].

4.2 Reactive Protocols for UAV-Network

On demand searching of route from one node to another is basically called reactive behavior. There are two main components which include route discovery and maintenance. AODV, DSR and TORA routing protocols are based on reactive nature [21].

4.3 Hybrid Protocols for UAV-Network

Hybrid routing protocols are the combination of proactive and reactive nature. Zone routing protocol is hybrid natured strategy which use to divide topology in different zones and clusters [22].

UAV-networks routing protocols are further classified on the basis of either position, topology or may be cluster [23]. Some of the routing mechanisms which can be deployed in flying ad hoc networks are as under:

- Topology aware routing
- Topology construction method
- Jamming Resilient Multipath routing
- Position aware routing
- Cluster based routing protocols.

5   Intrusion Detection System for UAV-Networks

Detection of cyber-attacks is considered a very serious problem. However, if the security attacks will be identified on specified then entire system can be safe-guarded. Else due to this issue might excess of flooding data packets can confuse the network due to which direct control goes to intruder/attacker. Therefore, giving proper solution intrusion detection system is designed which use to identify different cyber-attacks. Combination of aerial vehicles, 5G networks with satellite has advanced the quality of service. Although, security is one of the major issue which comes in parallel with every network system. Rakesh et al. demonstrated the concept of novel IDS using machine learning algorithms on UAVs and satellite pathways [24].

Suspected/unwanted data packets are broadcasted by intruders to steal legitimate information. KDD-Cup 99 or DARPA 1999 data sets are mostly utilized to improve the performance of intrusion detection system. Preventing high profile damage in networks, IDS use to monitor unauthorized activities. Security attacks like DoS/DDoS affect the performance and availability of authentic node. Therefore, IDS detect unwanted data packets and nodes and send alert message to base station. Due to this activity entire network or system is protected from malicious attacks [25].

Rapid growth of malware is a serious threat to UAV-networks; therefore contemporary systems are needed to secure the systems. A novel strategy is proposed called DynODet which identify self-modification patterns by comparing with operands. Dynamic optimization is utilized to overwrite the function. Altering the dynamics malware use to change the permission in read and write options [26].

Devices/Systems are interconnected through internet, due to that many incidents of attacks are increased. Data mining strategy are used to design hidden naïve bayes model to detect possible security attacks [27].

Wireless technology is utilized for short range communications which use to have many security vulnerabilities. Therefore, IEEE 802.11 is having security problems where signature based IDS detect attacks. AWID family datasets are well explained to evaluate normal and unwanted data traffic [28].

However, Amdal's law is used for string match-selection process which enhances the processing load for about 75%. A novel IDS is designed which mainly evaluate processing time, header processing time and payload processing to monitor security violations [29, 30].

As 4.0 industrial revolutions has changed the mode to wireless in many applications like water, electricity, ground vehicles etc. In addition, machine learning approach ADIN proposes optimal solution for data traffic management to minimize error [31]. Figure 2, shows different types of intrusion detection system.

Apart from that intrusion detection system is further divided to identify various types of attacks which are given below:

- Network intrusion detection system
- Host based intrusion detection system
- Signature based intrusion detection system

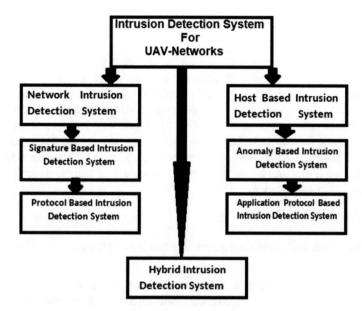

Fig. 2 Types of intrusion detection systems for UAV-networks

- Anomaly based intrusion detection system
- Hybrid intrusion detection system
- Protocol based intrusion detection system
- Application Protocol based intrusion detection system.

5.1 Network-Intrusion-Detection-System

Network traffic can be monitored through multiple locations to identify suspicious data packets can be made possible through network-IDS. Deep neural networks are utilized to improve IDS capabilities for detecting zero-day attacks. However, UNSW-NB15 dataset is used in the research study to check the adaptive behavior of novel network-IDS [32].

Network-IDS is usually connected with switch to monitor incoming and outgoing traffic and identify attacker's data packets. NIDS is very helpful while detecting cyber-attacks and can be deployed between firewall and switch. AI is used as tool in network-IDS which rely on three stages like data processing, training and testing [33].

5.2   Host-Based-Intrusion-Detection-System

IoT based intrusion detection system is designed to detect unwanted data which can be either anomaly like relying on behavior or signature. In the proposed approach Z-wave network is used to trace data collection, home controller process data packets and analysis system send an alert message to safeguard IoT networks [34]. Although, host-IDS observe the attributes of single node and monitor related events within host for unauthentic activity [35]. HIDS monitor unwanted data packets which might unbalance network energy level. Also, intruder sometimes edit or delete information therefore if any suspicious activity can be seen then alert is directly send to administrator or base station.

5.3   Signature-Based-Intrusion-Detection-System

Signature-IDS use to have data base to store history of authorized users which is helpful to identify intruders. Moreover, normal and abnormal data patterns are incorporated of the existing users to detect unauthorized nodes. Past history or traces can be utilized which need to matched while receiving in coming data packets [36]. Identification of virus and worms which might access to database of cyber-attacks signatures which need to be compare with existing data. In the proposed work snort is used as detection system [37].

5.4   Anomaly-Based-Intrusion-Detection-System

Formulation of threshold is helpful in the approach of anomaly based intrusion detection system to detect attacks. Therefore, behavior of data packets must be evaluated to specify unwanted activity if it crosses the threshold. Muaadh et al., describe deep learning techniques for anomaly based IDS to identify zero-day attacks [38]. Anomaly-IDS use to search abnormal data traffic which must have high rate of detecting false alarm [39].

5.5   Hybrid-Intrusion-Detection-System

Hybrid-IDS approach uses to integrate two or more different IDS features to make one hybrid. Host, anomaly and signature can formulate hybrid-IDS which is quite effective in comparison with other contemporary techniques. Interconnected network like cloud computing, various resources, IoT networks are vulnerable to attain high accuracy. Therefore, a combination of machine learning approaches are integrated which include K-means clustering and support vector machine utilized to form hybrid-IDS to identify unknown attacks [40]. Hybrid architecture is proposed to counter and detect malicious activities like data breaches, injection and DDoS attack. Also, cloud based network system is designed to monitor unwanted data packets and secure communication channels [41].

5.6 Protocol-Based-Intrusion-Detection-System

Protocol-IDS observe the working mechanism regarding server or network system. However, for online server, hyper-text-transfer protocol secure is implemented for optimal communication in between users. These protocols are widely used in web-applications for better security. Bayesiam, Naïve bayes, random forest and decision tree are utilized to mimic and eliminate false data packets [42]. Unsupervised technique is used to alter the numeric features where packet anomalies identify about seven possible attacks in training stage [43].

5.7 Application-Protocol-Based-Intrusion-Detection-System

Application-protocol IDS is specialized in monitoring data packets to detect false information within networks. This approach usually works as middle agent which resides within group of servers for identification unwanted nodes. Also, this technique is applied only on protocols and standards [44, 45].

6 Security Attacks on UAV-Networks

Unmanned aerial vehicles (UAVs) collectively make flying networks which is having security problems. Therefore, researchers either work on attack detection or prevention. Intruders attack on the UAV-network to unbalance the level of entire network. Data integrity and network availability are targeted for quick attack identification [46]. Aerial vehicles are mostly deployed on sensitive missions to collect data from ground and send to base station. Due to high mobility intruders directly attack on UAVs to hijack the entire UAV-network [47]. Figure 3, depicts security attacks on UAV-networks.

Some security attacks are categorized as follow:

Fig. 3 Security attacks on
UAV-networks

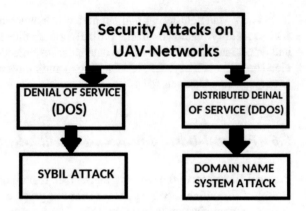

6.1   Denial of Service (Dos)

DoS is third party attack which means when two nodes are communicating with each other, third node pretend to be like them and send fake data packets. Due to DoS intruder send flooding information to reduce authentic node bandwidth. Three commonly utilized methods of DoS attacks are given below [48].

- Smurf Attack
- SYN Flood Attack
- UDP Flood Attack.

Dos attack usually disturb and unbalance internet data traffic. In addition DoS mislead the entire network due flooding. Apart from that on time access to legitimate information become quite tough because of denial-of-service [49].

6.2   Distributed Deinal of Service (DDOS)

DDoS attacks directly disrupt the entire network. DDoS means the concept of multiple or group of intruders which focus on single workstation. DDoS attacks mislead, devalue and unbalance entire network energy [50, 51]. Sub-categorized DDoS attacks are as under.

- DDoS zombie Attack
- Land Attack
- ICMP Flood Attack
- HTTP Attack
- Control layer DDoS
- Data layer DDoS.

DDoS attacks are deployed by intruders in the field which include smart grid, UAV-network, VANETs, SANETs, flying ad hoc networks, IoT, IoE and healthcare.

6.3 Sybil Attack

Sybil attack is most commonly used by intruders to take control of single legitimate user. In Sybil attack peer-to-peer or one-to-one correspondence is affected [52].
Sybil attacks usually cause/disrupt:

- Optimal Routes
- Voting Systems
- Create Collisions
- Unfair resource allocation
- Data aggregation.

6.4 Domain Name System (DNS) Attack

Utilizing DNS or IP-spoofing is very dangerous for single and multi-user networks [53]. DNS cache poisoning can be utilized positively for advancing countermeasures and also effect intruders by negative usage. Hijackers use to send many broadcasted messages to a random or specific node [54].

7 Mobility Models for UAV-Networks

FANETs or UAV-networks are having many applications due to high mobility patterns. Deployment of routing protocols in aerial-networks needs to have proper mobility model for effective communication. For UAVs random waypoint is widely utilized to optimize the performance. However, Manhattan grid uses probabilistic approach for selecting UAV node movements. For better node speed and direction boundless simulation area is the better choice to be utilized for UAV-network. Although, for randomness Gauss-Markov model is helpful for flying ad hoc networks [55]. Some other mobility patterns are as below:

- Smooth Turn model
- Semi random circular movement model
- Column mobility model.

8 Future Directions

This research study is comprehensive survey in the field of UAV-networks. FANET nodes are quite fast in movements and must have proper direction. Aerial vehicles are having many applications in civil and military fields. There exist a lot of limitations in UAV-networks which need to be optimized. Due to COVID-19, the use of UAVs is increased a lot in every field. Small and large size UAVs move in all three directions and congestion must be reduced. Privacy and security issues must be addressed properly in UAV-networks. Also, energy is considered one of the major problems in FANETs. Machine learning, deep learning, artificial neural networks, cryptography, evolutionary computing techniques like genetic algorithm, particle swarm optimization and ant colony optimization can improve the overall performance of UAV-networks. Apart from that novel techniques need to be formulated by researchers and practitioners to enhance quality-of-service metrics for UAV-networks [56].

9 Conclusion

UAV-networks use different mobility patterns for better communication in between nodes. FANETs dynamically change topological structures which creates many limitations. However, this research is the comprehensive study regarding UAV-networks which gives real-times solutions. Although, the above study include architectural design of UAVs, 5G networks integration with aerial vehicles, routing protocols, various types of intrusion detection algorithms, mobility patterns and widely used cyber-attacks. In addition, industrial and commercial use of UAVs has rapidly increased which will facilitate every human around the world.

References

1. I.U. Khan, I.M. Qureshi, M.A. Aziz, T.A. Cheema, S.B.H. Shah, Smart IoT control-based nature inspired energy efficient routing protocol for flying ad hoc network (FANET). IEEE Access **8**, 56371–56378 (2020). https://doi.org/10.1109/ACCESS.2020.2981531
2. K. Srivastava, P.C. Pandey, J.K. Sharma, An approach for route optimization in applications of precision agriculture using UAVs. Drones **4**, 58 (2020). https://doi.org/10.3390/drones4030058
3. J.-P. Aurambout, K. Gkoumas, B. Ciuffo, Last mile delivery by drones: an estimation of viable market potential and access to citizens across European cities. Eur. Transp. Res. Rev. **11**(1), 1–21 (2019)
4. I.U. Khan, S.B.H. Shah, L. Wang, M.A. Aziz, T. Stephan, N. Kumar, Routing protocols and unmanned aerial vehicles autonomous localization in flying networks. Int. J. Commun. Syst. e4885
5. I.U. Khan, M.A. Hassan, M. Fayaz, J. Gwak, M.A. Aziz, Improved sequencing heuristic DSDV protocol using nomadic mobility model for FANETS. CMC-Compu. Mater. Contin. **70**(2), 3653–3666 (2022)

6. I.U. Khan, S.Z.N. Zukhraf, A. Abdollahi, S.A. Imran, I.M. Qureshi, M.A. Aziz, S.B.H. Shah, Reinforce based optimization in wireless communication technologies and routing techniques using internet of flying vehicles, in *The 4th International Conference on Future Networks and Distributed Systems (ICFNDS)* (2020), pp. 1–6
7. A. Abdollahi, F. Mohammad Fathi, An intrusion detection system on ping of death attacks in IoT networks. Wirel. Pers. Commun. 1–14 (2020)
8. I.U. Khan, A. Abdollahi, R. Alturki, M.D. Alshehri, M.A. Ikram, H.J. Alyamani, S. Khan, Intelligent detection system enabled attack probability using Markov chain in aerial networks. Wirel. Commun. Mob. Comput. (2021)
9. S. Begum, Y. Nianmin, S.B.H. Shah, A. Abdollahi, I.U. Khan, L. Nawaf, Source routing for distributed big data-based cognitive internet of things (CIoT). Wirel. Commun. Mob. Comput. (2021)
10. I.U. Khan, M.A. Hassan, M.D. Alshehri, M.A. Ikram, H.J. Alyamani, R. Alturki, V.T. Hoang, Monitoring system-based flying IoT in public health and sports using ant-enabled energy-aware routing. J. Healthc. Eng. (2021)
11. A. Sher, M. Sohail, S.B.H. Shah, D. Koundal, M.A. Hassan, A. Abdollahi, I.U. Khan, New trends and advancement in next generation mobile wireless communication (6G): a survey. Wirel. Commun. Mob. Comput. (2021)
12. I.U. Khan, R. Alturki, H.J. Alyamani, M.A. Ikram, M.A. Aziz, V.T. Hoang, T. A. Cheema, RSSI-controlled long-range communication in secured IoT-enabled unmanned aerial vehicles. Mob. Inf. Syst. (2021)
13. M.A. Hassan, S.I. Ullah, I.U. Khan, S.B.H. Shah, A. Salam, A.W.U. Khan, Unmanned aerial vehicles routing formation using fisheye state routing for flying ad-hoc networks, in *The 4th International Conference on Future Networks and Distributed Systems (ICFNDS)* (2020), pp. 1–7
14. M.A. Hassan, S.I. Ullah, A. Salam, A.W. Ullah, M. Imad, F. Ullah, Energy efficient hierarchical based fish eye state routing protocol for flying Ad-hoc networks. Indones. J. Electr. Eng. Comput. Sci. **21**(1), 465–471 (2021)
15. J.-P. Condomines, R. Zhang, N. Larrieu, Network intrusion detection system for UAV ad-hoc communication (2018)
16. R.H. Jhaveri, S.J. Patel, D.C. Jinwala, DoS attacks in mobile ad hoc networks: a survey, in *Second International Conference on Advanced Computing and Communication Technologies* (IEEE, 2012), pp. 535–541
17. O.S. Oubbati, M. Atiquzzaman, P. Lorenz, M.H. Tareque, M.S. Hossain, Routing in flying ad hoc networks: survey, constraints, and future challenge perspectives. IEEE Access **7**, 81057–81105 (2019). https://doi.org/10.1109/ACCESS.2019.2923840
18. M.H. Tareque, M.S. Hossain, M. Atiquzzaman,On the routing in flying ad hoc networks, in *Federated Conference on Computer Science and Information Systems (FedCSIS)* (2015), pp. 1–9. https://doi.org/10.15439/2015F002
19. M.B. Yassein, N. Alhuda, Flying ad-hoc networks: routing protocols, mobility models, issues. Int. J. Adv. Comput. Sci. Appl. (IJACSA) **7**(6) (2016)
20. Y. Bai, Y. Mai, N. Wang, Performance comparison and evaluation of the proactive and reactive routing protocols for MANETs, in *Wireless Telecommunications Symposium (WTS)* (2017), pp. 1–5. https://doi.org/10.1109/WTS.2017.7943538
21. S. Mohseni, R. Hassan, A. Patel, R. Razali, Comparative review study of reactive and proactive routing protocols in MANETs, in *4th IEEE International Conference on Digital Ecosystems and Technologies* (2010), pp. 304–309. https://doi.org/10.1109/DEST.2010.5610631
22. Y. Kumar, Isha, A. Malik, Performance analysis of hybrid routing protocol, in *2nd International Conference on Intelligent Engineering and Management (ICIEM)* (2021), pp. 400–405. https://doi.org/10.1109/ICIEM51511.2021.9445324
23. Q. Sang, H. Wu, L. Xing, P. Xie, Review and comparison of emerging routing protocols in flying ad hoc networks. Symmetry **12**(6), 971 (2020)
24. R. Shrestha, A. Omidkar, S.A. Roudi, R. Abbas, S. Kim, Machine-learning-enabled intrusion detection system for cellular connected UAV networks. Electronics **10**(13), 1549 (2021)

25. A. Khraisat, I. Gondal, P. Vamplew, J. Kamruzzaman, Survey of intrusion detection systems: techniques, datasets and challenges. Cybersecurity 2(1), 1–22 (2019)
26. D. Kim, A. Majlesi-Kupaei, J. Roy, K. Anand, K. ElWazeer, D. Buettner, R. Barua, Dynodet: detecting dynamic obfuscation in malware, in *International Conference on Detection of Intrusions and Malware, and Vulnerability Assessment* (Springer, Cham, 2017), pp. 97–118
27. L. Koc, T.A. Mazzuchi, S. Sarkani, A network intrusion detection system based on a Hidden Naïve Bayes multiclass classifier. Expert Syst. Appl. 39(18), 13492–13500 (2012)
28. C. Kolias, G. Kambourakis, A. Stavrou, S. Gritzalis, Intrusion detection in 802.11 networks: empirical evaluation of threats and a public dataset. IEEE Commun. Surv. Tutor. 18(1), 184–208 (2015)
29. H.-J. Liao, C.-H.R. Lin, Ying-Chih Lin, and Kuang-Yuan Tung, Intrusion detection system: a comprehensive review. J. Netw. Comput. Appl. 36(1), 16–24 (2013)
30. J.B.D. Cabrera, J. Gosar, W. Lee, R.K. Mehra, On the statistical distribution of processing times in network intrusion detection, in *43rd IEEE Conference on Decision and Control (CDC) (IEEE Cat. No.04CH37601)*, vol. 1 (2004), pp. 75–80. https://doi.org/10.1109/CDC.2004.1428609
31. A. Meshram, H. Christian, Anomaly detection in industrial networks using machine learning: a roadmap, in *Machine Learning for Cyber Physical Systems* (Springer Vieweg, Berlin, Heidelberg, 2017), pp. 65–72
32. L. Ashiku, C. Dagli, Network intrusion detection system using deep learning. Procedia Comput. Sci. 185, 239–247 (2021)
33. Z. Ahmad, A.S. Khan, C.W. Shiang, J. Abdullah, F. Ahmad, Network intrusion detection system: a systematic study of machine learning and deep learning approaches. Trans. Emerg. Telecommun. Technol. 32(1), e4150 (2021)
34. R. Gassais, N. Ezzati-Jivan, J.M. Fernandez, D. Aloise, M.R. Dagenais, Multi-level host-based intrusion detection system for Internet of things. J. Cloud Comput. 9(1), 1–16 (2020)
35. F. Sabahi, A. Movaghar,Intrusion detection: a survey, in *Third International Conference on Systems and Networks Communications* (2008), pp. 23–26. https://doi.org/10.1109/ICSNC.2008.44
36. S. Einy, C. Oz, Y.D. Navaei, The anomaly-and signature-based IDS for network security using hybrid inference systems. Math. Probl. Eng. (2021)
37. S.N. Shah, M.P. Singh, Signature-based network intrusion detection system using SNORT and WINPCAP. Int. J. Eng. Res. Technol. (IJERT) 1(10), 1–7 (2012)
38. M.A. Alsoufi, S. Razak, M.M. Siraj, I. Nafea, F.A. Ghaleb, F. Saeed, M. Nasser, Anomaly-based intrusion detection systems in iot using deep learning: a systematic literature review. Appl. Sci. 11(18), 8383 (2021)
39. J. Jabez, B. Muthukumar, Intrusion detection system (IDS): anomaly detection using outlier detection approach. Procedia Comput. Sci. 48, 338–346 (2015)
40. I. Aljamal, A. Tekeoğlu, K. Bekiroglu, S. Sengupta, Hybrid intrusion detection system using machine learning techniques in cloud computing environments, in *IEEE 17th International Conference on Software Engineering Research, Management and Applications (SERA)* (2019), pp. 84–89. https://doi.org/10.1109/SERA.2019.8886794
41. A. Meryem, B.E.L. Ouahidi, Hybrid intrusion detection system using machine learning. Netw. Secur. (5), 8–19 (2020)
42. S.A. Mandal, S. Sabitha, D. Mehrotra, Analysis on protocol-based intrusion detection system using artificial intelligence, in *Machine Intelligence and Smart Systems: Proceedings of MISS* (2020), p.131
43. M. Labonne, A. Olivereau, B. Polvé, D. Zeghlache, Unsupervised protocol-based intrusion detection for real-world networks, in *International Conference on Computing, Networking and Communications (ICNC)* (2020), pp. 299–303. https://doi.org/10.1109/ICNC47757.2020.9049796
44. A. Chung, On testing of implementation correctness of protocol based intrusion detection systems, in *Ninth International Conference on Software Engineering Research, Management and Applications* (2011), pp. 171–174. https://doi.org/10.1109/SERA.2011.26

45. M.F. Wu, Protocol-based classification for intrusion detection, in *Proceedings of the WSEAS International Conference on Mathematics and Computers in Science and Engineering*, vol. 7 (World Scientific and Engineering Academy and Society, 2008)
46. H. Sedjelmaci, S.M. Senouci, M. Messous,How to detect cyber-attacks in unmanned aerial vehicles network?, in *IEEE Global Communications Conference (GLOBECOM)* (2016), pp. 1–6. https://doi.org/10.1109/GLOCOM.2016.784187
47. N.M. Rodday, R.d.O. Schmidt, A. Pras, Exploring security vulnerabilities of unmanned aerial vehicles, in *IEEE/IFIP Network Operations and Management Symposium (NOMS)* (2016), pp. 993–994. https://doi.org/10.1109/NOMS.2016.7502939
48. F. Lau, S.H. Rubin, M.H. Smith, L. Trajkovic, Distributed denial of service attacks, in *Proceedings of the IEEE International Conference on Systems, Man and Cybernetics (Smc): Cybernetics Evolving to Systems, Humans, Organizations, and Their Complex Interactions* (cat. no.0), vol. 3 (2000), pp. 2275–2280. https://doi.org/10.1109/ICSMC.2000.886455
49. A. Huseinović, S. Mrdović, K. Bicakci, S. Uludag, A survey of denial-of-service attacks and solutions in the smart grid. IEEE Access **8**, 177447–177470 (2020). https://doi.org/10.1109/ACCESS.2020.3026923
50. S. Dong, K. Abbas, R. Jain, A Survey on distributed denial of service (DDoS) attacks in SDN and cloud computing environments. IEEE Access **7**, 80813–80828 (2019). https://doi.org/10.1109/ACCESS.2019.2922196
51. R.R. Brooks, l. Ozcelik, L. Yu, J. Oakley, N. Tusing, Distributed denial of service (DDoS): a history, in *IEEE Annals of the History of Computing,* doi: https://doi.org/10.1109/MAHC.2021.3072582
52. R. John, J.P. Cherian, J.J. Kizhakkethottam, A survey of techniques to prevent sybil attacks, in *International Conference on Soft-Computing and Networks Security (ICSNS)* (2015), pp. 1–6. https://doi.org/10.1109/ICSNS.2015.7292385
53. S. Saharan, V. Gupta, Prevention and mitigation of DNS based DDoS attacks in SDN environment, in *11th International Conference on Communication Systems and Networks (COMSNETS)* (2019), pp. 571–573. https://doi.org/10.1109/COMSNETS.2019.8711258
54. I.M.M. Dissanayake,DNS cache poisoning: a review on its technique and countermeasures, in *National Information Technology Conference (NITC)* (2018), pp. 1–6. https://doi.org/10.1109/NITC.2018.8550085
55. A. Bujari, C.T. Calafate, J.-C. Cano, P. Manzoni, C.E. Palazzi, D. Ronzani. Flying ad-hoc network application scenarios and mobility models. Int. J. Distrib. Sens. Netw. **13**(10) (2017). 1550147717738192
56. S. Poikonen, J.F. Campbell, Future directions in drone routing research. Networks **77**(1), 116–126 (2021)

Computational Intelligence
for Cybersecurity

Role of Computational Intelligence in Cybersecurity

Muhammad Yaseen Ayub, Mohammad Ammar Mehdi, Syeda Ghanwa Tawaseem, Syeda Zillay Nain Zukhraf, and Zupash

Abstract In this era of connectivity where billions of devices are interconnected for various purposes. As the network expands new threats are emerging requiring researchers to work for an effective Intrusion Detection System (IDS) that can mitigate those attacks. In this paper, we discuss common cyber-attacks and IDS used to cope with them. We discuss the use of Computational Intelligence in designing an IDS which can effectively work on unseen data. Because in today's dynamically changing world, it is essential to have IDS that can work on unseen data (Schatz et al. in Journal of Digital Forensics, Security and Law 12:8, 2017; Kott in Towards fundamental science of cyber security. Springer, New York, NY, pp. 1–13).

Keywords Cybersecurity · Attacks · Threats · Computational intelligence

M. Y. Ayub · Zupash
Department of Computer Science, COMSATS University Islamabad, Attock, Pakistan
e-mail: SP19-BCS-025@cuiatk.edu.pk

Zupash
e-mail: SP19-BCS-049@cuiatk.edu.pk

M. A. Mehdi
Department of Electrical Engineering, The Islamia University of Bahawalpur, Bahawalpur, Pakistan
e-mail: engr.ammar72@gmail.com

S. G. Tawaseem
Department of Computer Science, HITEC University, Taxila, Pakistan
e-mail: 18-CS-080@student.hitecuni.edu.pk

S. Z. N. Zukhraf (✉)
Department of Electrical and Computer Engineering, KIOS Research and Innovation Centre of Excellence, University of Cyprus, Nicosia, Cyprus
e-mail: zukhraf.syeda-zillay@ucy.ac.cy

© The Author(s), under exclusive license to Springer Nature Switzerland AG 2022   127
M. Ouaissa et al. (eds.), *Big Data Analytics and Computational Intelligence for Cybersecurity*, Studies in Big Data 111,
https://doi.org/10.1007/978-3-031-05752-6_8

1 Introduction

As we see the connectivity is increasing exponentially with the ubiquitous availability of internet connectivity around us. With the further development of Wireless Sensor Networks (WSNs), we are seeing a world where everything is interconnected to provide a more personalized experience to the user. While we have reached the point where different computational intelligence-based smart devices are working together using WSNs, there are also security strings attached to it as there is a web of devices connected with reference to Fig. 1. WSNs have distributed highly scalable networks, are dynamic, and do not require any predefined infrastructure to function. While these are pros of WSNs these things make it difficult to maintain them. These things make the network more vulnerable to a wide variety of attacks [3]. Our mobile phones, handheld devices, and wearable gadgets all are connected through wireless networks. For Example, smart glasses and smartwatches are connected to our smartphones and share information wirelessly to provide us with a more personalized experience. Many smart devices require access to different services like GPS or require internet connectivity to work properly. Since Wireless Sensor Networks are resource-limited and are prone to cyber threats like GPS Spoofing, DoS attack, Ping of Death Attack, etc. It is essential to have a strong Intrusion Detection System (IDS) that can detect abnormal behavior in the network effectively, warn and prevent data loss. Attacks can occur against power stations or smart grids, Industrial systems, military systems (radars for detecting unidentified objects) [4], or any other thing that is connected to the internet. It is said that by 2022 the total number of networked devices will be 28.5 billion [5, 6].

Fig. 1 Introduction to computational intelligence in cybersecurity

Thus, an effective IDS is essential to mitigate these threats. Our world is dynamic. It is changing rapidly and each day new challenges occur. To design an IDS which is smart enough to work in this dynamic environment we make use of Computational Intelligence [7]. Computational intelligence is a field that uses different bio-inspired approaches to develop such a system that can solve complex problems in a dynamically changing environment [8]. Using computational intelligence techniques, we design adaptive IDS which can work in a dynamic environment and learn from data and experiments to cope with the changing environment [9, 10].

2 Motivation

Nowadays computers are being embedded into everything. From handheld devices to wearables everything is getting an embedded microcomputer [11]. In a world where everything is getting digitalized and connected to a network, we cannot look away from the security risks it brings along it. Making such a sophisticated network safe and secure is a challenging task and it has gained the attention of many researchers recently. The latest research and development work in Wireless Network technologies has enabled us to have such low power required wireless network which in return enables us to firm such complex hybrid networks of smart gadgets interconnected to each other. WSNs have enabled us to form such complex networks but due to the inherent nature of WSNs, it comes along with its security risks. Due to WSNs being resource-limited and being broadcasting in nature [3, 12]. Cyber-attacks are not limited to WNSs. Different devices and systems have a variety of attacks. Cyber Attacks do not only happen against Networks. The end devices can be the target too. By end devices, it can be anything from a mobile or laptop to smart grid, smart fridge, smart TV, Car, or any other digital system which is connected to the network [5, 13]. To deal with this evolving security situation researchers have worked on different security mechanisms, commonly a smart Intrusion Detection System (IDS) is used. IDS works by mapping ongoing network activity to known attacks. This way it can detect any intrusion in the network [13]. As nowadays there are adaptive systems that can work in a rapidly changing environment while keeping on learning from data. This phenomenon is known as computational intelligence [9]. Such smart and adaptive systems used in Intrusion Detection Systems can prove to be game-changing in the security of wireless networks.

3 Literature Review

An effective IDS is essential to mitigate these threats. Our world is dynamic. It is changing rapidly and each day new challenges occur. To design an IDS which is smart enough to work in this dynamic environment we make use of Computational Intelligence. Computational intelligence is a field that uses different bio-inspired approaches to develop such a system that can solve complex problems in a dynamically changing environment [8]. Using computational intelligence techniques, we design adaptive IDS which can work in a dynamic environment and learn from data and experiments to cope with the changing environment [9, 14].

There are tons of research being done in the field of Cyber Security. Different systems have a different set of venerability to which researchers have proposed different methods. In the series of research done to come up with an effective Intrusion Detection System researchers have worked closely with Computational Intelligence techniques to develop a low cost, low resource demanding IDS. There are multiple layers of defense mechanisms against attacks as a single standalone method is not feasible and cannot defend against a wide range of attacks. As the first layer of security different encryption and authentication methods are used to ensure data confidentiality and integrity. Then as the second layer, we have Intrusion Detection. An effective Intrusion Detection System ensures these things that data is kept confidential, has integrity (is not changed or altered), and is Available (meaning it is accessible to authorized people when needed) [11].

4   Introduction to Cybersecurity

In the face of numerous explanations of cybersecurity in literature, we chose a pragmatic quality of research strategy to help the process of defining it that combines objective of quality of the research with subjective quality of our research. In effect, the outcome is kind of a conceptual definition based on objective as in an intrusion-detection system rather than speculation like the intentions of a hacker [15]. In a survey, identifying prevalent themes and differentiating features, and the formulation of a workable definition were all part of the definitional process. This definition was then presented to the interdisciplinary group discussions for additional research, development, and refinement to arrive at the proposed definition [6, 16, 17].

Cyber Security has evolved as a frequently used word in recent years, with greater usage by practitioners and politicians alike. However, as with so much trendy jargon, there appears to be little comprehension of what the phrase means. Although this may not be a problem when the phrase is used casually, it can present significant issues when used in the context of corporate strategy, business objectives, or international agreements. In this paper, we examine the current literature to find the key meanings of the word "Cyber Security" supplied by reputable sources. We next employ a variety of lexical and semantic analysis approaches to better comprehend the breadth and context of these definitions, as well as their importance. Finally, based on the analysis, we propose a better definition, which then demonstrates a more representative definition using the same lexical and semantic analytic methodologies [2]. According to some analysis, there are the following few definitions of Cybersecurity as shown in Table 1.

Table 1 Definition of cybersecurity

	Definition	Country of origin
1	Cyberspace consists of all the virtual and physical ICT devices	USA, UK, France, India, Saudi Arab, and Turkey
2	Cyberspace is referred to as the "Internet" and internet-connected ICT devices	New Zealand, Australia, Germany, Spain, and Canada
3	The term "Cyber domain" has been used instead of cyberspace	Finland

5   CIA Triad

CIA Triad is a model formed to guide policymakers to form efficient cyber security-related policies. CIA stands for Confidentiality, Integrity, and Availability. It can be represented in hierarchal form as shown in Fig. 2

5.1   *Confidentiality*

Most individuals, even those who are not in the security profession, are familiar with the term confidentiality. Confidentiality is defined by the NIAG as "confidence that information is not divulged to unauthorized persons, procedures, or devices". The confidentiality of digital information needs real-world measures as well. Shoulder surfing or peering over another person's shoulder when he or she is using a computer is an unskilled approach for an attacker to collect personal information. Physical dangers, like theft, may threaten confidentiality. The consequence of a violation of confidentiality differs depending on the severity of the protected material. A credit card number breach, such as the one that occurred in 2008 with the Heartland Payment Systems processing system, might result in litigation with settlements in the millions of dollars [18].

Fig. 2   CIA triad pyramid

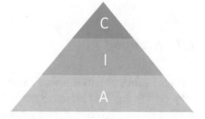

5.2   Integrity

Taking into consideration the context of information security, it is often related to data integrity or making sure that the stored data is correct and free of unauthorized alterations. Trying to interfere with the integrity of data at rest or in transit can have significant implications. An attacker might take advantage of the ability to change a message about the money transfer going between a user and his/her online banking website. By changing the account number of the receiver of the funds indicated in the message to the bank account that is owned by the attacker, he/she might hijack the transaction and steal the transmitted amounts. It is critical for any security system to ensure the integrity of this sort of message [18].

5.3   Availability

Users must be able to access information systems for them to be useful. A system that is offline or responding too slowly cannot offer the service that it should. The under-standing of the components of the CIA triad, as well as the principles underlying how to safeguard these ideals, is critical for all security professionals. Every component functions as a pillar, supporting a system's security. If an attacker compromises any of the pillars, the system's security will deteriorate. System designers can utilize authentication, authorization, and nonrepudiation technologies to keep these pillars in place. To apply these notions successfully, you must first understand how they interact with one another [18].

6   Threats/Attacks in Cybersecurity

The threat spectrum in digital social media is shown in Fig. 3. There are many threats in cyber security but a few of them are listed below:

6.1   Spoofing

GPS Spoofing attacks are also known as integrity attacks. As many devices heavily rely on Global Navigation Satellite Systems, more specifically on Global Positioning System (GPS) for their proper working. GPS-based devices require accurate and uninterrupted positioning information for their smooth operations, but different research has shown that GPS communications are very easy to be jammed or spoofed as they generally have no encryption or rigorous safety mechanism. As GPS positioning systems work on very low power signals, thus they can be easily jammed

Fig. 3 CIA and digital
social media

by higher power jamming signals transmitted towards the target system. Also, GPS is commonly used in our day-to-day life, it is very easy for attackers to replicate or spoof GPS signals to feed false positioning information to a target system which can lead to dangerous outcomes [19, 20].

6.2 DOS

DoS or Denial of Service attacks are attacks that prevent legitimate or authorized users from accessing the service. DoS attacks are carried out by sending an overwhelming amount of data packets to the target system and since there is no easy and simple way to differentiate between a legitimate or malicious packet it becomes hard to avoid such types of attacks which results in phenomena that are commonly known as signal jamming. DoS attacks generally exploit the weakness of TCP Protocol's three-way handshake mechanism [21, 22].

6.3 Ping of Death (POD)

Ping of Death attack is a type of DoS attack in which an attacker attempts to crash the target system by sending an oversized packet that the system cannot handle. Generally, in IPv4 protocol the packet size limit is 65,535 bytes, packets having a size greater than 65,535 bytes cannot be handled by most of the systems that do not have any effective PoD prevention mechanism. Now as in IPv4 protocol one cannot send packets larger than said size, the attacker accomplishes a PoD attack by sending a packet in smaller fragments which when the system tries to reassemble results in a packet that exceeds 65,535 bytes. This results in a system crash. Ping of Death attacks was more common because they only require attackers to know the IP address of the target system [23, 24].

7 Monitoring and Assessing Vulnerabilities

Monitoring and assessing vulnerabilities and hazards is a key component of the US government's cyber security strategy [24]. Continuous data gathering via automatic feeds includes network traffic information as well as host information from host-based agents: vulnerability information and patch status on network hosts; scan results from tools such as Nessus; TCP net flow data; DNS trees, and so on [8, 25, 26]. To determine the risks, these data are subjected to computerized analysis. The evaluation may entail identifying particularly serious vulnerabilities and exposures, as well as calculating points that offer an overall picture of the risk level at the network level. Risk measurements are frequently simply sums of the vulnerabilities and missing fixes in current practice [27, 28].

There are significant advantages to automating risk quantification that is, giving scores off the risk or other numerical measurements to the whole network, its subsets, and even individual assets [8]. This opens the door for the decision-making of actual risk management, which has the potential to be exceedingly rigorous and insightful [16, 26]. The Employees at all levels, from leaders of the senior level to the system administrators, will be aware of the constantly updated risk distribution across the Network and will utilize this information to prioritize the allocation of resources to the most effective corrective measures. Risk quantification can also help with the speedy, automated, or semi-automatic implementation of remedial programmers [27, 28].

According to a report, different keywords searched on google have the following analytics as shown in Fig. 4.

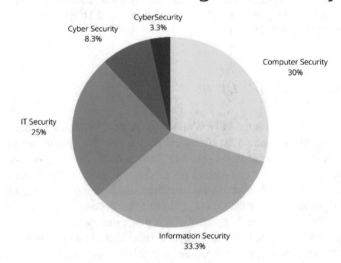

Fig. 4 Search trend percentage on google about security

8   Intrusion Detection Systems

Intrusion detection is the most prominent area of study that is widely acknowledged as belonging under the purview of cyber security. Much intrusion detection research is on developing innovative algorithms as well as structures of the tools to detect intrusion. A related subject is the evaluation of such tools' efficacy, as in the rate of detecting genuine intrusions or the false alarm rate of a given tool or algorithm in contrast to prior work [29].

8.1   Signature-Based Detection

Signature-based IDS are also known as rule-based IDS because they have a collection or list of signatures of all previously known attacks. It compared the data collected in the current network with the previously known attacks to find any matching patterns. It has high accuracy in detecting the known attacks which are in its dictionary, but it cannot work well against new attacks which are not in its dictionary.

8.2   Anomaly-Based Detection

In Anomaly-based IDS sudden changes in network behavior are detected. Over time the normal working of the network is monitored and is profiled as normal behavior. When any intrusion happens and there is a behavior change, it is classified as an anomaly and is reported. It can detect new attacks too, but it is prone to high false-positive alerts.

8.3   Hybrid Based Detection

Hybrid IDS are also known as Specification-based IDS, is a hybrid of the above-mentioned Signature-based and Anomaly-based Intrusion Detection Systems. Both mechanisms are combined, it also maintains a list of known attacks and detects anomalies based on changes in the normal behavior of the network. As it also has the signature of all previously known attacks, hence chances of false positives are low.

9 Conclusion

Cyber-security is a major factor to be considered while designing systems that are intended to be connected to some sort of network. Computational intelligence is increasingly employed for various applications because of its adaptive behavior. Computational intelligence-based IDS can easily adapt to changing environments and learn new normal patterns for a network based on which it can further classify irregularities and anomalies. It can be concluded that the use of Computational Intelligence techniques is the best-suited way for developing an IDS which can work in such expanding dynamic environment. Where many new attacks arise daily. Computational intelligence-based Intrusion Detection Systems can provide real-time threat monitoring and warning. Also, it can prevent intrusion and data loss. Computational Intelligence-based Intrusion Detection Systems are also cost-effective when compared to other IDS and generally perform better in mitigating the cyber threat. They can cover a wide spectrum of Cyber threats and enhance their abilities by continuously learning about new evolving threats through data and experiments. [29, 30].

References

1. D. Schatz, R. Bashroush, J. Wall, Towards a more representative definition of cyber security. J. Digit. Forensics, Secur. Law **12**(2), 8 (2017)
2. A. Kott, Towards fundamental science of cyber security, in *Network Science and Cybersecurity* (Springer, New York, NY, 2014), pp. 1–13
3. C.D. McDermott, A. Petrovski, Investigation of computational intelligence techniques for intrusion detection in wireless sensor networks. Int. J. Comput. Netw. Commun. 9(4) (2017)
4. M.G. Ball, B. Qela, S. Wesolkowski, A review of the use of computational intelligence in the design of military surveillance networks. Recent Adv. Comput. Intell. Def. Secur. 663–693 (2016)
5. M.Z. Gunduz, R. Das, Cyber-security on the smart grid: threats and potential solutions. Comput. Netw. **169**, 107094 (2020)
6. D. Craigen, N. Diakun-Thibault, R. Purse, Defining cybersecurity. Technol. Innov. Manag. Rev. **4**(10) (2014)
7. D. Staheli, T. Yu, R.J. Crouser, S. Damodaran, K. Nam, D. O'Gwynn, S. McKenna, L. Harrison, Visualization evaluation for cyber security: trends and future directions, in *Proceedings of the Eleventh Workshop on Visualization for Cyber Security* (2014), pp. 49–56
8. M.R. Jabbarpour, H. Zarrabi, R.H. Khokhar, S. Shamshirband, K.K.R. Choo, Applications of computational intelligence in vehicle traffic congestion problem: a survey. Soft. Comput. **22**(7), 2299–2320 (2018)
9. L. Thames, D. Schaefer, *Cybersecurity for Industry 4.0* (Heidelberg, Springer, 2017)
10. G. Apruzzese, M. Colajanni, L. Ferretti, A. Guido, M. Marchetti, On the effectiveness of machine and deep learning for cyber security, in *10th International Conference on Cyber Conflict (CyCon)* (IEEE, 2018), pp. 371–390
11. A. Gupta, O.J. Pandey, M. Shukla, A. Dadhich, S. Mathur, A. Ingle,Computational intelligence based intrusion detection systems for wireless communication and pervasive computing networks, in *IEEE International Conference on Computational Intelligence and Computing Research* (2013), pp. 1–7. https://doi.org/10.1109/ICCIC.2013.6724156

12. L.A. Maglaras, K.H. Kim, H. Janicke, M.A. Ferrag, S. Rallis, P. Fragkou, A. Maglaras, T.J. Cruz, Cyber security of critical infrastructures. Ict Express **4**(1), 42–45 (2018)
13. N. Ben-Asher, C. Gonzalez, Effects of cyber security knowledge on attack detection. Comput. Hum. Behav. **48**, 51–61 (2015)
14. M. Tavallaee, E. Bagheri, W. Lu, A.A. Ghorbani, A detailed analysis of the KDD CUP 99 data set. In: 2009 IEEE symposium on computational intelligence for security and defense applications (pp. 1–6). Ieee (2019)
15. S. Balusamy, A.N. Dudin, M. Graña, A.K. Mohideen, N.K. Sreelaja, B. Malar, Cyber Security and Computational Models
16. K. Demertzis, L. Iliadis, Computational intelligence anti-malware framework for android OS. Vietnam J. Comput. Sci. **4**(4), 245–259 (2017)
17. J. Graham, R. Olson, R. Howard (eds.), *Cyber Security Essentials* (CRC Press, 2016)
18. K.L. Dempsey, L.A. Johnson, M.A. Scholl, K.M. Stine, A.C. Jones, A. Orebaugh, N.S. Chawla, R. Johnston, Information security continuous monitoring (ISCM) for federal information systems and organizations (2011), p. 19
19. J. Kacprzyk, W. Pedrycz (eds.), *Springer Handbook of Computational Intelligence* (Springer, 2015)
20. J.S. Raj, A comprehensive survey on the computational intelligence techniques and its applications. J. ISMAC **1**(03), 147–159 (2019)
21. G.B. Huang, E. Cambria, K.A. Toh, B. Widrow, Z. Xu, New trends of learning in computational intelligence [guest editorial]. IEEE Comput. Intell. Mag. **10**(2), 16–17 (2015)
22. B. Xing, W.J. Gao, Innovative computational intelligence: a rough guide to 134 clever algorithms (2014)
23. M.Z. Zgurovsky, Y.P. Zaychenko, *The Fundamentals of Computational Intelligence: System Approach* (Springer International Publishing, 2017)
24. H. He, C. Maple, T. Watson, A. Tiwari, J. Mehnen, Y. Jin, B. Gabrys, The security challenges in the IoT enabled cyber-physical systems and opportunities for evolutionary computing and other computational intelligence, in *IEEE Congress on Evolutionary Computation (CEC)* (IEEE, 2016), pp. 1015–1021
25. D.S. Punithavathani, K. Sujatha, J.M. Jain, Surveillance of anomaly and misuse in critical networks to counter insider threats using computational intelligence. Clust. Comput. **18**(1), 435–451 (2015)
26. S. Shamshirband, M. Fathi, A.T. Chronopoulos, A. Montieri, F. Palumbo, A. Pescapè, Computational intelligence intrusion detection techniques in mobile cloud computing environments: review, taxonomy, and open research issues. J. Inform. Secur. Appl. **55**, 102582 (2020)
27. A. Kott, C. Arnold, The promises and challenges of continuous monitoring and risk scoring. IEEE Secur. Priv. **11**(1), 90–93 (2013)
28. W.J. Nistir, W. Jansen, P.D. Gallagher, Directions in Security Metrics Research (2009), 21
29. R.P. Lippmann, J.F. Riordan, T.H. Yu, K.K. Watson, Continuous security metrics for prevalent network threats: introduction and first four metrics. Mass. Inst. Tech. Lexingt. Linc. Lab. **23** (2012)
30. N. Shafqat, A. Masood, Comparative analysis of various national cyber security strategies. Int. J. Comput. Sci. Inform. Secur. **14**(1), 129 (2016)

Computational Intelligence Techniques for Cyberspace Intrusion Detection System

Abbas Ikram, Syeda Ghanwa Tawaseem, Muhammad Yaseen Ayub, and Syeda Zillay Nain Zukhraf

Abstract The main purpose of this study is to present ways and methods which could be used against malicious cyber activities that cause trouble and harm to our system and networks. Ways that can help us in detecting these attacks at the right moment and taking the necessary actions against them. Also, how AI can design autonomic solutions against these intrusions.

Keywords Computational intelligence · Cyberspace · Cybersecurity

1 Overview

Basically, Intrusion in an environment is an unavoidable problem as attackers continue to attack your system no matter how strong security measures are deployed in the system. To compete against these cyber security intrusions, the best Organizations are those who are always ready to act against such attacks and overcome them with their strategies against intrusion problems. Not being up to update against such attacks means vulnerability and losses to an individual, an organization, and the nation [1]. Physical devices are installed in our systems when sensors are not

A. Ikram
Department of Electrical Engineering, National University of Computer & Emerging Sciences, Peshawar, Pakistan
e-mail: p166384@nu.edu.pk

S. G. Tawaseem
Department of Computer Science, HITEC University, Taxila, Pakistan
e-mail: 18-CS-080@student.hitecuni.edu.pk

M. Y. Ayub
Department of Computer Science, COMSATS University Islamabad, Attock, Pakistan
e-mail: SP19-BCS-025@cuiatk.edu.pk

S. Z. N. Zukhraf (✉)
Department of Electrical and Computer Engineering, KIOS Research and Innovation Centre of Excellence, University of Cyprus, Nicosia, Cyprus
e-mail: zukhraf.syeda-zillay@ucy.ac.cy

sufficient to monitor, detect, and protect these new intrusions and threats; therefore, there is a need for more intelligent and advanced IT that can take serious actions against abnormal ones. Protect our network and system against these compromises. Cyber protection systems should be flexible, versatile, powerful, and able to detect malicious threats and to make intelligent and quick real-time decisions [2].

2 Introduction

To be safe and secure from cyber-attacks and intrusions we take security measures such as IDS and IPS. These tools and technologies minimize the threats of cyber intrusion attacks [1]. These systems are used to monitor the behavior of the system. If abnormal behavior is seen or detected, it is a sign of a possible cyber-attack, and actions must be taken. We can classify networks in this regard into two types which are signature-based and anomaly-based.

A signature-based IDS goals and focuses to track down an attack by comparing its signatures with packet payloads. It is very much effective in identifying the known cyber threats however, ineffective in detecting unknown malicious exploits and variants of known threats. Whereas an anomaly-based IDS detects and discovers unknown attacks. It compares the established normal profiles and models against newly observed events to identify significant fluctuations in the system, if there is an abnormal or suspicious behavior contradicting the model, alerts will be triggered. However, these systems do not give accurate results because of the complexity of the profile generation [3]. Also, in anomaly-based IDS we get many false alarms resulting in the confusion of what is normal and what is an abnormal alarm, along with the consumption of computer resources as per shown in Fig. 1. Therefore, in order to achieve maximum accurate results both case techniques need to be used and deployed in our cyber network systems to achieve a low false alarm rate [1, 4].

Artificial intelligence is another innovative approach that can be helpful against cyber-attacks. A great many computing methods of AI have been progressively doing a vital role in detection as well as prevention. AI helps us to design and create autonomous computing systems which are very intelligent and can adapt to their context of use. Implementing such methods which are self-management, self-tuning, self-configuration, self-diagnosis, and self-healing can protect our network and systems from malicious cyber intrusions. In the future, AI techniques seem a very likely looking area of research that emphasizes improving the security measures against cyber intrusions [2].

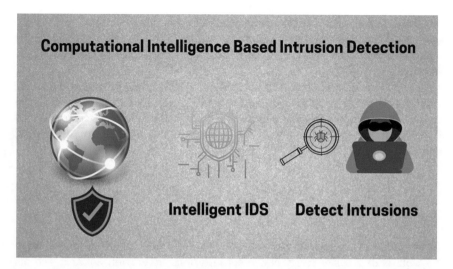

Fig. 1 Computational Intelligence based IDS

3 Motivation

In today's world, computer networking is the most common means of sending, receiving, and sharing information. Countries, organizations, businesses, or even a single person are dependent upon computerization and systems. Increased use of systems means data breaching techniques being developed hence, giving rise to cyberattacks upon our network and systems. Therefore, we need to have ways to protect from such intrusion attacks. An anomaly-based system is one of the common methods used against the protection of cyberattacks. Anomaly-based IDS detects and discovers attacks that are unknown. It compares the established normal profile model against new events to identify abnormal or suspicious behavior. However, these methods fail to report or trigger alerts against malicious content such as SQL injection, cross-site scripting, and shellcode. Hence, we need methods that could protect us against unrecognized malicious content and attacks. Deep learning algorithms have proved extraordinary against such threats and gave phenomenal results in many fields, e.g., Computer Vision (CV), Natural Language Processing (NLP), and Automatic Speech Recognition (ASR). They are proven to be proficient to extract primary features from unstructured data [5, 6]. Deep neural networks (DNN) based methods have shown outstanding accomplishment against the previous machine learning-based Cyber intrusion detection systems and showed robustness in the presence of dynamic IP addresses [7, 8].

As per Cybersecurity Ventures, IT security services and products spending has exceeded about $1 trillion over the past 5 years, from 2017. The losses against cybercriminal activities have increased $3 trillion in 2015 to $6 trillion in 2021 and so on in the future. By looking at these facts, it is necessary to look for new, advanced, and efficient methods of detection of cyberattacks in intrusion detection systems. Another solution to this growing problem is the use of machine learning algorithms which are unconventional methods, however, some of them are highly affected [9].

As computerization is increasing every second of the minute, therefore we should be prepared for more advanced and threatening intrusions to our systems. This could only be done by using advanced and efficient techniques against data exploitation. Detection Systems can prove to be game-changing in the security of wireless networks [10].

4   Literature Review

In this world of the Internet and networks, the most important implementation of a system is its security. Every day, cyber intrusions are growing stronger, and our network is vulnerable to these attacks. The use of the IoT and distributed devices has given us a new approach to look into the matter of the implementation of intrusion detection systems [11]. Therefore, we need intrusion detection techniques as stronger as these attacks, which are accurate and could protect our systems. One of the techniques is cyber intrusion systems, which are based on machine learning algorithms with high performance and accuracy in the detection of such attacks [12, 13]. An intrusion Detection System (IDS) with the most recent and updated methods can help to protect IoT-based systems. In the last two decades, IDS has become an extremely important area for a lot of researchers. Most of the methods have become outdated as new attacks by hackers are difficult to identify hence strenuous to act on them. The complete research is done on IDS using previous examinations and methods in mind and proposed the new techniques which will help protect against these attacks in IoT [14, 15]. Another solution using IDS is recursive feature elimination is performed using a random dataset and the most important values of the features are calculated. Using this method malicious intrusions are detected with maximum accuracy [16].

Anomaly-based detection systems lead to a high level of security and long-lasting protection against cyber-attacks [17]. This is due to the vast improvement and development in big data solutions [18]. Anomaly detection is also known as behavior-based detection systems was developed using deep learning and machine learning algorithms which compare the established normal profiles and models against newly observed events to identify significant fluctuations in the system. The standard network datasets were used to assess the suggested model which is built and designed for enhancing the cybersecurity system [17]. An anomaly or an intrusion over the system comes under the network forensic branch. If a cyber intrusion takes place on a network how to collect, analyze and investigate using the evidence available is called Digital forensic analysis and investigation [19, 20].

Detection Banking Trojans are one of the challenging and confusing tasks. This is due to continuous advancement in methods used to obscure and evade the existing intrusion detection systems. cyber kill chain-based taxonomy is a technique that can help protect the systems and networks against banking Trojans [21]. Wireless sensor networks (WSN) most challenging issue is to face the problems of cyber-attacks and intruders, in order to make a sensor network secure and protected from such attacks a technique is used known as Artificial Intelligence System (AIS) [22]. Network infrastructure devices help significantly in managing numerous enterprises like bank accounts, online transactions, social insurance, protection, and computer networks systems; however, simultaneously may suffer from unauthorized intrusion and several attacks which prevent assurance of continuous safety to authorize users of the network. For more accuracy and protection of our systems, we need to have Network intrusion detection (NID) [23, 24]. Deep learning methodologies can achieve a higher accuracy rate with a very low false alarm rate to detect a cyberattack or an intrusion over a system. Implementation of the hybrid algorithm provided improved results against intrusion detection. This kind of bidirectional algorithm showed the highest known accuracy, decreased false alarm rate, and false-negative rate [25, 26]. Big Data technologies for Cyber Intrusion Detection can help in resolving Big Heterogeneous Data challenges [27].

Cyberattacks risks are immensely high on smart grid security systems. These malicious viruses can spread from one meter to another to take out the power. In order to be safe from such harmful attacks, the American Reliability Corporation has developed some repetitions to be safe from such damaging intrusion attacks [28].

5 Different Algorithms for Computational Intelligence

In this paper, we will list down the computational methods and ways that will fight against Cyberattacks and protect our systems against malicious intrusions. it is a system with the ability to prevent attacks. An IDS needs to act in real-time when the malicious activity has just begun on the system. It needs to give indications and alarms after the detection of abnormal behavior in the system [29].

5.1 Method Based on Anomaly

Detection of a malicious activity follows different steps that lead to intrusion detection. One of the methods which lead to intrusion detection is dependent on signature and dependent on Anomaly.

Independent on signature we have a dataset available on our system. If the activity matches any of the harmful activities mentioned in the stored set of data, immediately the system will take appropriate action and remove those threats. What if a cyber intrusion has been made on your system or network, and it is not available on your dataset? In that scenario, we use methods based on Anomaly.

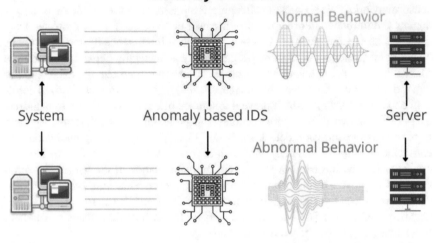

Fig. 2 Anomaly-based IDS

If any of the harmful packets differentiate from the model available and abnormal behavior is sensed or detected, alarms must be sent to distinguish it as malicious. This method must have the ability to pick out any anomaly from normal behavior by using normal signature packet loads [30].

However, these methods are not efficient enough since, most of the time we have to deal with false alarms generated through anomaly-based methods hence making it difficult and confusing for a system to understand which alarm is for malicious activity, Fig. 2 shows the context in pictorial form. Therefore, we need to find ways that are more accurate and have the ability to detect intrusion in the first place and act immediately.

5.2   Method Based on Artificial Intelligence

AI methodologies can be utilized to get good results in the Wireless sensor Network (WSN) systems. The possible threats that can occur are very challenging, therefore, using a system discussed in the above section, a system is designed using an artificial technique that will prevent serious security threats against WSN. Artificial immune intelligence-based systems can be implemented very easily as compared to the other types of networks. One of the most recent introduced malicious threats is the Interest Cache Poisoning attack. This type of attack targets the network layer of a system and has the ability to disordering the routing packets. Therefore, two main methods the Direct Diffusion and the Dendritic Cell Algorithm, both are applied for detecting any abnormality on the nodes. Direct diffusion handles two main tasks: data cache and

Artificial Intelligence based IDS

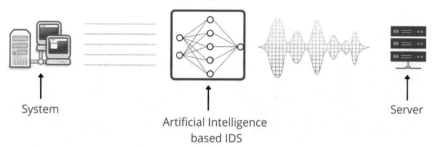

System   Server

Artificial Intelligence
based IDS

Fig. 3 AI-based IDS

interest cache. It manages two types of routing packets one of them is data packets and the other one is interest packets [31].

Whereas the DCA algorithm keeps track of two types of signals that are danger signals and safe signals. A cyber-Intrusion adaptive detection technique used for prevention and prediction can also be designed by the usage of immune systems. It is implemented on the MAC/Data link layer, in which first a gene is identified for the purpose of detection. For the purpose of monitoring the normal and abnormal behavior of the traffic in the neighboring nodes, a natural selection algorithm is being installed in every node. The artificial immune system monitors the incoming traffic, checks incoming and outgoing internal logs and databases throughout. If there is an intrusion attack, it immediately acts and generated alarms, notifications, and reports [22, 32] as referred in Fig. 3.

5.3 Method Based on Machine Learning Methods

Support vector machine (SVM), k-Nearest Neighbor (kNN), Decision Trees, and other traditional machine-learning methods are examples.

Deep learning-based techniques are gaining popularity as data sets grow larger and larger because they can learn computational processes in-depth and may lead to improved generalization capabilities. Autoencoder and more approaches are available. The concept of integrating several data classification algorithms to build a hybrid classifier has been investigated in order to boost recognition accuracy even more. For specific data sets, a considerable number of trials have proven that hybrid-based approaches perform better. They can achieve improved precision and detection thanks to a unique classifier and merging process [13, 33].

6　Conclusion

As the pool of networked devices is increasing daily and so are the attacks against those networked devices. Prevention against those attacks and possible intrusion is a major research topic nowadays. There are several Intrusion Detection and Intrusion Prevention methods in use around the world. We discussed some rule-based and Artificial intelligence-based methods. It can be concluded that Artificial Intelligence or Machine learning-based methods of intrusion detection are very effective and accurate in detecting and preventing intrusions. As the data is growing exponentially, several deep learning methods are emerging which are proving to be very useful in designing IDS. They are often used in coordination with data classification algorithms to further enhance their accuracy.

References

1. S.U. Otor, B.O. Akinyemi, T.A. Aladesanmi, G.A. Aderounmu, An improved bio-inspired based intrusion detection model for cyberspace. Cogent Eng. **8**(1), 1859667 (2021). https://doi.org/10.1080/23311916.2020.1859667
2. S. Dilek, H. Çakır, M. Aydın, Applications of artificial intelligence techniques to combating cyber crimes: a review. https://doi.org/10.5121/ijaia.2015.6102
3. X.K. Li, W. Chen, Q. Zhang, L., Wu, Building auto-encoder intrusion detection system based on random forest feature selection. Comput. Secur. (2020). https://doi.org/10.1016/j.cose.2020.101851
4. Y. Meng, L.F. Kwok, Enhancing false alarm reduction using voted ensemble selection in intrusion detection. Int. J. Comput. Intell. Syst. **6**(4), 626–638 (2013). https://doi.org/10.1080/18756891.2013.802114
5. E. Min, J. Long, Q. Liu, J. Cui, W. Chen, TR-IDS: anomaly-based intrusion detection through text-convolutional neural network and random forest. Secur. Commun. Netw. 9 (2018) (Article ID 4943509). https://doi.org/10.1155/2018/4943509
6. W. Lian, G. Nie, B. Jia, D. Shi, Q. Fan, Y. Liang, An intrusion detection method based on decision tree-recursive feature elimination in ensemble learning. Math. Probl. Eng. **2020**, 15 (2020) (Article ID 2835023). https://doi.org/10.1155/2020/2835023
7. H. Zhang, J.L. Li, X.M. Liu, C. Dong, Multi-dimensional feature fusion and stacking ensemble mechanism for network intrusion detection. https://doi.org/10.1016/j.future.2021.03.024
8. Y. Gao, Y. Liu, Y. Jin, J. Chen, H. Wu, A novel semi-supervised learning approach for network intrusion detection on cloud-based robotic system. https://doi.org/10.1109/ACCESS.2018.2868171
9. A. Gerka, Searching for optimal machine learning algorithm for network traffic classification in intrusion detection system. https://doi.org/10.1051/itmconf/20182100027
10. Y. Shen, K. Zheng , C. Wu, M. Zhang, X. Niu, Y. Yang, An ensemble method based on selection using bat algorithm for intrusion detection. https://doi.org/10.1093/comjnl/bxx101
11. M. Al-Omari, M. Rawashdeh, F. Qutaishat, M. Alshira'H, N. Ababneh, An intelligent tree-based intrusion detection model for cyber security. https://doi.org/10.1007/s10922-021-09591-y
12. X. Larriva-Novo, V.A. Villagrá, M. Vega-Barbas, D. Rivera, M. Sanz Rodrigo, An IoT-focused intrusion detection system approach based on preprocessing characterization for cybersecurity datasets. Sensors **21**, 656 (2021). https://doi.org/10.3390/s21020656

13. J., Zhang, Y. Ling, X. Fu, G. Xiong, R. Zhang, Model of the intrusion detection system based on the integration of spatial temporal features. Comput. Secur. (2019). https://doi.org/10.1016/j.cose.2019.101681

14. A.A. Gendreau, M. Moorman, Survey of intrusion detection systems towards an end to end secure internet of things. https://doi.org/10.1109/FiCloud.2016.20

15. A.G. Bombatkar, T.J. Parvat, Efficient method for intrusion detection and classification and compression of data. 978-1-5090-0076-0/15 $31.00 © 2015 IEEE. https://doi.org/10.1109/CICN.2015.213

16. S. Ustebay, Z. Turgut, M.A. Aydin, Intrusion detection system with recursive feature elimination by using random forest and deep learning classifier. https://doi.org/10.1109/IBIGDELFT.2018.8625318

17. H. Alkahtani, T.H.H. Aldhyani, M. Al-Yaari, Adaptive anomaly detection framework model objects in cyberspace. https://doi.org/10.1155/2020/6660489

18. H. Al Najada, I. Mahgoub, I. Mohammed, Cyber intrusion prediction and taxonomy system using deep learning and distributed big data processing. https://doi.org/10.1109/SSCI.2018.8628685

19. K.S. Sangher, A. Singh, A systematic review—intrusion detection algorithms optimisation for network forensic analysis and investigation. 978-1-5386-8010-0/19/$31.00 ©2019 IEEE

20. S. Shamshirband, A.T. Chronopoulos, A new malware detection system using a high performance-ELM method. https://doi.org/10.1145/3331076.3331119

21. D. Kiwiaa, A. Dehghantanhaa, K.K.R. Choob, J. Slaughtera, A cyber kill chain based taxonomy of banking Trojans for evolutionary computational intelligence. https://doi.org/10.1016/j.jocs.2017.10.0201877-7503/© 2017 Elsevier B.V. All rights reserved

22. G. Kalnoor, J. Agarkhed, Artificial intelligence-based technique for intrusion detection in wireless sensor networks. https://doi.org/10.1007/978-981-10-3174-8_69

23. N.B. Nilesh, A. Parikh, Classification and technical analysis of network intrusion detection systems

24. A.C. Enache, V. Sgârciu, M. Togan, Comparative study on feature selection methods rooted in swarm intelligence for intrusion detection. 2379-0482/17 $31.00 © 2017 IEEE. https://doi.org/10.1109/CSCS.2017.40

25. S. Khan, K. Kifayat, A. Kashif Bashir, A. Gurtov, M. Hassan, Intelligent intrusion detection system in smart grid using computational intelligence and machine learning. https://doi.org/10.1002/ett.4062

26. M. Ahsan, K.E. Nygard, Convolutional neural networks with LSTM for intrusion detection. Proceedings of 35th International conference on computers and their applications. EPiC Series in Computing, vol. 69, 69 (2020)

27. R. Zuech, T.M. Khoshgoftaar, R. Wald, Intrusion detection and big heterogeneous data: a survey. Zuech et al. J. Big Data **2**, 3 (2015). https://doi.org/10.1186/s40537-015-0013-4

28. Y. Wang, D. Ruan, J. Xu, M. Wen, L. Deng, Computational intelligence algorithms analysis for smart grid cyber security. ICSI 2010, Part II, LNCS 6146, pp. 77–84 (2010)

29. P. Shirani, M.A. Azgomi, S. Alrabaee, A method for intrusion detection in web services based on time series. Proceeding of the IEEE 28th Canadian conference on electrical and computer engineering Halifax, Canada, 3–6 May 2015

30. S.R. Alkhaldi, S.M.Alzahrani, Intrusion detection systems based on artificial intelligence techniques. Acad. J. Res. Sci. Publish **2**(21) (2021)

31. J. Shifflet, A technique independent fusion model for network intrusion detection. Proceedings of the mid states conference on undergraduate research in computer science and mathematics, vol. 3, no. 1, pp. 13–19

32. S.W. Lee, M. Mohammed sidqi, M. Mohammadi, S. Rashidi, A.M. Rahmani, M. Masdari, M. Hosseinzadeh, Towards secure intrusion detection systems using deep learning techniques: comprehensive analysis and review. https://doi.org/10.1016/j.jnca.2021.103111

33. Z. Wu, J. Wang, L. Hu, Z. Zhang, H. Wu, A network intrusion detection method based on semantic Re-encoding and deep learning. https://doi.org/10.1016/j.jnca.2020.102688

A Comparative Analysis of Intrusion Detection in IoT Network Using Machine Learning

Muhammad Imad, Muhammad Abul Hassan, Shah Hussain Bangash, and Naimullah

Abstract Recent innovations and advanced technology encourage users to implement solutions against harmful attacks. This is provided new capabilities of dynamic provisioning, monitoring, and management to reduce the IT barriers.IDS is one of the challenging tasks where attackers always change their tools and techniques. Several techniques have been implemented to secure the IoT network, but a few problems are expanding, and their results are not well defined. According to this study, machine learning techniques have been used to detect and classify the problem into the anomaly and normal from the Network Intrusion Detection dataset. First, the data is preprocessed and make it standardize by standard scaler function. The random forest technique has been used to extract the significant features from the dataset. Furthermore, five different classification technique has been used based on the performance measure and compared. The outcome represents that the Decision tree model accomplished the highest accuracy of 100% among other classifiers.

Keywords Intrusion detection · IoT · Machine learning · KNN · SVM · Decision tree · ANN · Logistic regression

1 Introduction

The Internet of Things (IoT) is probably the most advanced technology associated with the internet that collects and shares data to support human lives, careers, and cultures. In 2018, the quantity of IoT devices was estimated to be 28 billion, which will increase to 49.1 billion in 2022. IoT is a recognized mechanism connected through servers, sensors, and numerous software [1]. Recently, the Internet of Things prototype has been recycled to create intelligent systems, such as smart cities and smart homes, with various applications and associated services. Developing a smart environment aims to make human life further accessible and complacent, linking to living, energy consumption, and industrial need [2].

M. Imad (✉) · M. Abul Hassan · S. Hussain Bangash · Naimullah
Department of Computing and Technology, Abasyn University, Peshawar 25000, Pakistan
e-mail: Imadk28@gmail.com

© The Author(s), under exclusive license to Springer Nature Switzerland AG 2022
M. Ouaissa et al. (eds.), *Big Data Analytics and Computational Intelligence for Cybersecurity*, Studies in Big Data 111,
https://doi.org/10.1007/978-3-031-05752-6_10

An Intrusion Detection System (IDS) is a device or software component that monitors and detects harmful activity on computer systems or networks for a vengeful activity or policy negligence. Intrusion Detection Systems (HIDS) are directed at a single computer, whereas Network-based Intrusion Detection Systems (NIDS) are required at an entire network. NIDS are network-attached hardware or software components that evaluate the traffic generated by hosts and devices [3]. IDS monitors the network assets to detect incorrect activity or abnormalities. IDS are classified based on whether it monitors the existence of a host, which might be a single or several computers on a network. As shown in Fig. 1, The IDS is split into two primary grouping; Network-based intrusion detection systems and host-based intrusion detection systems. The Host-based IDS uses application logs in the inquiry while the network-based operates to capture and evaluate the packet accepted from network traffic.

The three phases required to build IDS [4]: Data/information resources for event logs, The prevalence of intrusion detection using analytic devices, and feedback are formed in reaction to the outcomes of the preceding stage. As shown in Fig. 2, IDS may perform the following tasks [4]: data collecting, preprocessing data, intrusion introduction, and remedial actions. Intrusion Detection Systems (IDS) are developing systems that can detect intrusion successfully in terms of time and accuracy [4].

The IDS is critical to identify the network threats. However, IoT has numerous key features, alike as sensor nodes that are often restricted in power, memory space, and wireless channel capacity [5]. According to the authors [6], the IDS controls harmful attacks while the attackers use various tools and techniques. It's a very challenging task to solve a problem as a consequence of machine learning classifiers.

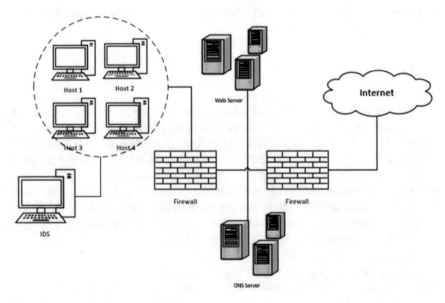

Fig.1 Intrusion detection system (IDS)

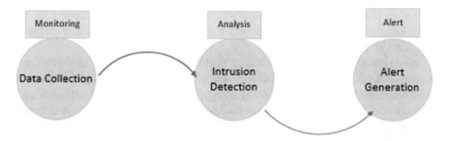

Fig. 2 Three phases to build IDS

The detection rate of IDS performance is a false positive and false negative. The algorithm finds out the minimum rate of false-negative and random forest gained the highest average accuracy rate. KDD datasets were first publicized, some IDS behaviours statistical measurements of value. The metrics performance determines various classifiers testing like Decision Table, MLP, Random Forest, J48, Naïve Bayes. The testing phase identifies the precision, accuracy rate, true negative and false negative, true positive [6].

Many approaches and frameworks have been implemented to mitigate network intrusion [7]. Machine learning and deep learning classifiers can assist in observing the large network which can reach millions in a day. Furthermore, some researchers used network attack-based datasets such as KDD cup 1999, UNSW-NB 15, IDS 2018 Intrusion CSVs (CSE-CIC-IDS2018), Network Intrusion Detection dataset from the Kaggle repository [8, 9].

The main contributions of this research can be reviewed as follows:

- A comprehensive analysis of IDS, dataset, and ML algorithms to predict optimal attacks in IoT systems.
- The achievement of some shallow machine learning classifiers has been calculated and compared that can be tested at particular levels of IoT-based IDS.

This article is coordinated as pursued. Section 2 discussed the previous literature on intrusion detection, attacks, & datasets mentioned in the table. The method and flow chart for intrusion detection are mentioned in Sect. 3. The result analysis and the dataset description is shown up in Sect. 4. Finally, in Sect. 5, current the results.

2 Related Work

Hongyu Liu and Bo Lang, research study about how deep learning and machine learning can solve the challenging task, improving the accuracy of detecting the false alarm rate and identifying the unknown attacks of IDS. Many researchers used various methods to improve normal and abnormal data accuracy. Deep learning performance is incredible and remarkable for a taxonomy of IDS data to detect the main objects and dimensions to classify with the help of taxonomy framework best of cybersecurity. The attack of cybersecurity internal and external include firewalls, IDS, antivirus [10].

In this study, researchers used a deep neural network (DNN) on intrusion detections system with the help of artificial intelligence (AI). The KDD cup 99 datasets are perfect for the classification of ever-evolving network attacks. DDN first performed preprocessed on the data transformation and normalization the accuracy detection rate and calculated efficacy of DNN model [11]. The authors deeply researched the probe and analysis of IDS attacks with machine learning methods. Instruction detection is challenging because it's impossible to determine all kinds of attacks today in the cyber world. They identified low-frequency attacks and detected different attack data analyses [12].

The authors further studied IDS problems to improve the IoT's security performance as a result of deep learning and machine learning classification. The researchers developed an algorithm to solve the problems and detect Denial-of-service (DoS) attacks to protect deep learning. The model accuracy increasing researchers used important python packages seaborn, TensorFlow, scikit-learn to get the target accuracy [13]. Additionally, security-related exploration in IoT and intrusion the strategy of rigorous state-of-the-art techniques applied on machine learning to detection the network security. The survey collected and analyzed 95 works on the subject and addressed security matters in IoT surroundings [14].

This study of research, feature selection method using supervised machine learning Intrusion detection to arrange network traffic. The network traffic utilizes the techniques of ANN, SVM, and machine learning on the NSL-KDD dataset to manage the network traffic. The model assessment is planned to achieve the accuracy rate and break down occurrence in the network traffic issues indication observed [15]. IoT collects a large amount of data, and this is a sizable consideration to handling the security and protection of the data and monitoring to control the device's cyberattacks. The experiments were conducted in the unauthorized network access to detect the DDoS attacks on these devices [16].

The authors studied the Hybrid model to detect Intrusion systems using deep learning and machine learning approaches to access resources and monitor the existing applications immediately. Distributed denial of service (DDoS) attacks are a risk of data injection vulnerabilities, abusive features, breaches, and account compromises. Cloud security is a major challenge to reduce the trust in cloud services [17]. Two new approaches applied to feature selection operation determine the algorithms according to the attack types. This studied experiment detected the performance on DR, F-Measure, MCC, TP Rate on NSL KDD datasets [18].

Most researchers proposed IoT-based machine learning applied to intrusion detection networks. This paper implements access control, encryption, and security measures authentication for IoT devices. The research emphasizes identifying different IoT threats that apply various algorithms like decision trees, random forest, SVM, DNN, LSTM, and DBN [19]. In A comprehensive study, machine learning algorithms are implemented to improve IoT-based applications accuracy. The classifiers are implemented on metrics and validation methods (Table 1). The main goal

Table 1 Machine learning-based summarized work

References	Dataset	Attack/intrusion category	Algorithm	Result
Hanif et al. [21]	UNSW-NB 15	Generic, exploits, shellcode, analysis, worms, DoS	ANN supervised	Precision = 84% FPR = < 8%(tenfold corss validaation)
Imad et al. [22]	KDD Cup 99 & AWID	ARP Flooding, DOS, U2R, R2L, Probe	Supervised Active Learning	Active learning improves performance and response faster
Cassales et al. [23]	Kyoto 2006+	Eight different types of Honeypots	DT and KNN	Speed up to 450 kpbs with less device resources
Wu et al.[24]	CAN intrusion	DoS and fuzzy	Deep Learning, SVM, ANN	The method is effective against attacks on automobiles
Karatas et al. [25]	CSE-CIC-IDS2018	SQL injection, brute force, DoS	Machine Learning-based SMOTE implementation	The imbalances in the dataset reduced due to which attack reduces
Hu et al. [26], Chen et al. [27]	DARPA- 98	Penetrate traffic-profiling-based network idss	ANN, SVM	ANN = 96% SVM = 99% Detection Rate
Shafi et al. [28]	KDDCUP- 99	Payload-based attacks	SMO classifier	97% Detection Rate
Ustebay et al. [29]	CICIDS-2017	IDS classification	Deep Multilayer Perceptron (MLP) and Payload Classifier	95.2% Accuracy Detected
Koroniotis et al. [30]	Bot-IoT	Probing attacks, Denial of Service, Keylogging, Data theft	SVM model	98% Detection Rate

(continued)

Table 1 (continued)

References	Dataset	Attack/intrusion category	Algorithm	Result
Creech [31]	ADFA-WD	Doppelganger attack, Chimera Attack	Hidden Markov Model(HMM), Extreme Learning Machine(ELM),SVM	HMM = 74.3% ELM = 98.57% SVM = 99%
Adebowale et al. [32]	NSL-KDD	IDS classification	K-Nearest Neighbor (k-NN)	94% Accuracy

of researchers is to use ensemble learning and statistical measurement strategies to improve the performance of the classifiers [20].

3 Methodology

This section presents the flowchart, dataset description, and ML techniques involved in our proposed model. The model is consists of Preprocessing, feature extraction, and classification. Furthermore, we used five different ML techniques based on the network intrusion dataset and compared them in classification.

3.1 Machine-Learning Algorithms for Anomaly Detection

The statistical categorization of measures used in the implementation of ML is used to investigate attack detection problems. In cyber security, the spam filter isolates spam from various communication services. Spam appears to be the most widely used ML approach in information security. Three steps involve preprocessing, Feature extraction, and classification to find the intrusion detection which are presented in Fig. 3. For classification, the supervised learning labelled data approach is commonly employed. This study discussed ML techniques such as DT, Logistic Regression, RF, KNN, Naïve Bayes, and ANN for intrusion detection.

3.2 Preprocessing

We applied standard scaler and label encoder function in the preprocessing to make the data inappropriate. StandardScaler function standardizes a feature by eliminating the mean and scaling to group fluctuation. Unit variance is determined by parting all values by the standard deviation. The formal concept of scale that I provided before

Fig. 3 Flowchart of the intrusion detection using machine learning

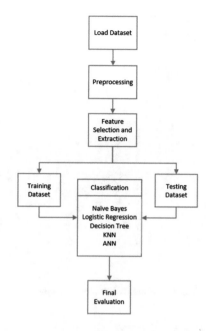

does not apply to StandardScaler. StandardScaler produces a distribution with a standard deviation of one. And the variance is also 1. The mean of the distribution is approximated to be 0 using StandardScaler.while label encoding converted all values in a columns number with sequences.

3.3 Feature Selection

A random forest Technique was used to extract the important feature of IDS.for feature extraction; it combines the filter and wrapper methods. Random forests are comprised of 4–12 hundred decision trees, each of which is constructed using a random extraction of measurement from the dataset. The trees are tested and hence have less chance of over-fitting. Any tree consists of binary based on multiple attributes [33]. The Selected feature from the intrusion detection dataset is presented below in Fig. 4 with feature importance.

3.4 Classification

Five different classifications were used, which are discussed in detail.

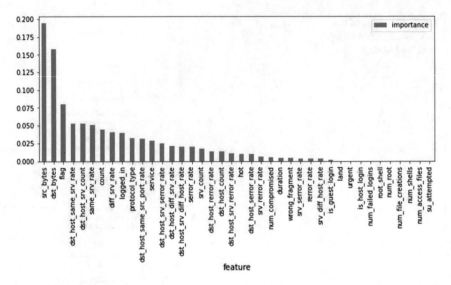

Fig. 4 Feature of intrusion detection

3.4.1 Naïve Bayes

Naïve Bayes is one of the supervised algorithm types in data science. It has built on the Bayes theorem [22]. For an individual data entity, the approach of Naïve Bayes expects a certain level of autonomous naivety amidst features. Naïve Bayes works on an input vector, and it takes some numbers such as variant, average and uses an input vector for which class is fit for it [5, 34].

3.4.2 Decision Tree (DT)

DT is a non-parametric supervised learning technique applied for classification and regression. The decision tree split the data numerous times based on particular feature cutoff values. The DT splits the dataset into small subsets where each instance belongs to one subset. DT consists of the internal node, branch node, and leaf node. The internal node presents the attribute, the branch node shows the value of the attribute node, and the leaf node presents the class distribution [35].

3.4.3 K-Nearest Neighbor

KNN is a supervised non-parametric classifier that is applied to the data with several classes to predict the classification of a new point.KNN stores all the classes and classifies them based on a distance measure's similarity. For the first time, KNN was used for statistical estimation and pattern recognition [34, 36].

3.4.4 Logistic Regression

Logistic Regression is a classification algorithm that anticipates a dependent variable's category based on the independent variable's values. Its output is either 0 or 1. LR is a predictive analysis algorithm based on probability. The input data in Logistic Regression is classified, which implies that various input values translate to the same output values [35, 37].

3.4.5 Artificial Neural Networks (ANN)

Artificial neural networks are the main algorithm mostly used in ML. Neural networks consist of the input, hidden, and output layers, which are attached. The first layer includes neurons that forward the data to the second layer, and the output is sent to the third layer.ANN is a non_linear statistical technique where the relationships between input and output are modelled.

4 Experiments and Results

This field of the study explains the dataset and the performance of five machine learning algorithms in detail.

4.1 Dataset Description

The Network Intrusion Detection dataset has been applied in this research. The dataset is available on the Kaggle site. The dataset consists of a wide range of intrusion, simulated in a military network status. Furthermore, each link is identified as either normal or an attack with a single attack type. Each connection record is around 100 bytes long. Forty-one quantitative and qualitative characteristics are extracted from normal and attack data for each TCP/IP connection. There are two types of class variables: Normal and Anomalous.80% of the dataset has been applied for training purposes and 20% for testing from the intrusion dataset.

4.2 Performance Metrics

The suggested framework used five different performance metrics for intrusion detection. The performance strength of the classifier are accuracy, precision, recall, F1-score, and support. We also calculated the Cross-Validation Mean Score and Model

Table 2 Performance metrics for five classifiers

Performance measure	Formula
Accuracy	$\frac{TP+TN}{TP+FP+FN+TN}$
Precision	$\frac{TP}{TP+FP}$
Recall	$\frac{TP}{TP+FN}$
F1-Score	$2 * \frac{Precision*Recall}{Precision+Recall}$

TP = True Positive, TN = True Negative, FP = False Positive, FN = False Negative

Table 3 Confusion matrix and cross-validation of five classifier

Classification techniques	Confusion matrix				Cross-validation Mean Score
	TP	FP	TN	FN	
Naive Baye classifier	8007	1403	10,299	444	0.90
Decision tree classifier	9410	0	10,743	0	0.99
KNN classifier	9335	75	10,708	35	0.99
Logistic regression	8837	573	10,414	329	0.95
ANN model	9294	116	10,652	91	0.98

accuracy. A confusion matrix arises where the performance measure is determined for every classifier. Table 2 shows the performance metrics of the classifier [36, 37].

4.3   Confusion Matrix

Table 3 and Fig. 5 show the confusion matrix of five classifiers (Naïve Bayes, Decision Tree, KNN, Logistic Regression, and ANN classifier). From the matrix, we discover that the Decision Tree can allocate data accurately in TP, FP, FN, TN and achieve the cross-validation mean score of 99% [33, 38].

4.4   Performance of ML Technique

Table 4 shows the performance of five different classifiers. From the table, the decision tree has been succeeded the highest accuracy of 100% in other classifier models. The KNN and ANN classifier has performed well, while the Naïve Bayes classifier has achieved 91% accuracy. The accuracy, precision, recall, and F1-Score of Decision Tree are 100%, while others are presented in Table 4.

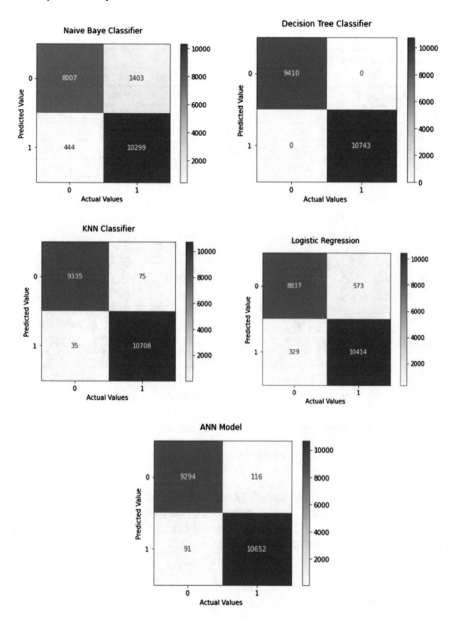

Fig. 5 Confusion matrix for Naïve Bayes, decision tree, KNN, logistic regression, and ANN classifier

Table 4 Classification report of five ML techniques in terms of accuracy, precision, recall, F1-Score and Support

Classification techniques	Classification report				
	Accuracy	Precision	Recall	F1-Score	Support
Naive Baye classifier	0.90	0.91	0.91	0.91	20,153
Decision tree classifier	1.00	1.00	1.00	1.00	20,153
KNN classifier	0.99	0.99	0.99	0.99	20,153
Logistic regression	0.96	0.96	0.96	0.96	20,153
ANN model	0.99	0.99	0.99	0.99	20,153

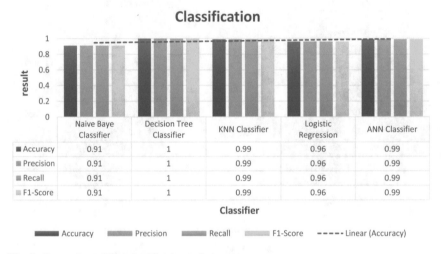

Fig. 6 Comparison of five classification techniques

These results indicate the feasibility and efficiency of our approach using five different ML algorithms to detect anomalous and normal data from the intrusion dataset. Furthermore, Fig. 6 shows that the decision tree achieved the highest detection rate among other ML techniques.

5 Conclusion

Machine learning plays a vital role to aids attackers and improving the performance of IDS. This study used five disparate machine learning approaches like Naïve Bayes, Decision tree, KNN, Logistic Regression, and ANN. The dataset(Network Intrusion Detection Dataset) has been downloaded from the Kaggle site. The dataset has been counterfeit in a military network surrounding and includes normal and anomalous classes. The main steps involve preprocessing, feature extraction and classification.

We applied a standard scaler and label encoder function to standardize the feature and eliminate the mean scaling to group fluctuation in preprocessing. In feature extraction, we used a random forest technique to extract the important features from the network intrusion dataset. Finally, in classification, five different ML techniques (NB, DT, KNN, Logistic regression and ANN) were used to detect and classify intrusion detection efficiently. The result analysis shows that the DT algorithm performs well with 100% accuracy among other techniques, while others are NB achieved 90%, KNN and ANN achieved 99%, and LR with 96% accuracy. In future, we aim to use CNN and the deep learning technique on the dataset to check the performance accuracy.

References

1. I.U. Khan et al., Monitoring system-based flying IoT in public health and sports using ant-enabled energy-aware routing. J. Healthcare Eng. **2021** (2021)
2. V.P. Kafle, Y. Fukushima, H. Harai, Internet of things standardization in ITU and prospective networking technologies. IEEE Commun. Mag. **54**(9), 43–49 (2016)
3. A. Lakhina, M. Crovella, C. Diot, Diagnosing network-wide traffic anomalies. ACM SIGCOMM Comput. Commun. Rev. **34**(4), 219–230 (2004)
4. J.R. Vacca, *Computer and Information Security Handbook* (Newnes, 2012)
5. K. Yang, J. Ren, Y. Zhu, W. Zhang, Active learning for wireless IoT intrusion detection. IEEE Wirel. Commun. **25**(6), 19–25 (2018)
6. *Evolution of Machine Learning Algorithms for Instruction Detection System with Multiple Classifiers*, eds. by M. Alzubi, S. Kovacs et al. (University of Miskolc Hungry, IEEE, 2017). Electronic ISSN: 1949-0488
7. M.A. Hassan, S.I. Ullah, A. Salam, A.W. Ullah, M. Imad, F. Ullah, Energy efficient hierarchical based fish eye state routing protocol for flying Ad-hoc networks. Indones. J. Electr. Eng. Comput. Sci. **21**(1), 465–471 (2021)
8. Find Open Datasets and Machine Learning Projects|Kaggle, Kaggle.com (2022). https://www.kaggle.com/datasets?search=intrusion+detection
9. S.I. Ullah, A.W. Ullah, A. Salam, M. Imad, F. Ullah, Performance Analysis of POX and RYU Based on Dijkstra's Algorithm for Software Defined Networking, in *European, Asian, Middle Eastern, North African Conference on Management & Information Systems*. (Springer, 2021), pp. 24–35
10. H. Liu, B. Lang, Machine learning and deep learning methods for intrusion detection systems: a survey. Appl. Sci. **9**(20), 4396 (2019)
11. J. Kim, N. Shin, S.Y. Jo, S.H. Kim, Method of intrusion detection using deep neural network. Paper presented at IEEE international conference on big data and smart computing (BigComp) (IEEE, 2017), pp. 313–316
12. P. Mishra, V. Varadharajan, U. Tupakula, E.S. Pilli, A detailed investigation and analysis of using machine learning techniques for intrusion detection. IEEE Commun. Surv. Tutor. **21**(1), 686–728 (2018)
13. B. Susilo, R.F. Sari, Intrusion detection in IoT networks using deep learning algorithm. Information **11**(5), 279 (2020)
14. K.A. da Costa, J.P. Papa, C.O. Lisboa, R. Munoz, V.H.C. de Albuquerque, Internet of Things: A survey on machine learning-based intrusion detection approaches. Comput. Netw. **151**, 147–157 (2019)
15. K.A. Taher, B.M.Y. Jisan, M.M. Rahman, Network intrusion detection using supervised machine learning technique with feature selection. Paper presented at 2019 International

conference on robotics, electrical and signal processing techniques (ICREST) (IEEE, 2019), pp. 643–646

16. K.S. Kiran, R.K. Devisetty, N.P. Kalyan, K. Mukundini, R. Karthi, Building a intrusion detection system for IoT environment using machine learning techniques. Proc. Comput. Sci. **171**, 2372–2379 (2020)

17. A. Meryem, B.E. Ouahidi, Hybrid intrusion detection system using machine learning. Netw. Secur. **2020**(5), 8–19 (2020)

18. Ü. Çavuşoğlu, A new hybrid approach for intrusion detection using machine learning methods. Appl. Intell. **49**(7), 2735–2761 (2019)

19. N. Islam et al., Towards machine learning based intrusion detection in IoT networks. Comput. Mater. Contin **69**, 1801–1821 (2021)

20. A. Verma, V. Ranga, Machine learning based intrusion detection systems for IoT applications. Wireless Pers. Commun. **111**(4), 2287–2310 (2020)

21. S. Hanif, T. Ilyas, M. Zeeshan, Intrusion detection in IoT using artificial neural networks on UNSW-15 dataset. Paper presented 2019 IEEE 16th international conference on smart cities: improving quality of life using ICT & IoT and AI (HONET-ICT) (IEEE, 2019), pp. 152–156

22. M. Imad, F. Ullah, M.A. Hassan, Pakistani currency recognition to assist blind person based on convolutional neural network. J. Comput. Sci. Technol. Stud. **2**(2), 12–19 (2020)

23. G.W. Cassales, H. Senger, E.R. de Faria, A. Bifet, IDSA-IoT: an intrusion detection system architecture for IoT networks. Paper presented at 2019 IEEE symposium on computers and communications (ISCC) (IEEE, 2019), pp. 1–7

24. W. Wu et al., A survey of intrusion detection for in-vehicle networks. IEEE Trans. Intell. Transp. Syst. **21**(3), 919–933 (2019)

25. G. Karatas, O. Demir, O.K. Sahingoz, Increasing the performance of machine learning-based IDSs on an imbalanced and up-to-date dataset. IEEE Access **8**, 32150–32162 (2020)

26. J. Hu, X. Yu, D. Qiu, H.-H. Chen, A simple and efficient hidden Markov model scheme for host-based anomaly intrusion detection. IEEE Netw. **23**(1), 42–47 (2009)

27. W.-H. Chen, S.-H. Hsu, H.-P. Shen, Application of SVM and ANN for intrusion detection. Comput. Oper. Res. **32**(10), 2617–2634 (2005)

28. K. Shafi, H.A. Abbass, Evaluation of an adaptive genetic-based signature extraction system for network intrusion detection. Pattern Anal. Appl. **16**(4), 549–566 (2013)

29. S. Ustebay, Z. Turgut, M.A. Aydin, Intrusion detection system with recursive feature elimination by using random forest and deep learning classifier. Paper presented at 2018 international congress on big data, deep learning and fighting cyber terrorism (IBIGDELFT) (IEEE, 2018), pp. 71–76

30. N. Koroniotis, N. Moustafa, E. Sitnikova, B. Turnbull, Towards the development of realistic botnet dataset in the internet of things for network forensic analytics: Bot-iot dataset. Futur. Gener. Comput. Syst. **100**, 779–796 (2019)

31. G. Creech, *Developing a high-accuracy cross platform Host-Based Intrusion Detection System capable of reliably detecting zero-day attacks* (University of New South Wales, Canberra, Australia, 2014)

32. A. Adebowale, S. Idowu, A. Amarachi, Comparative study of selected data mining algorithms used for intrusion detection. Int. J. Soft Comput. Eng. (IJSCE) **3**(3), 237–241 (2013)

33. M. Imad, A. Hussain, M. Hassan, Z. Butt, N. Sahar, IoT based machine learning and deep learning platform for Covid-19 prevention and control: a systematic review. In: *AI and IoT for Sustainable Development in Emerging Countries*, pp. 523–536 (2022). https://doi.org/10.1007/978-3-030-90618-4_26

34. M. Imad, S.I. Ullah, A. Salam, W.U. Khan, F. Ullah, M.A. Hassan, Automatic detection of bullet in human body based on X-Ray images using machine learning techniques. Int. J. Comput. Sci. Inf. Secur. (IJCSIS) **18**(6) (2020)

35. M. Imad, N. Khan, F. Ullah, M.A. Hassan, A. Hussain, COVID-19 classification based on Chest X-Ray images using machine learning techniques. J. Comput. Sci. Technol. Stud. **2**(2), 01–11 (2020)

36. S.I. Ullah, A. Salam, W. Ullah, M. Imad, COVID-19 Lung image classification based on logistic regression and support vector machine. Paper presented at European, Asian, Middle Eastern, North African conference on management & information systems. (Springer, 2021), pp. 13–23
37. A. Salam, F. Ullah, M. Imad, M.A. Hassan, Diagnosing of dermoscopic images using machine learning approaches for melanoma detection. Paper presented at 2020 IEEE 23rd international multitopic conference (INMIC) (IEEE, 2020), pp. 1–5
38. A. Hussain, M. Imad, A. Khan, B. Ullah, Multi-class classification for the identification of Covid-19 in X-ray images using customized efficient neural network. In: *AI and IoT for Sustainable Development in Emerging Countries*, pp. 473–486 (2022). https://doi.org/10.1007/978-3-030-90618-4_23

Blockchain Enabled Artificial Intelligence for Cybersecurity Systems

Keshav Kaushik (iD)

Abstract Blockchain is a prominent technology having a wide range of applications in the domain of cybersecurity, Artificial Intelligence, Internet of Things and many more. Blockchains have the benefits of being immutable, data not being damaged by the host machine, and strong confidentiality. The drawbacks are that they need a lot more energy to store the same amount of data and are prone to 51 percent attacks. Blockchains have the potential to greatly improve cybersecurity. Blockchain is an unchangeable distributed ledger that allows data to be stored without the involvement of a third party. The implementation of blockchain technology in artificial intelligence for cybersecurity has piqued researchers' curiosity. The emphasis of this chapter is on blockchain's applicability in cybersecurity. This chapter envisages the role of blockchain and AI in Cybersecurity. Moreover, the open challenges for blockchain enabled AI solutions for Cybersecurity systems are also discussed in this chapter.

Keywords Artificial intelligence · Blockchain · Cybersecurity · Smart contracts · IoT

1 Introduction

People utilize an increasing range of social media sites, and weak and insecure credentials secure the majority of them. Large volumes of data are captured during social media interactions, and if hackers get access to this data, they may create havoc. Blockchain technology may be used to develop a consistent security standard since it is a highly secure method to end-to-end encryption. It could also be used to safeguard private conversations by creating a standardized API architecture that supports cross-messenger communication. Attackers have already been able to get access to broad networks via edge devices in the past. With the increasing demand for smart home gadgets, attackers may be able to gain access to home automation via edge devices like smart switches if the security mechanisms on these IoT devices

K. Kaushik (✉)
School of Computer Science, University of Petroleum and Energy Studies, Dehradun, India
e-mail: officialkeshavkaushik@gmail.com

© The Author(s), under exclusive license to Springer Nature Switzerland AG 2022   165
M. Ouaissa et al. (eds.), *Big Data Analytics and Computational Intelligence for Cybersecurity*, Studies in Big Data 111,
https://doi.org/10.1007/978-3-031-05752-6_11

are poor. Blockchain technology [1] may be used to secure similar connections or individual devices by decentralizing their administration. Blockchain is a game-changing technology that has the potential to transform business, especially in the emerging business sector's worldwide collaborations. Blockchain technology is a distributed and decentralized record-keeping system that can track transactions across several computers. Along with its capabilities to disclose any injustice and provide certainty in the dependability of trades, the innovation is regarded as a robust internet protection convention. The blockchain is yet another innovative technology that is now being used by organizations to manage records.

The blockchain technology provides a number of benefits, including decentralization, transparency, traceability, and data integrity. This article [2] describes the blockchain infrastructure and discusses the idea, features, and importance of blockchain in security, as well as how Bitcoin works and how to improve Internet of Things (IoT) security. It tries to emphasize the importance of Blockchain in defining the direction of cyber security, cryptocurrencies, and IoT adoption. This article discusses the importance of blockchain technology in a variety of technological sectors, as well as its benefits over traditional systems.

Decentralized alternatives to trust administration are gaining traction as connectivity grows, cloud services become more popular, and the IoT becomes more prevalent. Blockchain technologies, which offer a distributed ledger, are attracting a lot of interest from the scientific community in a variety of sectors. Unfortunately, this technique does not guarantee security on its own. Among the most serious issues in computer system cybersecurity is deciding where to put trust. When a given application necessitates the administration of sensitive data, it's usually managed by protecting a particular node or equipment.

Blockchain technology has already been used in a variety of situations. Bitcoin and several other cryptocurrencies, for example, make use of blockchains in such a manner that each economic transaction is added to the blockchain as a new record. Blockchain is a distributed, autonomous, and digital ledger in which transactions are recorded in the block form. Because of its immutability and limited access, this ledger aids in the storage of information in a transparent manner. Purchases, transactions, accounts, manufacturing, and much more may all be tracked via a blockchain network. Permissioned members gain trust and confidence in their dealings with other firms, as well as new efficiency and possibilities, since they share a single vision of the truth.

Cybersecurity is the activity of safeguarding systems and networks against digital assaults aimed at gaining access to, altering, or destroying digital data in order to steal money or sensitive information. As people grow more dependent upon technology and data, it's more critical than ever to strengthen security safeguards to safeguard digital transaction records. Viruses, Trojans, Rootkits, and other types of malware may be used in cyberattacks. Phishing [3], Distributed Denial of Service (DDoS) assaults, Man in the Middle (MITM) attacks, SQL injection attacks, and Ransomware attacks are all prevalent forms of cyberattacks. Insider-attacks are also prominent type of cyber-attacks in which AI and Blockchain can be used to provide the solution. Insiders [4] who may be aware of the weaknesses of the networks and corporate

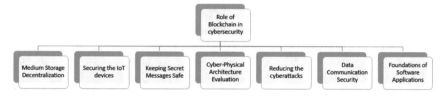

Fig. 1 Role of blockchain in cybersecurity

types presented have allowed clients with legitimate access to sensitive and secret information. Many cyber-attacks perpetrated by hostile insiders are more difficult to detect than those perpetrated by outside attackers, whose fingerprints are more eagerly hidden. The role of blockchain in cybersecurity is explained with the help of Fig. 1.

1.1 Medium Storage Decentralization

Businesses are increasingly concerned about data breaches and exploitation in the workplace. A centralized storage solution is still used by the majority of enterprises. A hacker just has to exploit one flaw to have entry to all the information stored on these sites. As a consequence of such an attack, a criminal has access to the sensitive and confidential data, including corporate financial information.

1.2 Securing the IoT Devices

Hackers to obtain access to larger networks are increasingly leveraging Smart objects, such as heaters and gateways. Only because of the current AI frenzy has it become difficult for hackers to get control to whole systems like home automation via edge devices like "smart" switches. In most cases [5], the overwhelming majority of these IoT devices have poor security procedures. In this case, blockchain might be used to decentralize and secure the administration of such large-scale equipment and systems. The approach gives the device the power to make security decisions on its own.

1.3 Keeping Secret Messages Safe

As the technology reduces the world into a worldwide society, people are increasingly turning to social media. Furthermore, the number of social media outlets continues to increase. Every day, new social apps are being created as verbal commerce grows more popular. Massive amounts of metadata are recorded throughout these conversations. To protect their accounts and data, the millions of customers of social networking platforms choose weak, insecure passwords.

1.4 Cyber-Physical Architecture Evaluation

Data tampering, configuration management difficulties, and component failure have all harmed the integrity of information provided by cyber-physical systems. Nonetheless, the information coherence and validation characteristics of blockchain technology might be utilized to validate the status of any cyber-physical infrastructure. Throughout the whole chain of custody, the data created by blockchain on the infrastructure's components may be more trustworthy.

1.5 Reducing the Cyber-Attacks

The blockchain's immutability is one of its most prominent features. Due to the use of consecutive hashing and cryptography, as well as the decentralized structure, it is almost difficult for any entity to arbitrarily modify data on the ledger. Companies that handle confidential material might use this to ensure data integrity and to deter and detect manipulation.

1.6 Data Communication Security

In the long run, blockchain may have been used to prevent unauthorized access to data in motion. Transmission of data may be protected by encrypting it using the technology's full encryption feature, which prevents bad actors, whether people or organizations, from obtaining access to it. The total integrity and validity of information conveyed through blockchain would be improved as a consequence of this technique. Hackers with criminal intent capture data in transit with the purpose of altering or erasing it. Due to inefficient communication, channels, such as emails, leave a big gap in the system.

1.7 Foundations of Software Applications

To avoid foreign interference, blockchain may be used to secure the authenticity of software installation. Blockchain, like MD5 hashes, may be used to validate actions like firmware upgrades, installers, and fixes to prohibit malicious software from infiltrating machines. Latest software identities are matched against hashes accessible on vendor webpages in the MD5 paradigm. Because the hashes provided on the company's system could already be corrupted, this approach is not fully failsafe.

The decentralized feature of blockchain lends itself to a range of applications, including cybersecurity. Since system components instantly cross-reference one another to detect the node with incorrect data, data on public blockchains cannot be changed. Blockchain technology allows for the highest levels of data transparency and integrity. Since blockchain technology arranges data storage, one of the most common causes of data leaks is human error. The greatest major threat to organizations is cybercrime, which blockchain technology may be able to assist tackle.

2 Introduction to Artificial Intelligence for Cybersecurity Systems

AI uses computer systems, data, and occasionally robots to replicate the human brain's reasoning, problem-solving and judgment skills. It also includes the subfields of deep learning and machine learning, which use AI algorithms that are trained on data to produce predictions or categorization and improve over time. Automation of monotonous activities, enhanced decision-making, and a positive customer experience are all advantages of AI. The blockchain's digital record enables insight into AI's architecture as well as the source of the data it utilizes, addressing the difficulty of understandable AI. This increases confidence in the accuracy of data and, as a result, in the AI suggestions. When using blockchain to maintain and disseminate AI models, an audit trail is created, and combining blockchain with AI may improve data security. The authors [6] looked at how blockchain and artificial intelligence may help safeguard and secure personal information in this article. Users have choice over what, when, and with whom their personally identifiable information may be shared using decentralized and federated identity systems. Such technologies may also help to decrease cyber-threats. Artificial intelligence improves blockchain-based security choices by giving consumers greater control over their data and ensuring that the information and algorithms generated from it are more exact, fair, and reliable.

AI can read, comprehend, and correlate data quickly and exhaustively, adding a new degree of intellect to blockchain-based corporate network. Blockchain helps AI to grow by giving access to massive volumes of data from both inside and outside the enterprise, allowing it to provide more actionable insights, control bandwidth utilization and model exchange, and create a trustworthy and transparent data industry. By lowering friction, increasing speed, and increasing efficiency, AI, robots, and blockchain may provide added features to multi-party commercial processes. For instance, AI algorithms embedded in smart contracts on the blockchain might suggest recalled things, conduct trades depending on predetermined criteria and events,

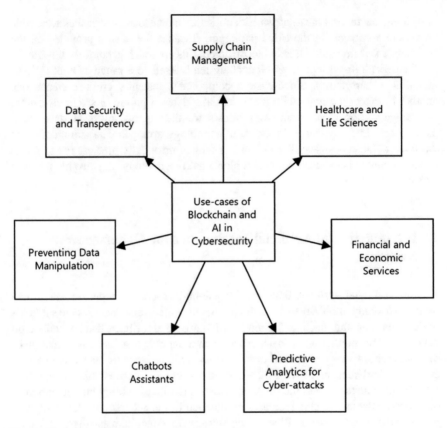

Fig. 2 Blockchain and AI use-cases in cybersecurity

resolve disputes, and pick the most environmentally friendly transport network. The prominent use-cases of Blockchain and AI in Cybersecurity is described in Fig. 2.

2.1 Supply-Chain Management

AI and blockchain [7] are revolutionizing supply chains throughout sectors and providing new possibilities by automating a mostly paper-based system, making data accessible and reliable, and adding knowledge and robotics to perform transactions. A company, for example, may monitor carbon pollution statistics at the good or service or component level, enhancing decarbonization efforts with precision and insight.

2.2 Healthcare and Life Sciences

AI can assist develop practically every discipline in healthcare, from exposing therapeutic ideas and supporting user demands to recognizing findings from patient information and showing trends. Companies may collaborate to enhance treatment while respecting patient privacy by storing patient data on blockchain, especially electronic health records. In the pharmacy sector, blockchain and AI may improve medication supply chain visibility and traceability while also drastically raising clinical trial success rates. Advanced data analysis combined with a distributed clinical trial infrastructure allows for data consistency, accessibility, patient monitoring, consent management, and trial involvement and data gathering automation.

2.3 Financial and Economic Services

By facilitating trust, reducing friction from multilateral interactions, and increasing system throughput, blockchain and AI are changing the financial services business. Examine the loan application procedure. Candidates provide their permission for personal data to be kept on the blockchain. Assurance in the data and automated processes for assessing the transaction help to speed up closings and increase client satisfaction.

2.4 Predictive Analytics for Cyber-Attacks

A data breach may be detected using predictive analytics before it occurs. These analytics, like a sensor that indicates where and when an opponent is arriving, decide when and where assaults may happen. This allows your corporation to sound the alert, raise the drawbridge, and get your troops ready. Predictive analytics allows you to outsmart hackers and emerge triumphant rather than detecting a vulnerability after the conflict has already been lost. The goal of this study [8] is to determine whether there is a link between both the COVID-19 outbreak and the rise in cybercrime. Blockchain can also be used in predicting the pandemic like COVID-19 [9].

2.5 Chatbots Assistants

Chatbots and virtual assistants [10] are programs that perform online conversations utilizing text or text-to-speech in lieu of actual operators. By using limited resource, firms may grow while retaining a "human-like" customer experience.

2.6   Preventing Data Manipulation

Among the most secure information security solutions is blockchain. Significant advancements [11] in digital technologies has promoted new attack vectors and data security issues. Companies must implement robust verification and cryptographic key vaulting techniques to safeguard their data. Notwithstanding previous methodologies, many Blockchain development organizations are being motivated by this innovation to re-architect and restructure security problems. Blockchain provides a genuine feeling of delivering data trust aspects.

2.7   Data Security and Transparency

Because AI algorithms demand a large quantity of data from as many locations on the Internet as feasible, data security is one of the most important problems for any network design. However [12], a more capable AI can defend data security at a greater level, since an upgraded AI can work out complex and intricate threats more quickly than a standard AI.

Data, algorithm, and computational power are three crucial parts of AI technology, in the meaning that large information is necessary for retraining the process to acquire a categorization method, and the training program necessitates optimal processing power. In practice, diverse entities design, train, and employ AI models. Customers may not completely trust the model they are employing since the training program is unclear to them. Furthermore, as AI algorithms get more complicated, it is becoming increasingly difficult for individuals to comprehend how the learning result is generated. As a result, there has been a recent trend toward moving away from centralized AI systems and toward decentralized AI approaches.

AI's most valuable resource is data. The efficiency [13] of AI categorization findings is directly influenced by the amount and quality of data. However, there are certain issues with the data exchange procedure. First, diverse stakeholders own the data required for training, and they do not rely on each other. The data is tough to approve or verify. Second, bad people may share harmful data for specific goals. Millions [14] of variables and compromises between confidentiality, efficiency, decentralization, and other factors go into the implementation and construction of a blockchain. AI can help with these choices, as well as automating and regulate blockchain for improved performance and governance. Furthermore, since all data on the blockchain is public, AI is critical in ensuring users' security and privacy. For better understanding of the role of Blockchain and AI in the domain of Cybersecurity, the comparative analysis of latest related papers are done in Table 1.

Table 1 Comparative analysis of related work in the area of Blockchain, AI and Cybersecurity

Author	Year	A	B	C	D	E	F	Major findings
Shinde et al. [15]	2021	✔	✔	✖	✔	✖	✖	This research examines how Blockchain offers a security blanket for AI-based systems using bibliometric and literary analysis. For this quantitative investigation and evaluation, two popular research resources i.e. Web of Science and Scopus were used. The study discovered that conference concept proposals and select journal publications make a significant impact. However, much more study is needed before actual and robust Blockchain-based AI systems can be implemented
Aguilar et al. [16]	2021	✔	✖	✔	✖	✖	✖	This study attempts to offer a complete overview of the strategies and aspects suggested for achieving cybersecurity in blockchain-based platforms. The study is aimed at local scholars, cybersecurity experts, and blockchain developers. The authors looked at 272 articles from 2013 to 2020, as well as 128 industrial applications, for this aim. We outline the lessons learnt and highlight a number of issues that need to be addressed in order to encourage additional work in this domain
Mengidis et al. [17]	2019	✔	✔	✔	✔	✖	✖	The authors have examined the increasing cybersecurity concerns and how advances in AI and blockchain technology might mitigate the resulting dangers in this paper

(continued)

Table 1 (continued)

Author	Year	A	B	C	D	E	F	Major findings
Taylor et al. [18]	2020	✔	✘	✔	✔	✔	✔	This study finds peer-reviewed publications that aims to use blockchain for cyber security and gives a comprehensive examination of the most widely used blockchain encryption methods. This relevant systematic analysis also illuminates future studies, teaching, and practice areas in the blockchain and cyber security sector, like blockchain security in IoT, blockchain security for AI applications, and sidechain security
Kim et al. [19]	2020	✔	✔	✔	✘	✘	✔	Through an analysis of cybersecurity attack patterns, the authors suggested the direction of constructing an information AI system, that is a blockchain-based knowledge data environment prototype to authenticate the trustworthiness of learning data, and the authors have introduced the orientation of learning information management before deep learning
Maleh et al. [20]	2021	✔	✔	✔	✔	✔	✔	This book covers the state-of-the-art as well as the most recent findings in the fields of AI and Blockchain approaches and techniques for upgrades applications. It provides new paradigms, real solutions, and technical advancements for Cybersecurity including Artificial Intelligence and Blockchain
Chentouf et al. [21]	2021	✔	✘	✔	✘	✘	✘	In this article, we examine Blockchain technology and how it might aid in the development of a safer IoT system, as well as the significant difficulties and security concerns that are impeding the concept's progress

A: Blockchain B: Artificial Intelligence C: Cybersecurity D: Smart Contracts E: Bitcoin F: IoT

3   Open Challenges for Blockchain Enabled AI Solutions for Cybersecurity

AI can exist on top of Blockchain and derive insights from the shared information created by the network, which can then be utilized to make forecasts. Blockchain [22] is an elevated cybersecurity system that uses binding understanding between nodes to construct chains that link old blocks recorded in nodes with new blocks in a chronological order. Finance, healthcare, cybersecurity, prediction, medical services, cryptocurrencies, and other businesses see their development accelerated by technological convergence. The more these sectors use digital systems and deliver services, the higher the danger of these networks being hacked.

Despite its benefits in terms of transparency and integrity, blockchain systems pose a variety of cybersecurity and data-security concerns. Users benefit from the decentralized access approach to public blockchains, which has rapidly become a feature abused for criminal activity, particularly with cryptocurrency. Complete decentralization [23] is however a major problem for this sort of system's governance. The absence of inbuilt governance standards is a flaw that makes the decision-making process difficult when making judgments that effect all users. In this case, the danger is that the decision is left to the programmers or a small number of individuals, undercutting all of decentralization's advantages. Examining at the consensus processes of contemporary blockchain systems, the same problem can be seen.

The author has examined and highlighted current problems for merging AI, blockchain technology and cybersecurity in Fig. 3. The following are some of the anticipated challenges associated with the harmonization and incorporation of AI, Blockchain and Cybersecurity.

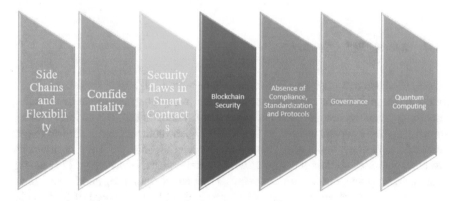

Fig. 3 Challenges for Blockchain enabled AI solutions in Cybersecurity

3.1 Side Chains and Flexibility

In current scenario, the scalability is one of the biggest problem that is being faced by blockchain network [24]. Bitcoin blockchain platforms can conduct a mean of 4 transactions per second, whereas Ethereum blockchain platforms can do an average of 12 operations per second. Especially contrasted to Facebook, which processes thousands of transactions every second, involving likes, comments, and replies, such speed is just inadequate.

3.2 Confidentiality

Nevertheless, acquired data is publicly open and usable for all readers on public blockchain ledgers, allowing for secure and legitimate processing of data. Furthermore, these secure blockchain systems [25] will stop access to and exposure of the vast volume of data that AI may need to digest and implement precise and accurate decision-making and analysis.

3.3 Security Flaws in Smart Contracts

It is critical to guarantee that a smart contract's architecture is free of defects and flaws, as well as safe against assaults. Because [26] the code and data on the networks may be subject to assaults, it is critical to keep them secure.

3.4 Blockchain Security

Cases of abuse and misuses of blockchain's decentralised ability have been documented. Despite the fact that blockchain offers comprehensive methods for safeguarding predictive analysis, IoT and blockchain systems are subject to cyber-attacks similar to the 51 percent assault. In this work [27], the authors have examined major blockchain systems and undertake a thorough investigation on the security risks to blockchain, as well as the similar to the case assaults. Researchers [28] give a complete assessment on security problems, blockchain topologies, and implementations from an industrial viewpoint in order to find the huge potential of blockchain for the IIoT.

3.5 Absence of Compliance, Standardization and Protocols

Standards for blockchain technology have yet to be developed. IEEE, NIST, ITU, and a number of other standards organizations are working to develop blockchain connectivity, administration, interaction, and architectural standards. Furthermore, governmental [29] and organizational regulations, rules, laws, regulatory requirements, and practices for blockchain implementation, arbitration, and dispute resolution in the perspective of AI applications, particularly for public blockchain financial transactions financing and automated transfer of funds using cryptocurrencies, must be established at the local and global levels.

3.6 Governance

It takes a lot of time and effort to deploy, build, and manage a blockchain platform with several players and stakeholders. Even with a corporate or cooperative blockchain, substantial challenges exist in terms of the kind of blockchain to install, as well as who maintains and diagnosing and resolving it.

3.7 Quantum Computing

It is expected that future quantum computers would be able to break public key encryption, allowing secret keys to be discovered. The existing blockchain system is based on digital signatures that employ public key encryption. Several experts anticipate that quantum computing will make blockchain's fundamental security unbreakable by 2027 [30].

4 Conclusion

This chapter highlights the blockchain enabled AI for its applications in cybersecurity systems. The chapter discusses the role of blockchain in cybersecurity and the various applications of artificial intelligence in cybersecurity systems. This domain has several open challenges like governance, absence of compliance, security, etc. A comparative study of the related work in the domain of blockchain, AI, and cybersecurity has also been highlighted in this chapter.

References

1. M. Liu, W. Yeoh, F. Jiang, K.-K. R. Choo, *Blockchain for Cybersecurity: Systematic Literature Review and Classification,* pp. 1–17 (2021). https://doi.org/10.1080/08874417.2021.1995914
2. P. Bansal, R. Panchal, S. Bassi, A. Kumar, Blockchain for cybersecurity: a comprehensive survey. Paper presented at Proceedings of the 2020 IEEE 9th international conference communication system network technology (CSNT 2020), pp. 260–265 (2020). https://doi.org/10.1109/CSNT48778.2020.9115738
3. K. Kaushik, S. Singh, S. Garg, S. Singhal, S. Pandey, Exploring the mechanisms of phishing. Comput. Fraud Secur. **2021**(11), 14–19 (2021). https://doi.org/10.1016/S1361-3723(21)00118-4
4. K. Kaushik, A systematic approach to develop an advanced insider attacks detection module. J. Eng. Appl. Sci. **8**(1), 33 (2021). https://doi.org/10.5455/JEAS.2021050104
5. A. Dhar Dwivedi, R. Singh, K. Kaushik, R. Rao Mukkamala, W.S. Alnumay, Blockchain and artificial intelligence for 5G-enabled Internet of Things: challenges, opportunities, and solutions. Trans. Emerg. Telecommun. Technol. e4329 (2021). https://doi.org/10.1002/ETT.4329
6. S. Heister, K. Yuthas, How blockchain and AI enable personal data privacy and support cybersecurity, in *Blockchain Potential AI [Working Title]*, (2021). https://doi.org/10.5772/INTECHOPEN.96999
7. Blockchain and artificial intelligence (AI)|IBM. https://www.ibm.com/topics/blockchain-ai.. Accessed 18 Dec 2021
8. E.F. Adeniran, H. Jahankhani, *A Descriptive Analytics of the Occurrence and Predictive Analytics of Cyber Attacks During the Pandemic*, pp. 123–159 (2021). https://doi.org/10.1007/978-3-030-87166-6_6
9. K. Kaushik, S. Dahiya, R. Singh, A.D. Dwivedi, Role of blockchain in forestalling pandemics. Paper presented at Proceedings of the 2020 IEEE 17th international conference on mobile ad hoc and smart systems (MASS 2020), pp. 32–37. https://doi.org/10.1109/MASS50613.2020.00014
10. Top Emerging Technologies in 2021| CompTIA. https://connect.comptia.org/content/infographic/list-of-emerging-technologies. Accessed 19 Dec 2021
11. Blockchain Technology Ensuring Data Security & Immutability|by Gautam Raturi|Towards Data Science. https://towardsdatascience.com/blockchain-technology-ensuring-data-security-immutability-7150d309352c. Accessed 20 Dec 2021
12. K. Wang, J. Dong, Y. Wang, H. Yin, Securing data with blockchain and AI. IEEE Access **7**, 77981–77989 (2019). https://doi.org/10.1109/ACCESS.2019.2921555
13. R. Wang, M. Luo, Y. Wen, L. Wang, K.K. Raymond Choo, D. He, The applications of blockchain in artificial intelligence. Secur. Commun. Netw. **2021** (2021). https://doi.org/10.1155/2021/6126247
14. T.N. Dinh, M.T. Thai, AI and blockchain: a disruptive integration. Computer (Long. Beach. Calif) **51**(9), 48–53 (2018). https://doi.org/10.1109/MC.2018.3620971
15. R. Shinde, S. Patil, K. Kotecha, K. Ruikar, Blockchain for securing AI applications and open innovations. J. Open Innov. Technol. Mark. Complex., **7**(3) (2021). https://doi.org/10.3390/joitmc7030189
16. M. Gimenez-Aguilar, J.M. de Fuentes, L. Gonzalez-Manzano, D. Arroyo, Achieving cybersecurity in blockchain-based systems: a survey. Futur. Gener. Comput. Syst. **124**, 91–118 (2021). https://doi.org/10.1016/J.FUTURE.2021.05.007
17. N. Mengidis, T. Tsikrika, Blockchain and AI for the next generation energy grids: cybersecurity challenges and opportunities. Vrochidis I. Kompatsiaris **43**(1), 21–33 (2019). https://doi.org/10.11610/isij.4302
18. P.J. Taylor, T. Dargahi, A. Dehghantanha, R.M. Parizi, K.K.R. Choo, A systematic literature review of blockchain cyber security. Digit. Commun. Netw. **6**(2), 147–156 (2020). https://doi.org/10.1016/J.DCAN.2019.01.005

19. J. Kim, N. Park, Blockchain-based data-preserving AI learning environment model for AI cybersecurity systems in IoT service environments. Appl. Sci. **10**(14), 4718 (2020). https://doi.org/10.3390/APP10144718
20. Y. Maleh, Y. Baddi, M. Alazab, L. Tawalbeh, I. Romdhani, *Artificial Intelligence and Blockchain for future Cybersecurity Applications*, p. 376
21. F.Z. Chentouf, S. Bouchkaren, *Blockchain for Cybersecurity in IoT* , pp. 61–83 (2021). https://doi.org/10.1007/978-3-030-74575-2_4
22. F. Muheidat, L. Tawalbeh, *Artificial intelligence and Blockchain for Cybersecurity Applications,* pp. 3–29 (2021). https://doi.org/10.1007/978-3-030-74575-2_1
23. J.L. Hernandez-Ramos, D. Geneiatakis, I. Kounelis, G. Steri, I.N. Fovino, Toward a data-driven society: a technological perspective on the development of cybersecurity and data-protection policies. IEEE Secur. Priv. **18**(1), 28–38 (2020). https://doi.org/10.1109/MSEC.2019.2939728
24. D. Khan, L.T. Jung, M.A. Hashmani, Systematic literature review of challenges in blockchain scalability. Appl. Sci. 11(20), 9372 (2021). https://doi.org/10.3390/APP11209372
25. W. Liang, N. Ji, Privacy challenges of IoT-based blockchain: a systematic review. Clust. Comput. **2021**, 1–19 (2021). https://doi.org/10.1007/S10586-021-03260-0
26. G. Destefanis, M. Marchesi, M. Ortu, R. Tonelli, A. Bracciali, R. Hierons, Smart contracts vulnerabilities: a call for blockchain software engineering? Paper presented at 2018 IEEE 1st International work blockchain oriented softw. eng. (IWBOSE 2018), vol 2018, pp. 19–25 (2018). https://doi.org/10.1109/IWBOSE.2018.8327567\
27. X. Li, P. Jiang, T. Chen, X. Luo, Q. Wen, A survey on the security of blockchain systems. Futur. Gener. Comput. Syst. **107**, 841–853 (2020). https://doi.org/10.1016/j.future.2017.08.020
28. S. Latif, Z. Idrees, Z.E Huma, J. Ahmad, Blockchain technology for the industrial Internet of Things: a comprehensive survey on security challenges, architectures, applications, and future research directions. Trans. Emerg. Telecommun. Technol. **32**(11), e4337 (2021). https://doi.org/10.1002/ETT.4337
29. A. Anjum, M. Sporny, A. Sill, Blockchain standards for compliance and trust. IEEE Cloud Comput. **4**(4), 84–90 (2017). https://doi.org/10.1109/MCC.2017.3791019
30. E.O. Kiktenko et al., Quantum-secured blockchain. Quantum Sci. Technol. **3**(3), 035004 (2018). https://doi.org/10.1088/2058-9565/AABC6B

Approaches for Visualizing Cybersecurity Dataset Using Social Network Analysis

Iytzaz Barkat and Obaidullah

Abstract In this paper, a comprehensive evaluation of social network analysis approaches performed with Cybersecurity prospect to analyze and visualize cyber-security information. this paper help to understand the supporting features and their relevancy to security. However, these approaches are open source and supporting to many Operating system so these are easy to access and can be used by individuals to get their desired output.

Keywords Social network analysis · Gephi · Tulip · Pajek · Cybersecurity · Security · Visualization of security

1 Introduction

Visualization in Cybersecurity is still a difficulty; numerous visualization tools give visualization for Cybersecurity, but there appear to be some shortcomings; it appears that there is a lack of a visual analytic technique that provides analysis of larger Cyber-security datasets. Common cybersecurity risks include phishing, adware, cross-site scripting, data breaches, DDOS/Dos attacks, and so on. The wide range of possibili-ties and the rise in Cybersecurity demand highlight the importance of analysis in the field of Cybersecurity. Our primary purpose is to evaluate various visualization tools for analyzing Cybersecurity datasets. Our primary purpose is to evaluate various visualization tools for analyzing Cybersecurity datasets. Cybersecurity is a crucial pillar in the new era of technology; hackers have been quite active in recent years, and many significant organizations have had data breaches. Information security is a challenge. The primary goal of this project is to investigate and address the visu-alization needs in cybersecurity. For this aim, a few visualization tools were picked and compared to each other based on their features and methods. The importance of visualization in cybersecurity is emphasized in this study. Few articles are relevant

I. Barkat (✉) · Obaidullah
Riphah Institute of System Engineering (RISE), Riphah International University, Islamabad, Pakistan
e-mail: Barkatiytzaz@gmail.com

© The Author(s), under exclusive license to Springer Nature Switzerland AG 2022   181
M. Ouaissa et al. (eds.), *Big Data Analytics and Computational Intelligence for Cybersecurity*, Studies in Big Data 111,
https://doi.org/10.1007/978-3-031-05752-6_12

to cybersecurity visualization, but none of them address the Evaluation of the visual analysis approach that provides a comparison view in visualization form. To visualize data, we conducted multiple tests on various approaches to determine which approach provides the most reliable visualization and which approach provides large-scale visualization. In this research, data is visualized in graph format to make it easy to understand for people who are unfamiliar with cyber security, even though there are multiple factors, some of which are paid and some of which have limited features.

1.1 Visualization

A graphical depiction of data is known as visualization. Visualization is a simple method for analyzing massive text data. Some people are not interested in reading enormous text files, or they are unsure which parts are necessary to read and which are not. Furthermore, visualization can be in a variety of formats such as photos, diagrams, graphs, or animations, among others. In our advanced technological era, visualization has ever-expanding applications in science, education, and practically all sectors of technology, and visualization is the application of computer graphics. Tamassia et al. [1] investigate the process of graph sketching for the purpose of security visualisation in 2009. The use of visualisation in the current environment benefits us in a variety of ways. Visualization is not a new technology; it has been utilised in maps and drawings for over a thousand years.

1.2 Visualization in Cybersecurity

Visualization in Cybersecurity is a field that aims to lessen the labour of security analysts by presenting data in visual analytics rather than text or characters. However, visualization in Cybersecurity does not meet the demand for visualization in Cybersecurity. Every day, we collect information in the form of log files; it is tough to assess the entire data set. Shiravi et al. (2012) proposed in 2012 that systems facilitate offline forensics analysis. Their studies' findings showed a dearth of research into real-time capabilities in cybersecurity visualization [2]. Data visualization is a data analytic approach used to assist the human brain in seeing patterns in data. IP information, server logs, and communication records are examples of cyber big data that can be difficult to grasp; nevertheless, converting cyber data into a graphical format can help us understand the pattern of events. Shiravi et al. [2] examined and filled this gap of limited work and employed a case- based approach to get a widespread review of network security through visualizations system. The result suggested that there exist a strong positive association between degree of awareness regarding system and analysis based on real time. The major contribution of the work suggested that such systems support offline forensics analysis. The findings of the study highlighted the lack of research for real-time capabilities in cybersecurity visualization and develop a provision of the necessary basis for the development of security data visualization system. However, the entire set of work in this area is at its primary level.

1.3 Problem Statement

There are several ways to visualisation in social network analysis, but none of them are particularly effective. Specifically, for data sets pertaining to information security. As a result, various methodologies must be reviewed in order to develop an appropriate way for information security visualisation. As a result, this study focuses on social network analysis methodologies tailored to Cybersecurity datasets, which give improved visualisation of datasets. Some techniques have limited capabilities, while others are commercial and still support specific algorithms and layouts that do not meet the need for visualisation in cybersecurity. Some of them are only compatible with specific operating systems. In this sense, we are looking for different ways that are more efficient than others. Additionally, it supports many operating systems and layouts, and it is versatile enough to allow us to tailor the output by changing property values.

1.4 Scope

The purpose of this research is to evaluate social network analysis methodologies for analysing cybersecurity datasets. There are numerous visualisation approaches available for the depiction of various types of datasets with various properties. Each strategy is distinct from the others. This dissertation will assess visual techniques to social network analysis in order to create a better approach for visual data analysis and to evaluate social network analysis approaches. First, related content is studied at Google Scholar, IEEE, and other internet platforms, and then each approach is installed and analysed. In this research, five social network analysis approaches are collected and then evaluated based on their features, and then some Cybersecurity related dataset is gathered and analysed in each approach.

2 Social Network Analysis Approaches

In this study, we will assess and depict Cyber Security data using social network analysis methodologies. In SNA, there are numerous ways. Which we will compare and determine which ways are superior. Several techniques, layouts, input and output formats are accessible in selected social network analysis methodologies. Nadeem Akhtar, as well as submitted a study on Social Network Analysis tools [3]. SNA is used by Samtani, S., and Chen to identify key hackers for keylogging approaches.

According to the findings of this study, multiple hackers are the most senior and longest-tenured members of their network [4].

2.1 Social Network Approaches and Datasets

Selection of Social Network Approaches is based on different parameters, although there are multiple approaches for visualization, but all those approaches have their own compatible requirements, pricing, user friendly and many more. In this thesis we have select five approaches on the multiple basis.

Following are the Criteria for selection of approaches:

- GUI and Python
- User Friendly
- Multiple OS supportive
- Free and Open Source.

In order to visualize Cyber data, this study choose some social network analysis techniques.

Below in this study approaches will review.

- Gephi [5] social network analysis approach capable to visualize dataset.
- Pajek [3] approach of social network analysis to analyze datasets.
- Tulip [6] visualization approach.

Selection of Dataset in this thesis is based on different parameters, finding security related dataset can be a complex task, but in this thesis, dataset is taken from open repository and for educational use. A Cybersecurity breaches dataset is taken and execute in the above-mentioned approaches. Cybersecurity expert analyzed and validate results, also point out how an unlabeled dataset can use to evaluation and to construct a network IDS [7].

- CybersecurityDatasetbreach   [8]."https://www.kaggle.com/alukosayoenoch/cyber-security-reaches-data?select=Cyber+Security+Breaches.csv".

2.2 Gephi

Gephi is a social network analysis approach which is capable to visualize different types of data various fields. Among multiple key features, one is it has the ability to show specialization process, which have ability to shift the entire set-up in the form of diagram, and the default algorithm layout is ForceAtlas2, latter it is developed by the Gephi team, the user of the Gephi technique provide elucidation to adopt a free of scale or a range of nodes between 10 to 10,000 for network analysis [5].

2.3 Pajek

Pajek is another approach of social network analysis to analyze datasets with different features, Pajek is also a strong approach with its multiple features, In special edition of Pajek consumption of memory is lower, it support multiple OS and large network visualization with hundreds of vertices, it support multiple algorithm and features for customization of graph, in some cases if the dataset is large and complex then customization of graph and differentiation of graph into different part is difficult in Pajek [3].

2.4 Tulip

Tulip is also a visualization approach, tulip the framework can be customized and extended according to the needs of users, its plug-in architecture accepts one to add any domain-specific routine, tulip also includes python scripting to easily experiment and even animate networks. In Tulip Visualization approach, four datasets will be visualized in tulip approach, each dataset have different size and different values. In tulip there are multiple panels at a same time which can visualize a dataset into multiple ways, tulip can handle large dataset. there is multiple option to visualize dataset different layouts and different parameter are given in these approaches, along with this it can visualize selective data in a dataset for better understanding of dataset [6].

   In regard to comparison, this study closely compares each one of the approaches by their features, pricing, and layouts. Gephi has a variety of features with the flexibility of add/remove plugging and it supports all kinds of operating systems as well as also supports multiple algorithms that help us to make different graphs for different types of data, as well as other techniques. In spite of all features, there have some deficiencies, some may support only the particular operating system and specific algorithm. In all these comparisons this study found Gephi is the better to approach to further visualizing Cyber Security.

   In Table 1 each approach is compared based on their features, each feature plays a vital role in visualization approaches to visualize datasets. All visualization approaches have their own supported features which is compatible to their version and support specific type of dataset. Some specific and crucial features selected in Table 1 and compared, details of features is given below.

- Degree

Degree represents the number of edges connected to a node or number of connections to nodes, in cybersecurity it helps to know the number of connections to effected node.

Table 1 Features
comparison of selected
approaches

Features	Gephi	Tulip	Pajek
Degree	✓	✓	✓
Assortivity	✓	☒	☒
Centrality	✓	✓	✓
Diameter of network	✓	☒	✓
Clustering	✓	✓	✓
Flow	✓	✓	✓
Connected components	✓	✓	✓
Clique	✓	✓	☒
Modularity	✓	✓	✓
BFS	☒	✓	☒
DFS	✓	✓	✓
Hits	✓	✓	✓
Density	✓	✓	☒

- Assortivity

Assortivity is inclination of node which is attach to other node that are same in some way. It describes the correlation between the same system in a network.

- Centrality

Centrality define the node in graph which act as the bridge between two nodes which are connected to each other like a proxy server which act as a bridge between client and actual server.

- Network Diameter

Network Diameter in graph shows the distance between vertices pair, it actually shows the shortest path between systems.

- Clustering

Clustering in graph differentiate the node by dividing the node into clusters, it makes group of similar nodes which belong to same network or similar to some properties.

- Flow

Flow graph is directed graph which shows the direction in security it shows the direction of attack or any other type of data it will make a flow of it which will be customizable.

- Communities

Communities in a graph define as a subset of node which are strongly connected to node and loosely connected to other community node.

- Clique

Clique is an undirected and subset of vertices in a graph such that two vertices are adjacent to each other.

- BFS

Breadth First Search traverse the graph from starting node and find all near node. It searches nodes in horizontal way by going through each node in row to destination node.

- DFS

Depth First Search Find the node in vertically down which make's distance between nodes far to each other.

- Density

Density of graph define the maximum number of edges or connection between systems or nodes in a graph.

- Core

Core is any complete graph is known as a core. Partitioning of graph to reduce the graph to smaller graph.

2.5 Cybersecurity Breaches Dataset

In this dataset breaches of some organizations in USA are listed, some states of USA are mentioned and along with state different organization is mentioned and their type of breach and location of breach like laptop, email, mobile etc. also date of breach is mentioned in it and no of individual effected with this breach is also given in this dataset [8].

Attributes in Tunnel Dataset.

- Nodes: 447
- Edges: 668
- Security Area: Data.

In Table 2 details of "**Cybersecurity data breaches dataset**" has been mentioned, State Column shows the state in which data breach occur and organization suffers by data loss. In Organization columns name of organization is mentioned. 3rd column mention the number of people got effected by attack, 4th column shows the date of breach when does the data of organization get breached, 5th column shows the type of breach either organization suffer by hacking or loss or any unauthorize access, in 6th column location of information is given from where information is lost.

Table 2 Cybersecurity data breach details

State	Organization	People Affected	Dateof Breach	Breach Type	Location Information
DC	N/A	3120	10/09/2009	Loss	Laptop
NC	Rick Lawson, Professional Computer Services	2000	12/08/2009	Hacking/IT Incident	Desktop Computer, Network Server, Electronic Medical Record
TX	N/A	3120	11/19/2009	Loss	Other Portable Electronic Device, Other
WY	N/A	9023	12/02/2009	Unauthorized Access/Disclosure	Network Server
CT	N/A	957	02/04/2010	Hacking/IT Incident	Network Server

3 Methodology

This study focuses on the evaluation of social network analysis. Methods for visualizing and analyzing cybersecurity data. This report is a thorough evaluation of the available literature on cyber security networks and visualization systems. The current investigation is based on Shiravi et al. [2].'s analysis. Furthermore, this research is based on the most recent cyber security visualization system analysis. The main approach of this research will be explained in this chapter, a flow chart of this research will be added, and the data collection procedure will be outlined in this chapter.

Figure 1 shows the methodology is defined where each step is mentioned that in 1st step social network analysis approaches is selected and in the 2nd step approaches

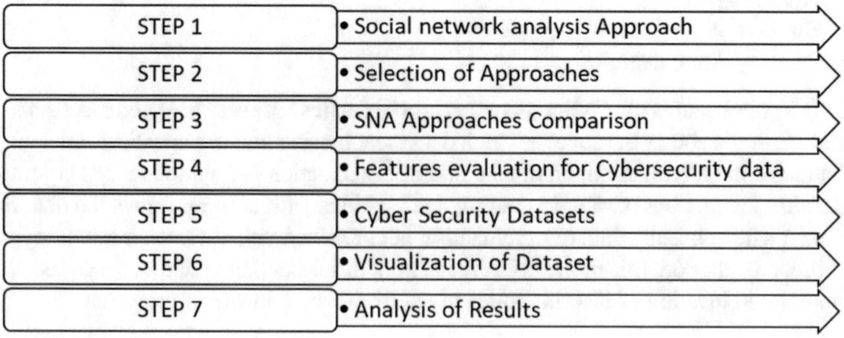

STEP 1	• Social network analysis Approach
STEP 2	• Selection of Approaches
STEP 3	• SNA Approaches Comparison
STEP 4	• Features evaluation for Cybersecurity data
STEP 5	• Cyber Security Datasets
STEP 6	• Visualization of Dataset
STEP 7	• Analysis of Results

Fig. 1 SNA and dataset

are selected by comparing basic features. In 3rd step a brief comparison of selected approach is done by comparing key features of approaches and in 4th step features are evaluated for cybersecurity datasets. In 5th step cybersecurity dataset is selected and in 6th step selected dataset is visualized in each approach with best offering features and in the last step an analysis has been done for each approach [2].

3.1 Collection of Data

In this study, data collection was personally observed and tested; to collect data, all methods were briefly studied and tested; the researcher discovered some deficiencies while analyzing the various types of datasets in these approaches; some have limited support to visualize datasets, some have limited support of the operating system, and some of the techniques were discovered to be outdated and limited to support the particular format. Some open-source repositories have collected datasets and obtained the most recent ones to analyze and visualize that dataset; some datasets have been discovered to be extremely fruitful with the most recent Cybersecurity data and supporting format for the selected approach.

This study uses certain well-known platforms, such as Google Scholar, IEEE, Research Gate, and others, to collect data on social network analysis aspects. This study examines some SNA methodologies and analyses them one by one before comparing them all. In this study, this study delves into numerous social network analysis approaches and examines these techniques for visualization, Cyber Security Datasets this study takes some recent datasets from various forums and then analyses each dataset one by one. Data were gathered to assess the behavior of visualization on various scales of datasets by adding or subtracting values, nodes, edges, degrees, and so on.

Figure 2 a complete methodology is defined for research cybersecurity datasets which is used in this research, multiple factors are in under consideration while choosing dataset, that it should be understandable by everyone, and different platform and keyword is used to find best dataset.

4 Visualization of Dataset

At this stage, datasets have been implemented into a social network, analysis approaches have been performed, and the output of those datasets has been compared At this stage, datasets have been implemented into a social network, analysis approaches have been performed, and the output of those datasets has been compared each approach has different input methods, a different parameter for visualization, some approaches require manual input, and some approaches take value based on dataset format. Each approach visualized each dataset separately and used different parameters to make it more enriched and easier to comprehend; nevertheless, some

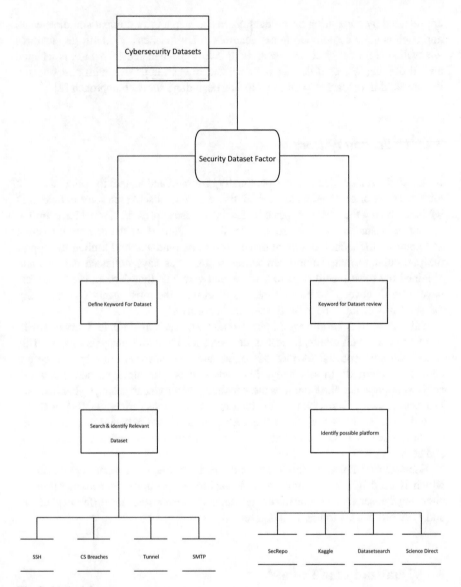

Fig. 2 Methodology

datasets are too large, thus some approaches did not cover all sections of a large dataset. The dataset's implementation is presented in the following step.

4.1   Gephi Visualization

In this phase visualization of selected datasets has been added. Also, description of datasets also described. Gephi has capabilities to visualize large and small datasets in customizable format [1]. Cybersecurity breaches dataset visualize breaches of different organization of USA in different states of USA, also it shows number of individuals get effected after that data breach and type of breach is also mentioned, this dataset contains different information regarding cyber breaches [5].

Figure 3 cybersecurity breaches dataset visualized with modularity and connected component feature. In this figure greater size of node represent the connection to them and the states in which breaches has been happened and the number of breaches per state. Cybersecurity breaches dataset is visualized with Fruchterman-Reingold layout algorithm. Fruchterman-Reingold algorithm determine how far a node can get away to other by single step. Greater Size of nodes represents most cause of breach. And nodes with smaller size and same color to edges represents the organization which suffer with data breach, how many people get affected with that breach, and date of breach [5].

Above Cybersecurity dataset graph visualization shows multiple nodes, In Gephi users can customize size and appearance of nodes and edges, in this case some node

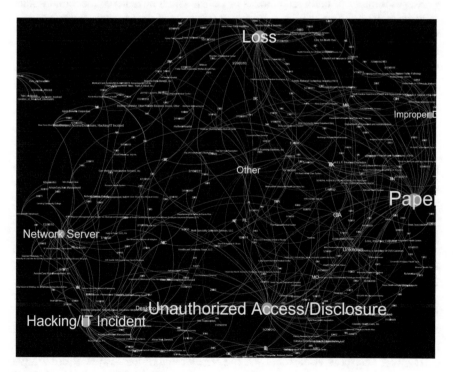

Fig. 3 Security breaches modularity

with larger size then other shows the type of loss and smaller nodes shows other information, along with node size label size is also customizable with the size of node which help to understand it better, multiple nodes are connected with larger node which represent how much nodes are affected with that type of loss.

- Average Degree: 0.228
- Average number of connections to and from each node
- Weakly Connected Components: 159
- Weekly connected because this dataset shows each state separately because each state has its own kind of data breach and thus states show the attached nodes
- Strongly Connected Components: 206
- Strongly connected are those they are receiving and sending information connected in both ways. Organization suffering by data breach and showing link to their state
- Network Diameter: 1
- Network Diameter shows the distance between two nodes, how many edges are there to reach the destination node
- Centrality: 1.0
- Average Path tells the average number of edges are connected to each node
- Density: 0.047
- Shows number of undirected nodes or states, which have no connection to other states
- Modularity: 0.951
- Number of nodes which are connected to other nodes with in same network. Dense connection in a group of modules where nodes are strongly connected to each other with in the module.

4.2 Pajek Visualization

Pajek is another approach of social network analysis to analyze datasets with different features, Pajek is also a strong approach with its multiple features, In special edition of Pajek consumption of memory is lower, it support multiple OS and large network visualization with hundreds of vertices, it support multiple algorithm and features for customization of graph, in some cases if the dataset is large and complex then customization of graph and differentiation of graph into different part is difficult in Pajek. In Pajek to visualize Cybersecurity Breaches dataset first it needs to convert Cybersecurity Breaches dataset file into.net file in order to visualize in Pajek for this purpose it uses txt2pajek approach to convert our dataset. Entries of dataset is in.net format. After converting into.net format it will import our dataset in network block and then it will set some value and define some parameters to visualize dataset and then it will view the graph by clicking on the draw button [3].

Figure 4 dataset is visualized by using circular algorithem. Pajek offer multiple features but its user interface is complex, normal user found it hard to use

Fig. 4 Security breaches modularity

- Average Degree: 0.228
- Average number of connections to and from each node
- Weakly Connected Components: 159
- Weekly connected because this dataset shows each state separately because each state has its own kind of data breach and thus states show the attached nodes
- Strongly Connected Components: 206
- Strongly connected are those they are receiving and sending information connected in both ways. Organization which suffering from data breach and showing link to their state
- Network Diameter: 1
- Network Diameter shows the distance between two nodes, how many edges are there to reach the destination node
- Centrality: 1.0
- Average Path tells the average number of edges are connected to each node
- Density: 0.047
- Shows number of undirected nodes or states which have no connection to other states
- Modularity: 0.951
- Number of nodes which are connected to other nodes with in same network. Dense connection in a group of modules where nodes are strongly connected to each other with in the module [3].

4.3 Tulip Visualization

In this dataset breaches of some organization in USA are listed, some states of USA are mentioned and along with state different organization is mentioned and their type of breach and location of breach like laptop, email, mobile etc. also date of breach is mentioned in it and no of individual effected with this breach is also given in this dataset [6].

Here is a parallel coordinate view of Cybersecurity breach in which information of cybersecurity breach is given in different columns.

Figure 5 cybersecurity breaches dataset is visualized in tulip with 'multi-level' and 'spectral' algorithm. In this dataset organization names and type of loss is shown.

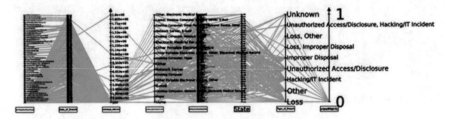

Fig. 5 Cybersecurity breaches dataset visualization in tulip with degree

The degree of Cybersecurity breaches dataset is calculated in the last column 'view metrices' at right side with 1 and 0. [6].

- Average Degree: 0.228
- Average number of connections to and from each node
- Weakly Connected Components: 159
- Weekly connected because this dataset shows each state separately because each state has its own kind of data breach and thus states show the attached nodes
- Strongly Connected Components: 206
- Strongly connected are those they are receiving and sending information connected in both ways. Organization which suffering from data breach and showing link to their state
- Network Diameter: 1
- Network Diameter shows the distance between two nodes, how many edges are there to reach the destination node
- Centrality: 1.0
- Average Path tells the average number of edges are connected to each node
- Density: 0.047
- Shows number of undirected nodes or states which have no connection to other states
- Modularity: 0.951
- Number of nodes which are connected to other nodes with in same network. Dense connection in a group of modules where nodes are strongly connected to each other with in the module.

5 Analysis

5.1 Gephi

Gephi is a social network analysis approach which is capable to visualize different types of data various fields. Among multiple key features, one is it has the ability to show specialization process, which have ability to shift the entire set-up in the form of diagram, and the default algorithm layout is ForceAtlas2, latter it is developed by the Gephi team, the user of the Gephi technique provide elucidation to adopt a free of scale or a range of nodes between 10 to 10,000 for network analysis. Among multiple key features, one is it has the ability to show specialization process, which have ability to shift the entire set-up in the form of diagram, and the default algorithm layout is ForceAtlas2. Gephi has interactive interface which make it efficient and easy to use, Gephi shown runtime process of visualization of dataset [5].

5.2 Results

Gephi is one of better approach among other approaches, Gephi has multiple features, in all aspects, in layout it has different algorithm and in customization it offers multiple features, best thing in Gephi it has real-time visualization of datasets, with running visualization of datasets different type of customization option can be selected and in layout algorithm by changing the properties of different values changing can be seen real-time in graph [9].

5.3 Analysis

If Gephi compares with other approaches it is the best approach with its user-friendly environment and attractive user interactive interface and offers multiple features, multiple input and output formats, with variety of algorithm and different type of customizable option, all these things make it better approach and while other approaches also do visualization but those approaches have their own deficiencies and different limitations [10].

5.4 Limitation

Following are the limitations of Gephi.
- Clustering of complex dataset
- Need to improve export options.

5.5 Pajek

Pajek is another approach of social network analysis to analyze datasets with different features, Pajek is also a strong approach with its multiple features, In special edition of Pajek consumption of memory is lower, it support multiple OS and large network visualization with hundreds of vertices, it support multiple algorithm and features for customization of graph, in some cases if the dataset is large and complex then customization of graph and differentiation of graph into different part is difficult in Pajek Pajek support specific input format.net etc. its interface is not interactive, its complex to use Pajek although it has variety of options and parameters but due to complex interface its challenge to use them. Also, in Pajek to visualize dataset first it should insert data into.net file with specific format or convert those datasets with an approach in a.net format and then visualize. Pajek is a Social network analysis approach which analyze datasets but if it compare with others approaches except Gephi it is also one of the best approach with multiple option to visualize dataset, but it is not much user friendly as Gephi is also it didn't support real-time customization while in Gephi it can customize output in real-time and in Pajek limited layouts and if it analyze large dataset the data in graphs become messy and difficult to understand and there is also lack of Clustering in Pajek [3].

5.6   Results

Pajek is a Social network analysis approach which analyze datasets but if it compare with others approaches except Gephi it is also one of the best approach with multiple option to visualize dataset, but it is not much user friendly as Gephi is also it didn't support real-time customization while in Gephi it can customize output in real-time and in Pajek limited layouts and if it analyze large dataset the data in graphs become messy and difficult to understand and there is also lack of Clustering in Pajek [11].

5.7   Analysis

Pajek is not user friendly, it is difficult to use if user need to insert multiple option for customization and for other layout algorithm then it values can be changes with different section and to find those section is also a complex task along.

5.8   Limitation

Following are the limitations of Pajek.

- Improper format can cause a hang
- Mismatch in the vertices can causes error
- Keyword are case sensitive
- No space between the sections
- Complex to cluster large dataset.

5.9 Tulip

Tulip is also a visualization approach, tulip the framework can be customized and extended according to the needs of users, its plug-in architecture accepts one to add any domain-specific routine, tulip also includes python scripting to easily experiment and even animate networks. In Tulip Visualization approach, four datasets will be visualized in tulip approach, each dataset have different size and different values. In tulip there are multiple panels at a same time which can visualize a dataset into multiple ways, tulip can handle large dataset [12]. there is multiple option to visualize dataset different layouts and different parameter are given in these approaches, along with this it can visualize selective data in a dataset for better understanding of dataset. In Tulip Visualization approach, dataset is visualized in tulip there are multiple panels at a same time it can visualize a dataset into multiple ways, tulip can handle large dataset. there are multiple options to visualize dataset different layouts and different parameter which is given in this approach, along with this it can visualize selective data in a dataset for better understanding of dataset [6].

5.10 Results

Tulip is also a standalone software which support multiple layouts and input and output formats, in tulip there are multiple layouts with different panels, in this approach multiple layouts with multiple datasets can be opened and, in this approach, desired data can be selected in the form of columns. In tulip there is variety of options which can be used, but real-time visualization in this approach is not supported [13].

5.11 Analysis

Tulip has multiple features and different layout and support multiple formats but it is not much user friendly. In tulip large dataset graph can be messy difficult to understand, in graph view of dataset there is lack of clustering to differentiate data to other information [14].

5.12 Limitation

Following are the limitations of Tulip.

- Lack of advance network metrics
- Lack of multiple input and output format support
- Complex to cluster large dataset

Table 3 Analysis table

	Gephi	Pajek	Tulip
Package	Gui	Gui	Gui
Usability	M-H	M-L	M-H
Quality Graphs	H	M	M
Large Network	M-H	H	M
User Interaction	M-H	M-L	M
Import/Export	Multiple	Specific	Multiple
Effective	H	M	M-H
Operating System	Linux, Windows, Mac	Linux, Windows, Mac	Windows/Linux/Mac
Orientation	Education, Business	Business, Education	Education
Support	Online, Books	Manual, Books	Online, Books, Manual

L = Low; M = Medium; H = High

- Analysis of Evaluated Approaches

    In this study, three approaches are evaluated and different security related datasets are visualized in them, analyzing different datasets in selected approaches is such a complex task, because each approach have different working, some approaches are easy to use and easy to implement different feature while some are very difficult to use. After evaluation and visualizing datasets this study found Gephi is a best approach to visualize datasets of Cybersecurity [15].

    Table for an Analysis of SNA Approaches.

    In Table 3 represent analysis of define approaches with different aspects for users to choose best approach for their work. In this table a general analysis is driven, for a normal user it is easy to understand its effectiveness, also it could be understanding either the tool meet the requirement also it define the usage complexity, either a basic user can use it only or it is only for pro users [16].

Conclusion

Evaluation of visualisation approaches for analysing cybersecurity datasets, in this study, social network analysis approaches are compared to one another on several characteristics. Each approach has its own set of pros and weaknesses, as well as its own set of characteristics [17].Approaches to social network analysis contain a variety of characteristics for visualising various types of datasets. These systems visualise datasets based on their specifications and supporting formats; nevertheless, these approaches require a specific input format to import datasets in order to visualise them. In each approaches dataset is visualized using different algorithm and customization option, to visualize dataset in social network analysis approaches first dataset is converted to supported format for that approach in which dataset is visualized and in this study each approach visualization result is appended with description [18].

This study found Gephi is best approach to analyze Cybersecurity dataset. because in Gephi there are multiple features which are very fruitful to analyze Cybersecurity information, especially degree feature which let us know about the number of connection to each node, and in Gephi network is clustered into different layout with different colour which makes it easy to understand which node is sending/receiving most request, how may nodes in network [19]. There are multiple limitations in approaches. And each approach has advantages and disadvantages, in this chapter this study explain the analysis and limitation of approaches which is experienced during visualization of datasets, after examining all approaches this chapter define analysis in tabular form, and using evaluation of approaches this research found Gephi is an effective and easy to use approach with its vast variety of features and have effective usability with its effective interface and users have an easy access of all features of this approach, in this research [20].

References

1. R. Tamassia, B. Palazzi, C. Papamanthou, Graph drawing for security visualization, in *Graph drawing*. ed. by I.G. Tollis, M. Patrignani. Lecture Notes in Computer Science number 5417 (Springer, Berlin Heidelberg, 2009), pp. 2–13
2. H. Shiravi, A. Shiravi, A. Ghorbani, A survey of visualization systems for network security. IEEE Trans. Vis. Comput. Graph. **18**(8), 1313–1329 (2012). ISSN 104-2626. https://doi.org/10.1109/TVCG.2011.144
3. V. Batagelj, A. Mrvar, Pajek-program for large network analysis. Home page http://vlado.fmf.uni-lj.si/pub/networks/pajek/
4. S. Samtani, H. Chen, Using social network analysis to identify key hackers for keylogging approaches in hacker forums. Paper presented at IEEE conference on intelligence and security informatics (ISI) (2016), 319-321
5. M. Bastian, S. Heymann, M. Jacomy, Gephi: an open source software for exploring and manipulating networks. Paper presented at International AAAI conference on weblogs and social media (2009)
6. D. Auber, D. Archambault, R. Bourqui, M. Delest, J. Dubois, A. Lambert, P. Mary, M. Mathiaut, G. Melançon, B. Pinaud, B. Renoust, J. Vallet, *Encyclopedia of Social Network Analysis and Mining* (Springer, 2017)
7. AlukoSayo, Cyber security breaches data https://www.kaggle.com/alukosayoenoch/cyber-security-breaches-data(2018)
8. N. Akhtar, Social network analysis approaches. Paper presented at Fourth international conference on communication systems and network technologies, pp. 388–392 (2014)
9. M.N. Habibi, Analysis of Indonesia politics polarization before 2019 president election using sentiment analysis and social network analysis. Int. J. Mod. Educ. Comput. Sci. (IJMECS) **11**(11), 22–30 (2019). https://doi.org/10.5815/ijmecs.2019.11.04
10. L.M. Romero-Moreno, Methodology with Python technology and social network analysis approaches to analyze the work of students collaborating in facebook groups. Paper presented at 2019 14th Iberian conference on information systems and technologies (CISTI), pp. 1–6 (2019). https://doi.org/10.23919/CISTI.2019.8760969
11. C.C. Yang, N. Liu, M. Sageman, Analyzing the terrorist social networks with visualization approaches, in *Intelligence and Security Informatics (ISI 2006)*. ed. by S. Mehrotra, D.D. Zeng, H. Chen, B. Thuraisingham, F.Y. Wang. Lecture Notes in Computer Science, vol 3975 (Springer, Berlin, Heidelberg, 2006). https://doi.org/10.1007/11760146_29

12. W. Ahmed, S. Lugovic, Social media analytics: analysis and visualisation of news diffusion using NodeXL. Online Inf. Rev. **43**(1), 149–160 (2019). https://doi.org/10.1108/OIR-03-2018-0093

13. O.E. Llantos, M.R.J.E. Estuar, Characterizing instructional leader interactions in a social learning management system using social network analysis. Proc. Comput. Sci. **160**, 149–156 (2019). ISSN 1877-0509. https://doi.org/10.1016/j.procs.2019.09.455

14. J. Klein, S. Bhulai, M. Hoogendoorn, R. Mei. R. Hinfelaar, Detecting network intrusion beyond 1999: applying machine learning techniques to a partially labeled cybersecurity dataset. Paper presented at 2018 IEEE/WIC/ACM international conference on web intelligence (WI), pp. 784–787 (2018)

15. J. Schmidt, Usage of visualization techniques in data science workflows. VISIGRAPP (2020)

16. T. Spiliotopoulos, I. Oakley, Applications of social network analysis for user modelling (2012)

17. Y. Kondakçı, S. Bedenlier, O. Zawacki-Richter, Social network analysis of international student mobility: uncovering the rise of regional hubs. High. Educ. **75**, 517–535 (2018)

18. Mike Sconzo, SecRepo samples of security related data, http://www.secrepo.com/#network(2021)

19. A. Ficara, L. Cavallaro, P.D. Meo, G. Fiumara, S. Catanese, O. Bagdasar, A. Liotta, Social network analysis of Sicilian Mafia interconnections. In: *Complex Networks* (2019)

20. K. Zhang, S. Bhattacharyya, S. Ram, Large-scale network analysis for online social brand advertising. MIS Q. **40**, 849–868 (2016)

Big Data Analytics and Applications

Data Footprinting in Big Data

Sathana Venkadasubbiah, D. Yuvaraj, Subair Ali,
and Mohamed Uvaze Ahamed Ayoobkhan

Abstract Today the internet is an integral part of human life to access and share any information in the societal media. This allows the users to create online communities of their own interest and share a variety of personal information in the form of text, video, image or audio. It has been extended gradually in various other applications such as facebook, whatsapp, twitter etc. Some of these applications were not originally intended to collect information from the user. However, the users may not be aware that their digital information is being collected. Devices such as smart televisions, smart cars and even smart grids are collecting massive quantities of user data without the user's knowledge. Users of social media and the internet in general, leave fragments of their activities and intentions behind them across an increasing range of technologies. These fragments collectively and passively create a hidden identity built up from metadata of which the user is mostly unaware. Given that the user builds this hidden identity during the normal course of their day, without editing elements that the user may not wish to share with others, might the passive digital footprint more accurately reveal the individual's genuine or authentic self than the individual realizes? This chapter briefs and reviews about the passive and active data footprinting in unstructured data termed BigData.

Keywords Footprinting · Big Data · Semantic Data · Passive footprinting · Active footprinting

S. Venkadasubbiah (✉)
K Ramakrishnan College of Engineering, Tiruchirapalli, India
e-mail: anusathana@gmail.com

D. Yuvaraj
Cihan University, Duhok, Kuridsitan Region, Iraq
e-mail: d.yuvaraj@duhokcihan.edu.krd

S. Ali
Westminster International University in Tashkent, Tashkent, Uzbekistan
e-mail: sliayakath@wiut.uz

M. Uvaze Ahamed Ayoobkhan
Westminster International University in Tashkent, Tashkent, Uzbekistan
e-mail: m.u.ahamed@wiut.uz

© The Author(s), under exclusive license to Springer Nature Switzerland AG 2022
M. Ouaissa et al. (eds.), *Big Data Analytics and Computational Intelligence for Cybersecurity*, Studies in Big Data 111,
https://doi.org/10.1007/978-3-031-05752-6_13

1 Introduction

Data mining is fresh and popular in reflecting the effort of discovering knowledge from data [1]. It provides the techniques that allow managers to acquire managerial information from their legacy systems. Data mining used to identify valid, novel, potentially useful, and understandable correlation and patterns in data. Data mining is made of large databases. While data mining often refers to the process of discovering useful knowledge from data, data mining focuses on the application of algorithms for separate patterns from data. Knowledge discovery tries to find patterns in data and to infer rules (that is, to discover new information) that queries and reports do not reveal effectively.

Data mining is a basis of knowledge discovery. It is a collection of algorithms that are used to find some novel, useful and interesting knowledge in databases. Data mining algorithms are based on applied fields of mathematics and informatics, such as mathematical statistics, probability theory, information theory, neural networks. Some methods of these fields can be used to find hidden relations between data, which can be used to create models that predict some behavior or describe some common properties of analyzed objects. It will combine methods of data mining with tools of reliability analysis to investigate the importance of individual database attributes. Results of such investigation can be used in database optimization because it allows identifying attributes that are not important for purposes for which the database is used [2].

Data mining also refers to web mining [3] and related techniques that are used to automatically discover and extract information from web documents and services. When used in a business context and applied to some type of personal data, it helps companies to build detailed customer profiles, and gain marketing intelligence. Data mining does, however, create a danger to some important ethical values like privacy and individuality. Web mining makes it difficult for an individual to autonomously control the unveiling and dissemination of data about his/her private life.

Data mining raises privacy concerns when web users are traced, and their actions are analyzed without their knowledge. Furthermore, both types of web mining are often used to create customer files with a strong tendency of judging and treating people on the basis of group characteristics instead of on their own individual features and merits. Although there are a variety of solutions to privacy-problems, none of these solutions offer sufficient protection. Only a combined solution package consisting of solutions at an individual as well as a collective level can contribute to releasing some of the tension between the advantages and the disadvantages of web mining. The values of privacy and individuality should be respected and protected to make sure that people are judged and treated fairly. Employees should be aware of the seriousness of the dangers and continuously discuss these ethical issues. This should be a joint responsibility shared by web miners (both adopters and developers), web users, and governments.

Application of data mining in health, finance, Financial Data Analysis, Retail Industry, Telecommunication Industry, Biological Data Analysis, Other Scientific

Applications, Intrusion Detection, Social media such as twitter and facebook [4]. In Medicine it's used to discover effective medical therapies for diverse illnesses. Techniques in data mining are classification, clustering, and regression. Data mining helps to unearth facts about customers from your database, which you previously didn't know about, including purchasing behavior. It lends automation benefits to existing hardware and software. Also used in Crediting/Banking helpful to financial institutions in such areas as loan information and credit reporting.

Data mining research makes the process of data analysis faster. It can assist law enforcers with keying out criminal suspects and taking them into custody, by looking into trends in various behavior patterns. And also helps to foretell the products which customers would like to buy. With the help of data mining knowingly or unknowingly, consumers disseminate personal data in daily activities. Credit and debit card transactions, ATM visits, Web site browsing and purchases even mobile phone use all generate data downloaded for analysis and customer profiling. Users may use this data to enhance customers' experience, but may also share information with marketers more focused on customer purchase.

In Medicine it's used to discover effective medical therapies for diverse illnesses. So data mining handles huge amounts of data such as sensitive information, financial details etc. If hackers hack the data through footprinting it will affect the company reputation and create loss to the business. There is a possible malware, cross site scripting and phishing attack in the social media data but stealing the data through footprinting creates large effects to the organizations such as loss of financial and affects' the reputation. It will ruin the business.

Footprinting is friendly to the hacker for gathering information [5] about the target organization. Hackers can collect the following information through footprinting such as IP address, Domain, Name Namespaces, Employee information; Phone numbers, Emails, Job Information. Types of footprinting passive footprinting means hackers gather publicly available information using open source tools. Here there is no direct contact with the target organizations. Active footprinting means a hacker gathers the information from employees of the target organization.

Hackers focus on the employee from social engineering, on site visits, interviews, questionnaires. Anonymous footprinting refers to collecting information from sources anonymously. Pseudonymous footprinting means collecting information publicly available on the internet but it is not directly linked to the author's name. There are many open tools there so hackers use these tools to footprint the data. The following method used to gather the information about organization or system or any person: Network enumeration, DNS Queries, Network Queries, Operating system identification, Organizational queries, and Ping sweeps, and World Wide Web spidering. Footprinting tools are Sam Spade, Super scan, Nmap, Tcpview.

There are many open source tools are available such as crawling (crawling is used to gathering required information about target organization such as target website, blogs, social networks), whois (whois tool is a large dataset is used get collect the information about administrator email address), trace route (trace route is also used to gather the information about path between the user and target system). Hacker uses availing tools to gather information about Target Company such as company details, employee details, and their email address, relation with other companies, company deals with other companies, project details and structure of the company. If hackers gathered enabling information about a person or organization then easily hack the target company and ruin their reputation. Footprinting is the first step to affect the security of any IT infrastructure; if an attacker gathers enough details about the target system then the attacker exploits it [6].

Footprinting in data mining: hackers use footprinting methods and tools to extract as much as possible information. Data mining handles large amounts of data about company details such as project details, employee details, account details, sensitive information passwords, and credit numbers so if hackers hack the data it means it affects the business and it creates loss of business and spoils the reputation of the company. Footprinting in data mining is the first step to attacker gathering information about target sensitive information.

2 Foot Printing

Footprinting is used to create a blueprint or map of an organization's network and systems. Footprinting is also known as reconnaissance. Reconnaissance lays the foundation for all other processes in information security. And also helps in gaining an insight into security incidents. Footprinting is used to collect as much as possible information about target organization. Footprinting is the first step for hackers to gather publicly available information about a system, organization. The following are footprinting terminology,

Passive information gathering: Passive footprinting means hackers gather the publicly available information using open source tools. Here there is no direct contact with the target organizations.

Active information gathering: Active footprinting means hackers gather the information from employees of the target organization. Hackers focus on the employee from social engineering, on site visits, interviews, questionnaires.

Anonymous footprinting: Anonymous footprinting refers to collecting information from sources anonymously.

Pseudonymous footprinting: Pseudonymous footprinting means collecting information publicly available on the internet but it is not directly linked to the author's name.

Private Footprinting: From organizations websites, emails, calendars etc.

Internet Footprinting: Collect information from Internet.

Attacker use footprinting to find information everywhere.

- Use public resources: Websites, directories, email addresses, job sites, and social networks.
- On the logical side: Network architecture, defense mechanisms, operating system, and applications.

Finally Hacker collects the valuable information about the organization, so it will affect the organization reputation and financial loss. Footprinting allows hackers to identify the security of target organizations and vulnerabilities to take perfect steps to exploit the system. It is also used to draw an outline network map for the target system. Reconnaissance is one of the three pre-attack phases (scanning, enumeration, footprinting).An attacker spends 90% of the time in profiling an organization and another 10% in launching the attack. Reconnaissance results for the organization profile with respect to networks (Internet/intranet/extranet/wireless) and systems involved. The following information of organization are collected by hackers [9]

Internet—Domain Names, Network blocks, IP addresses of reachable systems, TCP and UDP services running, System architecture, ACLs, IDSes running, System enumeration (user and group names, system banners, routing tables, and SNMP info) [7]

Intranet—Networking protocols used, Internal domain names, Network blocks, IP addresses of reachable systems, TCP and UDP services running, System architecture, ACLs, IDSes running, System enumeration.

Extranet—Connection origination and destination, Type of connection, Access control mechanism.

Remote access—Analog/digital telephone numbers, Remote system type, Authentication mechanisms.

The goal of footprinting is to identify computers and networks to attack and give detailed information about those computers and networks. Footprinting used to identify the people who work in the organization. The reasons for footprinting are *understand the Security point*: attackers use Footprinting to know the outer security point of the target organization. *Reduce Focus Area*: Footprinting reduces hacker's focus area to specific range of IP address, networks, domain names, remote access, etc. *Identify Vulnerabilities*: It allows attacker to identify vulnerabilities in the target systems in order to select appropriate exploits. *Draw Network Map*: "Reconnaissance allows drawing an outline the target organization's network infrastructure to know about the actual environment that they are going to break". Reconnaissance reveals the system vulnerability and weakness; attackers use this information to exploit the system. The attacker uses the following way to gather information.

1. Company Web page
2. Related Organization
3. Company employee
4. Current Affairs
5. Archive Information
6. Search Engines
7. Google Hacking
8. Metadata Analyzer (Extension of a File)
9. Internet Facing Devices
10. DNS Interrogation
11. Registrar queries
12. Network Reconnaissance.

2.1 Company Web Page

Starting your footprinting from the company's website would be a good start. Sometimes the website of the organization reveals much more information than they should do. In the website you may find some interesting information (about the developer, company which developed that site, company that maintains the site, etc.) hidden in the comment tag (<! and –) of the HTML code.

In this case, it would be a good idea to have an offline copy of that website and make the footprinting more efficient. Offline copy of the website can be achieved using the following tools:

- Wget (gnu.org/software/wget/wget.html) for UNIX/Linux
- Teleport Pro (tenmax.com) for windows

The website can have many hidden files and directories that are not linked but can be discovered using some brute-force techniques. The best tool to perform this operation is OWASP's DirBuster (owasp.org/index.php/Category:OWASP_DirBuster_Project). Brute-force techniques are very noisy and attract attention, for this reason a proxy feature is included in DirBuster. Some other sites are worth investigating such as "http://www" and "https://www". Hostnames such as www1, www2, web, web1, test, test1, etc., are great places to poke.

Many companies may have a remote access active to its internal resources via a client software like web browsers. Microsoft's Outlook Web Access is a very common example. It acts as a proxy to internal Microsoft Exchange servers. Common URLs are https://owa.example.com or https://outlook.example.com.

2.2 Related Organization

It was mentioned earlier that you may find the name of the developer or company that developed the target company's website. The target company may have a rock solid policy for their safety and security, but the partnered company may not. You could use social engineering techniques, on the partnered company and learn a lot of things about the target company. You could even get the website code if you are clever enough.

2.3 Company Employee

You could gather information about the company employee, who has a greater security clearance. The information would be much greater help when you would conduct a social engineering on the employee. You could find about the company employees, in the following networks:

- Social Networking Sites (SNS): Facebook, Twitter, Google+, MySpace, Reunion, Classmates etc.
- Professional networking: LinkedIn, Plaxo
- Career Management Sites: Monster.com, carrerbuilder.com, dice.com
- Family Ancestry: Ancestry.com
- Photo Management Sites: Instagram, Flickr, Photobucket, Tumblr
- Employee/Business directory service: JigSaw.com (Paid service)
- People and Contact search (works for some countries): phonenumber.com, 411.com, yellowpages.com, blackbookonline.info, peoplesearch.com, bigyellow.com, whowhere.com, ussearch.com, usafind.com.

2.4 Current Affairs

Mergers, acquisitions, scandals, layoffs, rapid hiring, reorganization, outsourcing, extensive use of temporary contractors, and other events may give attackers an opportunity to intrude the company and gain trust in their difficult situation. If the attacker is targeting a publicly traded company in the US, then for him/her the internet is all he/she needs. In fact, these companies need to submit periodic reports to the Securities and Exchange Commission (SEC). The attacker may find two reports that are the quarterly and annually reports in the EDGAR database found in sec.gov.

2.5 Archive Information

Some companies launch their website with security bugs and much revealing information. So during security auditing of the company's website, the company personnel would cleanse it. But there is a saying. The attacker may retrieve your sins from the archived copies in websites like WayBack Machine (archive.org) and the cached results you see under Google's cached results.

2.6 Search Engines

Well all of you know how kind the search engines are, they help us to find our answers. So, if they help us, then they can help the bad guys too. The bad guys just need to use the appropriate words and then, bang!, the search engine would return a pile of information they are looking for. There are some search engines that out-stand other search engines, because they have a capability to search other search engines. These are:

- dogpile.com
- search.disconnect.me (a few months back it would give results from google and may other search engines, but not anymore)

And some common but useful search engines are:

- Google, Bing, Yahoo, Lycos, Hotbot, AltaVista, Excite

2.7   Google Hacking

We are not going to hack Google; instead we are forcing Google to deliver results that we are interested in.

Here is a simple example: If you need to reveal Microsoft Windows servers with Remote Desktop Web Connection, then type in Google search.

Allinurl: tsweb/default.htm.

There are many keywords that may find interesting results on the web. You may find the keywords in Google Hacking Database (GHDB) which is with hackersfor-charity.org/ghdb. Unfortunately this site redirects to Google now, so there is another site that can be of use i.e., https://www.exploit-db.com/google-hacking-database/.

Some tools can also make this work much easier like.

- Athena by Steve, SiteDigger (foundstone.com), Wikto.

2.8   Metadata Analyzer

Some tools such as FOCA (originally found at informatic.com/foca.aspx, but now can be downloaded from https://www.elevenpaths.com/labstools/foca/index.html), are designed to identify and analyze the metadata stored within a file. FOCA utilizes some of the same concepts as Google hacking to identify common document extensions such as.pdf,.doc(x),.xls(x), and.ppt(x).

2.9   Internet Facing Devices

The attacker may also find interesting targets like the devices, such as routers, perimeter firewall, etc., that faces the internet. The attacker can enter the mac id, model number or name of the device and can search for the device he/she is looking for in Sentient Hyper-Optimized Data Access Network (SHODAN) at https://www.shodan.io/.

3 Objectives of Footprinting

Collect Network Information—The hacker collects information about the network such as Domain name, Internal domain names,Network blocks,IP addresses of the reachable systems,Rogue websites/private websites,TCP and UDP services running, Access control Mechanisms and ACL's,Networking protocols,VPN Points,IDSes running,Analog/digital telephone numbers,Authentication mechanisms,System Enumeration.

Collect System Information—The information related to the system about User and group names, System banners, Routing tables, SNMP information, System architecture, Remote system type, System names, Passwords.

Collect Organization's Information—The information related to Employee details, Organization's website, Company directory, Location details, Address and phone numbers.

4 Footprinting Methodologies

Footprinting through search engine—Hacker uses the search engine to extract the information about targets such as employee details, login pages, and intranet portal and technology platforms. It also provides sensitive information about the target. To find a company's public and restricted websites use a search engine such as Google and Bing. Using trial and error method to find company's restricted URLs. Netcraft tool used to determine operating systems in use by the target organization. And also use the Shodan search engine to find out specific computers such as routers, servers. With the help of Google earth tool to get the physical location of the target. The great source of personal and organization information collected from social networking sites. We can find individual information from various people's search websites. From a people search engine we can get information about a person or organization such as residential addresses and email addresses, contact numbers and date of birth, photos and operating environment. Another search engine such as financial services provides market value of a company's shares, competitor details, company profile. Hackers can gather company's infrastructure details from job postings.

Footprinting using advanced Google hacking techniques—The attacker uses advanced Google hacking techniques: 1. Query string-hacker creating complex search queries in order to extract sensitive or hidden information. 2. Vulnerable targets-to find vulnerable targets. Google advanced search option is used to extract information such as partners, vendors, clients.

Footprinting through social networking on social networking sites—Attackers use social engineering websites such as facebook, MySpace, LinkedIn, twitter, pinterest, to gather sensitive information. Attackers create a fake profile and false identity on social networking to gather sensitive information from employees.

Attacker also collects the information about the employee's interest and tricks the employee to reveal more information.

Website footprinting—Attackers use website footprinting to monitor and analyze the target organization's website for information. If an attacker browsing the target website means website footprinting provides what software used and its version, operating system used, sub-directories and parameters, filename, path, database field name, scripting platform, contact details and CMS details. Use web spiders tool to perform automated searches on the target website and collect the information about employee details such as employee names, email addresses, etc. Mirroring the whole website used to browse websites offline, finding directory structure and valuable information from the mirrored copy without multiple requests to the web server. Web mirroring tools allow downloading a website to local directory, html, images, flash, and videos from server to your computer.

Email footprinting—Collecting information from email header with help of email tracking tools such as email lookup, emailtrackerpro, trace email, read notify, yesware and zendio. Email tracking is used to monitor the delivery of emails to an intended recipient. Attackers track emails to gather information about a target recipient in order to perform social engineering and other attacks. Get recipient's system IP address. It will find the geolocation of the recipient. Attackers can know about when the email was received and read. Whether or not the recipient visited any links sent to them. Get recipient's browser and operating system information. Time spent on reading the emails.

Competitive intelligence—Attackers use competitive intelligence methodology to identify, analyze, verifying and gathering information about competitors from resources such as the internet. It is non-interfering and subtle in nature. Attackers can collect sources of competitive intelligence such as company websites and employment ads, search engines, internet, and online DB.

Whois footprinting—Regional internet registries maintain whois databases, including information about the domain of the owner. Whois tool displays information about target organization such as Domain name details, Contact details of the domain owner, Domain name servers, NetRange, When a Domain has been created, Expiry Records, Records last Updated. There are some WHOIS lookup tools such as lanwhois, batch IP converter, CallerIP, Whois Lookup Multiple, Addresses, Whois Analyzer Pro, Hot Whois, Active Whois, Whois This Domain, and Whois.

DNS footprinting—The attacker uses DNS footprinting to gather information about DNS to perform social engineering attacks. DNS footprinting tools are DIG, myDNStools, Professional Tools Set, DNS Records, DNS Watch, Domain Tools, DNS Query Utility, DNS Lookup. Dns footprint collecting fields are,

- **Domain Name**—Identifying the domain name or owner of the records
- **Record Types**—Specifying the type of data in the resource record
- **Record Class**—Identifying a class of network or protocol family in use
- **Time to Live (TTL)**—Specifying the amount of time a record can be stored in cache before discarded.

- **Record Data**—Providing the type and class dependent data to describe the resources.
- **A (address)**—Maps a hostname to an IP address
- **CNAME (canonical name)**—Provides additional names or aliases for the address record
- **MX (mail exchange)**—Identifies the mail server for the domain
- **SRV (service)**—SRV services used to identify directory services
- **PTR (pointer)**—Maps IP addresses to hostnames
- **NS (name server)**—Identifies other name servers for the domain
- **HINFO** = Host Information Records

Footprinting through social engineering—To extract confidential Information about human behavior Social Engineering is an art of exploiting. Social engineering depends on people who are unaware of their valuable information and who care less about protecting it. Eavesdropping, shoulder surfing, and dumpster diving are the techniques used for footprinting through social engineering. Eavesdropping is hearing unauthorized access to others communication such as video, audio, message, phone call to collect the information. From shoulder surfing techniques, an attacker can gather critical information about the target organization such as account numbers, credit card no, personal identification number, passwords. Dumpster diving is gathering information from someone else's trash and it involves collecting the information from phone bills, contact information, financial information, company trash bin, printer trash bin, user desktop.

5 Tools

Maltego—The hacker uses Maltego tools to find the relationships and organizations, websites, real world links between people, groups of people such as social networks, companies, internet infrastructure, phrases, documents, and files.

Recon-ng—Recon-ng tools work in Linux and it is full featured about web based open source reconnaissance. Recon-ng is similar to Metasploit but quite different. Metasploit framework used to exploit and recon-ng used to conduct reconnaissance.

Foca—Hackers use FOCA (Fingerprinting Organizations with Collected Archives) tool to find metadata and hidden information from documents. Foca acts as multiple attacks such as metadata extraction, network analysis, DNS snooping, proxies search, and fingerprinting, open directories search. Additional footprinting tools are prefix whois, netmask, netscan tools pro, binging, tctrace, SearchBug, Autonomous system scanner, DNS-Digger, TinEye, Robtex.

SearchDiggity—Google Hacking Diggity Project primary tool is SearchDiggity. SearchDiggity is a GUI application front end for both GoogleDiggity and BingDiggity.

SmartWhois—SmartWhois is a tool used to find out host name, domain, and IP address. And also provides information about the city, state, country, name of the registered owner, and contact information.

Google—Use advanced Google search to gather information about the target's website, web servers and vulnerable information. Attackers gathering information from jobs posted in the companies' websites reveals valuable information about the type of information technologies used in the target company.

The harvester—You can use it to catalogue email addresses and sub domains. It works with all the major search engines including Bing and Google. Attacker uses the harvester tool in Kali Linux.

WHOIS—To get information about domains, IP address, DNS you can run whois command from your Linux machine. Just type whois followed by the domain name: Whois yourdomain.com [8].

Netcraft—They have a free online tool to gather information about web servers including both the client and server side technologies. Visit http://toolbar.netcraft.com/site_report/ and type the domain name.

Nslookup—You can use it to query a DNS server in order to extract valuable information about the host machine. Nslookup tools are used in both Linux and Windows. In windows use commands prompt and type nslookup then domain name [9].

Dig—Another useful DNS lookup tool used in Linux machines. Type dig followed by the domain name.

MetaGoofil—It's a Meta data collection tool. Meta data means data about data. For instance, when you create a word document in Microsoft Word, some additional information is added to this word file such as file size, date of creation, the user name of the creator etc.-all this additional information is called metadata. MetaGoogle scours the Internet for metadata of your target. MetaGoofil used in both Linux (built in Kali Linux) and Windows [6].

Threatagent drone—Threatagent drone is a web based tool. To sign up at https://www.threatagent.com/ and type the domain name that you want to reconnaissance. Attackers use drones to extract the information then it will create a complete report about the target, which will include the IP address range, email address, point of contacts etc.

Social engineering—It is perhaps the easiest way to gather information about an organization. You can find lots of free information about social engineering on the Internet. Depending on the types of information you need about your target organization, you need to choose the appropriate technique. But remember that social engineering techniques need time to master and need to plan it very carefully; otherwise your activity can easily trigger an alert [10].

Traceroute tools—Traceroute is a packet-tracking tool that is ready to use for most operating systems. It operates by sending an Internet Control Message Protocol (ICMP) echo to each hop (router or gateway) along the path, until the destination address is reached. When ICMP messages are sent back from the router, the time to live (TTL) is decremented by one for each router along the path. Traceroute allows a hacker to determine how many hops a router is from the sender. When it encounters a firewall or a packet-filtering router, one problem with using the traceroute tool is that it times out (indicated by an asterisk). Traceroute tool used to alert an ethical hacker to the presence of a firewall; then, techniques for bypassing the firewall can be used.

Sam Spade, NeoTrace, and VisualRoute have many other hacking tools used for traceroute. The Windows operating systems use the *syntax* tracert hostname to perform a traceroute.

Spiderfoot Tool—One of the tools of footprinting that an attacker can use is spiderfoot. Spiderfoot is an open source tool that will help in the pen test and scan domains and web content. This will be automated and in a simple way. The information that you find with this tool is [11]:

- Target domain information
- Email addresses on the target
- IP addresses for the domain
- Web server version and the application
- Crawl the Web content
- All associate URL list
- Physical place
- Information about SSL certificate (if expired and more)
- URL that accept passwords
- Externally hosted javascript
- Add module to run a bruteforce
- Social Media Presence etc.

6 Footprinting Threats

The valuable information about target systems gathered by hackers such as account details, operating system, and database schema, server names, and network components with help of footprinting [12]. The types of threats are.

Social Engineering: With help of non-technical social engineering gathering the information.

System and Network Attack

Information Leakage

Privacy Loss: Once u access system and escalate privileges, their privacy is lost and can access tender etc.

Corporate Espionage: corporate espionage means competitors can spy an attempt to steal sensitive data. Competitors can launch similar products in the market causing loss to the initial company.

Business loss

Reconnaissance helps attacker to outfitter an attack plan to exploit known vulnerabilities configuration errors in the target network. Attackers use much information which is available on the internet like DNS lookups to find the name and IP address of the target network, WHOIS information to find the contact details, name server names etc.

Launch Attack: After performing Footprinting, hacker gains as much as information about the target network. The next step is to launch an attack on the target network based on the found vulnerabilities.

Escalate privileges: If the hacker obtains access as a normal unprivileged user to escalate the user account to gain administrator-level rights.

Jump to other servers and devices: Once the hacker is inside the internal network, then he can gain access to other devices inside. And also collect extra information like applications running, operating systems, user ids, password etc.

Install Back Doors: After compromising servers and collected the information he required, hacker then try to install and configure back door or remote-control hacking tools to gain access to the system in future. A backdoor application allows the hacker to future access to compromised machines.

Hide the Tracks: After performing attack and installed back door applications, next step is to hide the tracks so that the hacker can hide the attack from administrators. Hackers do perform many actions for this. For example, deleting the log files.

Leverage the compromised network: Finally, hackers start using the target network. They can steal or destroy the target network data, bring servers down, or attack another organization using the target network's systems.

Footprinting pen testing :

To determine organization's publicly available information footprinting pen testing is used [13]. The target organization from the Internet and other publicly accessible sources are attempts to gather by the tester.

Footprinting pen testing helps organization to:

o   Prevent DNS record retrieval from publically available servers
o   Prevent information leakage
o   Prevent social engineering attempts

Define the scope of the assessment and Get proper authorization. Footprints search engines such as Google, Yahoo! Search, Ask, Bing, Dogpile, etc. to gather target organization's information such as employee details, login pages, intranet portals, etc. that helps in performing social engineering and other types of advanced system attacks. GHDB, MetaGoofil, SiteDigger, etc. tools used to Perform Google hacking. Gather target organization employees information from their personal profiles on social networking sites such as Facebook, LinkedIn, Twitter, Google+ , Pinterest, etc. that assist to perform social engineering. HTTrack Web Site copier [14], Black-Widow, Webripper, etc. tools are used to perform website footprinting to build a detailed map of website's structure and architecture. EMailTrackerPro, PoliteMail, Email Lookup-Free Email Tracker, etc. tools are used to perform email footprinting. Social engineering used to gather information about the physical location of an individual to perform that in turn may help in mapping the target organization's network.

Hoovers, LexisNexis, Business Wire, etc. tools are used to gather competitive intelligence. Perform WHOIS footprinting using tools such as SmartWhois, Domain Dossier, Perform DNS footprinting using tools such as DNSstuff, DNS Records, etc. to determine key hosts in the network and perform social engineering attacks. Path Analyzer Pro, VisualRoute, Network Pinger, etc. tools are used to perform network footprinting to create a map of the target's network. Implement social engineering techniques such as eavesdropping, shoulder surfing, and dumpster diving that may help to gather more critical information about the target organization.

7 Conclusion

Data mining is used to extract meaningful patterns in online sites and social media such as facebook. Here data mining handles large amounts of data. With the help of footprinting, an attacker can hack the organization information and personnel information and employee details to ruin the business. To prevent data from footprinting, hackers use footprinting techniques to gather the weakness and provide security for the data and also we have to follow, the organization's network has to restrict the employees to access social networking sites. Design the web servers to avoid information leakage. To use pseudonyms on blogs, groups, and forums have to educate employees. The critical assets to avoid domain level cross-linking. To protect sensitive information use encryption. Apply a strong access control system. Directory listing of websites have to disable. You can protect yourself (and your organization) against footprinting by performing footprinting on yourself or appoint hacker employee to do. Locate the sensitive and disclosed information and try to limit, hide or delete it.

References

1. G. Piatetsky-Shapiro, *Advances in Knowledge Discovery and Data Mining*, vol. 21, ed by U.M. Fayyad, P. Smyth, R. Uthurusamy (AAAI press, Menlo Park, 1996)
2. N. Padhy, D. Mishra, R. Panigrahi, The survey of data mining applications and feature scope. ArXiv preprint arXiv **1211**, 5723 (2012)
3. B. Thuraisingham *Data Mining: Technologies, Techniques, Tools, and Trends* (CRC press, 2014)
4. G. Barbier, L. Huan, Data mining in social media, in *Social Network Data Analytics* (Springer, Boston, MA, 2011), pp. 327–352
5. M. Walker, *CEH Certified Ethical Hacker All-in-One Exam Guide* (McGraw-Hill Osborne Media, 2011)
6. S.-P. Oriyano, *CEH v9: Certified Ethical Hacker Version 9 Study Guide*, vol. 9. (Wiley, 2016)
7. S. McClure et al., Hacking exposed: network security secrets and solutions (2009)
8. K. Graves, *CEH Certified Ethical Hacker Study Guide* (Wiley, 2010)
9. K. Graves, *Ceh: Official Certified Ethical Hacker Review Guide: Exam 312–50* (Wiley, 2007)
10. Certified Ethical Hacker-Part 2-Footprinting and Reconnaissance by Riazul H. Rozen

11. ETHICAL HACKING: A TECHNIQUE TO ENHANCE INFORMATION SECURITY
    Gurpreet K. Juneja Lecturer, Department of Computer Science & Engineering, Guru Nanak
    Dev Engineering College, Ludhiana,india1
12. G.K. Juneja, Ethical hacking: a technique to enhance information security. Int. J. Innov. Res.
    Sci., Eng. Technol. **2**(12), 7575–7580 (2013)
13. J. Vierstra, Stamatoyannopoulos, genomic footprinting. Nat. Methods **13**(3), 213–221 (2016)
14. D. Rohr, Data processing and online reconstruction. arXiv preprint arXiv:1811.11485.

An Investigation of Unmanned Aerial Vehicle Surveillance Data Processing with Big Data Analytics

N. Vanitha⦿, G. Padmavathi⦿, V. Indu priya, and S. Lavanya

Abstract In recent days, Unmanned Aerial Vehicles (UAV) have played an important role in military and security applications. UAVs are essential for our IoT, security, defence and other business activities. These technologies are becoming very popular in this modern world. Drone image processing is professional as many industries, including mapping, surveying, urban planning, forestry, precision agriculture, oil, gas, and mining, use this technology. UAV surveillance uses big data analytics-based image processing methods that are more convenient than traditional processing methods and requires fewer resources. Rather than depending on a single computer for image processing, Hadoop Image Processing Interface (HIPI) uses a cluster of systems to implement parallel computing principles. HIPI provides an image processing library and computer vision applications in a MapReduce Framework. In this paper application of UAV data processing using big data analysis like wildfire prevention, management and Geo-mapping are discussed.

Keywords Unmanned aerial vehicle · Big data analytics · Management · Data processing · Surveillance

N. Vanitha (✉)
Dr. N.G.P. Arts and Science College & Avinashilingam Institute for Home Science and Higher Education for Women, Coimbatore, India
e-mail: vanitha@drngpasc.ac.in

G. Padmavathi
Avinashilingam Institute for Home Science and Higher Education for Women, Coimbatore, India
e-mail: ganapathi.padmavathi@gmail.com

V. Indu priya · S. Lavanya
Dr. N.G.P. Arts and Science College, Coimbatore, India
e-mail: Priyadharshini7020@gmail.com

S. Lavanya
e-mail: lavanyacs8648@gmail.com

© The Author(s), under exclusive license to Springer Nature Switzerland AG 2022   219
M. Ouaissa et al. (eds.), *Big Data Analytics and Computational Intelligence for Cybersecurity*, Studies in Big Data 111,
https://doi.org/10.1007/978-3-031-05752-6_14

1   Introduction

The Unmanned Aerial Vehicle (UAV), usually accepted as Drone, remain as one aerial organization operated through a drone operator or flew individually through one onboard system. Drone construction consists of two phases, movement system and control system. The main unit of the drone is a frame. It must be lightweight. The wings of drones are made by carbon fibre and aluminium. Batteries electrify drones. Drones weights range from 5–2000 kg. In recent days, UAV combined with nanotechnology with tiny microrobots of 3–5 mm are called smart dust. They are applied for monitoring light and climate conditions. Big data management is the procedure and method of handling big data collected from different sources such as social media, product reviews, satellite images, drone footage etc. Drone data can be processed using big data-based image analysis techniques like Hadoop Image Processing Library, including Apache Hadoop MapReduce Framework.

1.1   Applications of UAV

Drones have a huge variety of applications. They are capable of performing both indoor and outdoor surveillance operations. They are given below,

Surveillance

Drones are mainly used in the military for surveillance of sensitive and human risky border areas. Military use a huge variety of drones, mainly bio-drones fabricated by the dead skin of birds and animals. They are also used to study animals and birds. Macrobiotic drones are of 60 mg weight that is used for monitoring forest border areas due to their small size. They cannot be identified and destroyed by intruders [1].

- Dull missions are the mission that consume too much time for example; maritime reconnaissance sortie and online.
- Dirty missions include biological, chemical hazards. Humans cannot perform these types of assignments.
- Dangerous missions involve missions like hazardous gas leakage identification. Eg: Incidents like Bhopal incidents, forest fire fighting.
- Different missions are missions that is not suitable for manned aircraft. Eg. In risky border line and in terrorist areas mini and micro-robots are used in monitoring [2].

Space and Marine exploration

Drones can be used in any type of environment, unlike other robots. NASA developed and used drones to explore other planets and space missions. Drones play a vital role in marine organisms' study of underwater surveillance as they are not risky.US introduced Scan Eagle, Valens, etc. for marine research.

Mailing and delivery

Drones an anonymous flying vehicle used in mailing and delivery systems. Many big companies like Amazon, Uber trying this method of delivery options in some countries due to reduced road congestion, reduced time and very low transportation cost.

Surveying and mapping industry environments

The drones give the best and precise monitoring control over large industrial inspections. Day to day routine include the use of ladders, ropes, machine must be turned off each time. Drones calculate and identify temperature, corrosion in pipes, leakage by processing aerial images collected by drones.

1.2 Advantages of UAV

There is a rapid increase in drones for surveillance and data collection. The commercial market of drones is growing rapidly. Points that attract people towards drones are

Reduced risk

UAVs are used to attack terrorists. They can search and destroy fixed targets. Drones are also used to transport food, medicine, explosives weapons to soldiers in peaks and tough terrains.

Cut–down cost

Unmanned Aerial Vehicles are faster. It includes zero risk even if the plane crashes. It is cheaper as it does not require experienced pilots and expensive fuels. Unmanned Aerial Vehicles batteries can last more than 24 h and perform repetitive tasks efficiently, achieving the faster scanning of the particular area for days and nights in the deep blackness and in bad weather conditions like heavy snowfall.

Quality Aerial imaging

Drones are highly suitable for taking Aerial video and collecting large quantities of imaging data of a particular target. These images are processed to create 3D maps and 3D interactive models.

2 Application of Big Data

In current days there are a huge amount of data. Business organizations make use of this data for business growth. Big data may originate from social media, comments, feedback, satellite images, drone images and video, data collected by sensors etc. This data is processed to discover hidden patterns and obtain useful information.

Healthcare

The data from health care increases the application of big data management in medical field. Numerous health care devices have been introduced with basic functionality of collecting and transmitting medical information like body pressure, sugar level, heart beat rate etc. Big data is analyzed into four categories. They are explanatory, syndrome, predicting, and prescribed [3]. Descriptive analytics is used to find out what happened actually. Diagnostic analytics identifies factor and reason for incident happened. Predictive analytics helps to find future complications of patients health. Prescriptive analytics suggests or reject particular treatment based on side effects.

Marketing

To optimize campaigns, companies gather customer data like name, mobile number, and web searches to know about their target audience. In addition, feedback review of products is collected and the reach of a particular product is analyzed.

Crime detection

The main role of big data in crime detection is detecting crime patterns from large available data sets. The pattern detection algorithm is implemented with patterns of detected crimes from the database and starts with a particular crime as a seed. This algorithm process common characteristics of all crimes and returns result.

3 UAV Data Management

Collect data like images and videos and process it to return geo-referenced maps viewable in all gadgets. Videos and images collected from drones are processed to get 3D map and 3D models of the target. UAV image stitching software combines thousand of images. The following output is obtained from these 3D maps and models following output is obtained 3D building models, orthophotos, and parametric features like the height of building edges of the road. These topographic sheets are applied in the forest, coastline, and urban planning. In agriculture crops, damage is identified using drones. Normalized Difference Vegetation Index is the most commonly used index for plant health monitoring systems [4, 5].

3.1 UAV Drone Data Processing

The images gathered by the drone during the mission are sorted, organized and managed. Image processing tools help to produce orthomosaic maps, 3D points cloud, digital surface model, digital terrain model (DTM), 3D textual model mesh. Orthomosaic maps are created using correcting and stitching thousands of images together. Every pixel of orthomosaic maps has 2D geo-information (x, y). 3D points cloud, it is generated from a drone image search point, has 3D geospatial (x, y, z) information and color information. The digital surface model is created from drone images. Each pixel of DSM model contains 2D geospatial information (x, y) and altitude (z).

3.2 Process of UAV Data Collection and Analysis

UAV data analysis is a huge process with numerous steps. It has the following steps: order, dataflow, indices calculation, verification of calculations, and another variability.

3.3 Process Flow in UAV Data Processing

Process flow has five stages which are further categorized into several sub-processes. They are, 1. data collection, 2. data-preprocessing, 3. classification, 4. indices calculation, 5. result publication. Data collection has two segments flight and flight planning. The data pre-planning process is important phase if this is not performed properly, the accuracy of result may vary. It has an image selection process from video captured by UAV, automatically eliminating blur images. Many images are processed and merged in image merging, and geo referencing is done here. Classification is the next process. It requires geo-referenced images as inputs appropriate method of classification is very important. It has two types supervised and unsupervised classification. Evaluation of those classified results is unavoidable because this makes the process complete. The very next step is index calculation. The final step is result publication which includes visualization and interpretation of results.

3.4 Dataflow in UAV Data Processing

Data collection provide recorded video and this video is an input for pre-processing after this geo reference mosaic image of the target is created. Then that image is

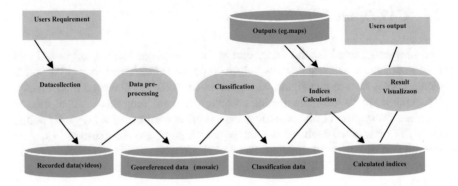

Fig. 1 Output of each process

classified and used for indices calculation. Indices are helpful to prepare research publications. Those results are stored as the output [6]. Figure 1 shows the UAV data processing.

3.5 Big Data Management Techniques in UAV Data Processing

UAV data management is enhanced by big data-based image analysis rather than depending on a single computer, a cluster of computers is used that implements the principles of parallel programming. Images captured by drones are copied instantly into Hadoop big data cluster where the Hadoop image processing library (HIPI) used.

UAV image transferring UAV images to Hadoop cluster

During mission UAV aerial images are saved in a micro-SD card. In order to transfer UAV data to cluster low-quality cached images, a small gadget(phone/tablet) is used. Smart gadget is connected with UAV with 4G connection through remote controller. To avoid loss of data UAV are connected to a remote controller through 58GHZ frequency channel. Frequency modulation FM signals are used for communication. The remote controller and smart gadget are connected with USB data cable helps to download data like flight status, camera angles and low-quality cached images from drone. Smart gadget is installed with flight planning and controlling software such as Litchi. Images are stored in image area of gadget later that album in gallery is uploaded to cloud spaces like Google Drive and iCloud Drive. Changes made in gallery adding images are synchronized automatically in cloud. This process requires 4G network connection. Once images are uploaded, cloud applications data is synchronized automatically and transferred into big data image processing interface (HIPI) is library used for image processing. HIPI is used with Apache Hadoop cluster as shown in Fig. 2.

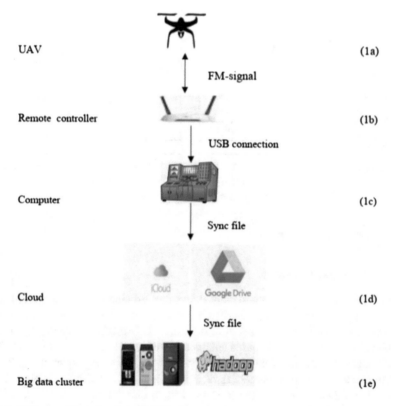

Fig. 2 Image transformation from UAV to cluster

Hadoop image processing interface

Hadoop image processing interface (HIPI) is a library used for image processing. HIPI is used with Apache Hadoop Map reduce framework. Map-reduce parallel programs use HIPI for storing huge quantities of images in the Hadoop Distributed File System (HDFS). The input of HIPI is a collection of images on a single file called HIPI Image Bundle (HIB). HIB images are allocated to map tasks. The records by map task are collected and passed to reduce task as by MapReduce shuffle algorithm [11]. HIPI framework does the following process.

- HIPI provides image processing library and computer vision applications in a MapReduce Framework
- Effective storage of images to make use in MapReduce application
- Filtering of a large set of images is made simple
- Offers interfaces for various image-based operations

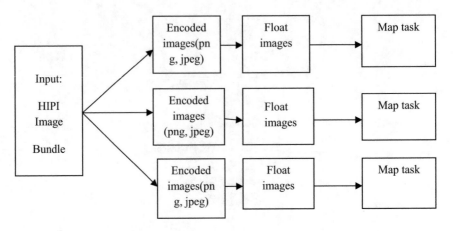

Fig. 3 Input distribution to map task

Image based MapReduce

Hadoop MapReduce programs handle data more efficiently. Images should pass as string type. A set of images is distributed to a set of map nodes later, images are decoded to access every pixel of information. This process of decoding images through process. Thus, image-based datatype is used to overcome this drawback. The set of images HIPI Image Bundle (HIB) is passed as an input as shown in Fig. 3 user have to create two classes, namely Input Format and Record Reader. These classes specify input distribution. On the process of input distribution to map task culling stage takes place. The culling stage allows filtering images based on properties. Users can set conditions for culling class example size lesser than 12 megapixels. HIPI offers an interactive interface for various image processing-based applications, which help effective MapReduce application creation. Images are distributed as float types to have easy access to pixel values.

4 Application of Big Data Management in UAV

Implementing UAV data processing with big data analytics involves various applications. Two applications are discussed below.

Forest fire prevention and Management

Unmanned aerial vehicles (UAV) are highly contributing in world fire management. For 30 years more than 45,000 forest fires are occurring every year. Satellite images and aerial images are not preferred more than UAVs because UAV data processing and capturing less time-consuming. Drones collect images from the forest and save it in SD card. Gadgets or computers are connected to UAV with 4G connection through a remote controller. To avoid loss of data UAV are connected to remote controllers through 58GHZ frequency channel. Frequency modulation FM signals are used for communication. The remote controller and smart gadget are connected with USB data cable that helps to download data like flight status, camera angles and low-quality cached images from drone. The smart gadget is installed with flight planning and controlling software such as Litchi. Images are stored in the image area of the gadget later that album in gallery is uploaded to cloud spaces like Google Drive and iCloud Drive. Changes made in gallery adding images are synchronized automatically in cloud. Once images are downloaded in cloud applications, they are transferred to the big data cluster. HIPI allows effective storage and retrieval of images to make use in MapReduce applications. Hadoop MapReduce programming model map task calculates the color pixel value of each single image collected by UAV. Reduce task calculate total average pixel colour of all images. Now these pixel color values are compared with flames of fire pixel value calculated predefine and forest fire detected.

Geo Information technology and mapping

The Implementation of UAV in surveillance has brought huge changes in the mapping and surveying industries. Hadoop file distributed system supports all forms of data like images, audio, videos and ect. A knowledge map is a Structure build according to graph data. Knowledge map consists of nodes and edges. Nodes specify real-world entity. Edges specify relationship between entity.

5   Background Study

Ivana cermakova et al. (2016) investigated the techniques of UAV data collection and data preprocessing techniques. Graphical representation of each process also included. A case study analyzing shoreline of small water body is done. Data is collected from water body by drones once in a month and the data is classified and result is published. Variance in shoreline for every month is displayed as result.

M. Hassanalian et al. (2017) investigated the various types of drones based on weight, size. Application of each Varity of drone, their design and fabrication challenger are discussed. Various methods and new approaches has been proposed for guidance navigation and control of drone (GNC) system. Some methods are using Google glass to control small drones by head movements and using Brain Computer interface for moving drones [1].

Jane Wynyard et al. (2019) investigated tools and data practices and come out with eight challenges to addressed in UAV data management faced by researchers in various sectors. This paper analyses the number of publications made on UAV [7].

Athanasis et al. (2019) investigated the enhancement of UAV data management by implementing big data image processing techniques. Hadoop Image processing library is used to process drone data. By using this method drone data is processed in wildfire prevention and control.

Miss. Alipta Anil Pawar et al., investigated battery because of uneven power supply. To overcome this problem, he has used 188-li-fo battery voltage tester. This

Table 1 Background study

Year	Author	Journal	Technique used
2016	Ivana cermakova JitkaKomarkova	Modeling a process of UAV data Collection and processing	UAV data collection and pre- Processing techniques
2017	M.Hassanalian A.Abdulkefi	Classification, Application, Design challenges of drone	Novel classification of drones
2019	Jane Wyngard Lindsay Barberi Andrea Thomer Don Sullivan	Emergent challenges for UAS Data management	Eight community distilled Challenge of Big data management
2019	Athanasis, Nikos, Themistocleous, Marinos Kalabokidis, Kostas Chatzitheodorou, Christ	Big Data Analysis in UAV Surveillance for Wildfire Prevention and Management	Proposing enhancement of UAV Data management by introducing Big data-based image processing Technique
2019	Alipta Anil Pawar, Dr. Sanjay L. Nalbalwar, Dr. Shankar B. Deosarkar, Dr. Sachin Singh	Surveillance Drone	BLDC (brushless DC) motor

tester repeatedly displays the battery's voltage for our battery. It falls cell rct voltage level battery for checking the round alarm. Table 1 Shows the techniques used by authors in processing UAV data.

6 Research Challenges in UAV Data Processing

Common challenges that need to be addressed in UAV data processing are

Guidelines for data and product levels

Guidelines on data and metadata formats will be helpful for UAV operators and technical developers of sensors. In some cases, additional metadata is required to get desired output. A simple UAV based project include sensors electrical platforms complex data transformations and various stakeholders. The output of the project is based upon interpretation of dataset. Suggesting product levels for multiple data types will help data archives in deciding the quality and resolution of data and associated metadata [7].

Filtering of unwanted data

More than 27,000 images are recorded in a drone within 15 s. The main challenge is filtering of those data. Removing unwanted data can be done manually and automatically. In automatic method, each image is given a score. Training the database will help to remove unwanted visual data set. At the stage of discrimination classification, large amount of big data is implemented on training to remove unwanted visible data set [8].

Lack of UAV data management education

Various tools exist for UAV data management and analytics. In addition, some tools are specific to some data types. Many effective steps and effort must be taken to overcome this drawback. Therefore, there is heavy demand for fresher's who effectively use the UAV system.

Security issues and attacks

The designing of lightweight dynamic cryptographic algorithms can secure drone communications, ensuring a higher level of confidentiality with low latency and resources overhead. This is the case of the approach in which uses a single-round function by relying on common channel parameters as there are several attacks on UAV network like false data dissemination attacks [9]. Physical and logical attacks on UAV is possible as drones breach people can attack public privacy drones, logical attacks are done by fake Wi-Fi nearby [10].

7 Conclusion

Unmanned Aerial Vehicles are used for various purposes. In this paper, big data management in UAV applications like Forest fire prevention and management, Geoinformation technology and mapping are investigated to know the processing of drone data using big data analytics. Various kinds of literature related to drone data processing have been studied and tabulated. Research challenges in UAV data processing are also explored. In future, drone data processing has been investigated with image processing techniques.

References

1. M. Hassanalian, A. Abdelkef, Classifications, applications, and design challenges of drones: a review Progress in Aerospace Sciences, 108–120 (2017)
2. J. Jayaraman, *Unmanned Aircraft Systems: A Global View* (Defence Research and Development Organization, New Delhi(2014-10-22)
3. N. El Aboudi, L. Benhlima, Big data management for healthcare systems: architecture, requirements, and implementation. Adv. Bioinform., pp.1–8 (2018)
4. N. Vanitha, V. Vinodhini, S. Rekha, A study on agriculture UAV for identifying the plant damage after plantation. Int. J. Eng. Manag. Res. (IJEMR) **6**(6), 310–313 (2016)
5. S. Huang, L. Tang, J.P. Hupy et al., A commentary review on the use of normalized difference vegetation index (NDVI) in the era of popular remote sensing. J. For. Res. **32**, 1–6 (2021). https://doi.org/10.1007/s11676-020-01155-1

6. I. Cermakova, J. Komarkova, Modelling a process of UAV data collection and processing, in 2016 *International Conference on Information Society (i-Society)* (2016). pp. 161–164, https://doi.org/10.1109/i-Society.2016.7854203.

7. J. Wyngaard, L. Barbieri, (Bar) & Thomer, Andrea & Adams, Josip & Sullivan, Don & Crosby, Christopher & Parr, Cynthia & Klump, Jens & Shrestha, Sudhir & T. Bell, Emergent Challenges for science sUAS data management: fairness through community engagement and best practices development. Remote. Sens. **11**, 1797 (2019). https://doi.org/10.3390/rs11151797

8. M. Kamari, Y. Ham, Automated filtering big visual data from drones for enhanced visual analytics in construction. ASCE Constr. Res. Congr., 400–402(2018). Retrieved from https://par.nsf.gov/biblio/10081679. https://doi.org/10.1061/9780784481264.039

9. N. Vanitha, G. Padmavathi, G., A comparative study on communication architecture of unmanned aerial vehicles and security analysis of false data dissemination attacks, in *2018 International Conference on Current Trends towards Converging Technologies (ICCTCT)* (IEEE, pp. 1–8, 2018)

10. J.-P. Yaacoub, H. Noura, O. Salman, Ali Chehab,Security analysis of drones systems: attacks, limitations, and recommendations, ISSN 2542–6605, pp. 15–17 (2020)

11. N. Athanasis, Themistocleous, Marinos & Kalabokidis, Kostas & Chatzitheodorou, Christos, Big Data Analysis in UAV Surveillance for wildfire Prevention and Management, in *15th European, Mediterranean, and Middle Eastern Conference, EMCIS 2018, Limassol, Cyprus, October 4–5, 2018* (2019) Proceedings, pp. 48–54. https://doi.org/10.1007/978-3-030-113 95-7_5

Big Data Mining Using K-Means and DBSCAN Clustering Techniques

Fawzia Omer Albasheer, Mohammed H. Ahmed, Awadallah M. Ahmed, Zia Khan, Said Ul Abrar, and Mian Dawood Shah

Abstract The World Wide Web industry generates big and complex data such as web server log files. Many data mining techniques can be used to analyze log files to extract knowledge and valuable information for both organizations and web developers. Large amounts of heterogeneous data are generated by websites, performing effective analysis on these data and transforming them into useful information using the existing traditional techniques is a challenging process. Therefore, this paper aims to analyze and cluster the log file data to get useful information that helps understand the users' behavior. A variety of data mining techniques were used to address the problem; three steps of data pre-processing were applied, namely the cleaning of data, the identification of users, and the identification of sessions. Results obtained after pre-processing phase showed that the data quality will improve when the number of records reduced by (51.45%). The density-based spatial clustering of applications with noise (DBSCAN) and the K-means algorithm were used to develop clustering algorithms. Density-based clustering with three clusters outperformed the K-Means algorithm with three clusters in terms of accuracy.

Fawzia Omer A. (✉) · Mohammed H. A. · Awadallah M. A.
Department of Computer Science, Faculty of Mathematical and Computer Sciences, University of Gezira, Wad Madani, Sudan
e-mail: fawzia.omer@uofg.edu.sd

Mohammed H. A.
e-mail: mohebir@uofg.edu.sd

Awadallah M. A.
e-mail: awadallah@uofg.edu.sd

Z. Khan
Department of Electrical and Electronic Engineering, Universiti Teknologi Petronas, Seri Iskandar, Malaysia

S. U. Abrar
Institute of Computer Science and Information Technology, University of Agriculture Peshawar, Peshawar, Pakistan
e-mail: saidulabrar@aup.edu.pk

M. D. Shah
Department of Computer Science, University of Peshawar, Peshawar, KPK, Pakistan

© The Author(s), under exclusive license to Springer Nature Switzerland AG 2022
M. Ouaissa et al. (eds.), *Big Data Analytics and Computational Intelligence for Cybersecurity*, Studies in Big Data 111,
https://doi.org/10.1007/978-3-031-05752-6_15

Keywords Big Data · Web server log file · Big data analytics · Clustering · K-Means · DBSCAN

1 Introduction

Data can be generated in the digital world from multiple sources, and the fast transition from digital technologies has led to the emergence of big data. Generally, big data can be defined as a large amount of dataset that needs modern techniques to be analyzed and mined, and The vast volume of data analyzed by Data Mining techniques is referred to as Big Data. Currently, Big data has no universally agreed definition [1], so researchers have outlined five critical aspects of big data beyond what our data processing technology can handle. These aspects are known as 5Vs, and Big Data can be defined according to the term of 5Vs as follow:

- Volume: huge amount of data is generated, in the case of this paper data contain 293,236 cases.
- Velocity: the speed of accumulation or generation of data. In the case of this paper, data entries of log files are generated every second.
- Variety: different types of data from various sources.
- Veracity: different accuracy and different qualities and uncertainty in the generated data; in the case of this paper, data quality is achieved by eliminating irrelevant data.
- Value: In this paper's case, to extract valuable data or measure the usefulness of data to support decision-making.

With the emergence of new technologies and all linked devices, significant volumes of data are expected to be created in the coming years—in fact, as much as 90% of current data was created in just the last two years—a trend that is expected to continue for an unforeseeable future [2]. A huge volume of data is available from many different online resources and services like social media websites, web server logs, cloud computing, etc. as we said, this kind of data faces many various problems due to its volume. Big Data is clustered in a covenant format to overcome and tackle such challenges. The clustering techniques are significant for data mining and there are several approaches to mine the data include genetic algorithms, neural networks, support vector machines. Many web pages and websites are created every day, hence the amount of data generated is dramatically growing on the web, and due to the increment of these data, the analysis process becomes a challenging task to extract useful and valuable information for the designers or organizations to enhance both the view and accessibility of the web page via search engines [3]. Web mining is a data mining approach for extracting meaningful information from large amounts of data on the web. The web servers use a weblog file to list actions on the web pages or files [4]. The weblogs file contains information about the user access to the data presented on the web pages, and there is an entry for every user request to be recognized. Therefore, this file records every event on the website [5] and is classified as semi-structured data. Currently, it is an exciting and necessary research area to satisfy customers' expectations. Web Usage Mining aims to examine the usage details and patterns of the visitors to a website and make the Website more effective. Therefore, websites' users have different interests but understanding the users' interests is becoming a fundamental need for websites owners and designers to better serve their users by making adaptive content, usage, and structure of the website according to user's preferences. Logfile containing irrelevant data should be pre-processed and analyzed to get valuable information about users' behavior. However, analyzing this kind of data for getting knowledge is a challenging process and it is very difficult to perform effective analysis using the existing traditional techniques. In this paper, the objectives are to understand the frequent access patterns of the users to allow the website owners to manage and improve their websites accordingly. Also, to analyze the web server log file data and improve the data quality when eliminating the irrelevant entries from data. Finally, to cluster data using clustering algorithms.

2 Resources and Tools

The data source used in this article is the web server log file of Sudan University of Science and Technology (SUST) from 7/Nov/2008 to 10/Dec/2009, containing 293,236 cases that satisfy big data characteristics. Jupiter notebook environment and ScikitLearn libraries (python-based toolkit for efficient and straightforward clustering) and visual studio framework are used to perform the experiments.

3 Big Data Analytics

Big data contain many different fields such as data storage, data processing, data visualization, and data analysis. Big data analytics is the process that focuses on examining large datasets containing a variety of data types also big data analytics is where advanced techniques of analytics are applied to big datasets [6]. It allows organizations to gain valuable information from a mix of structured, semi-structured, and unstructured data, using techniques such as data mining. Data Mining (DM) techniques are used to extract useful information and discover knowledge from Web documents [3]. Web Mining, which is a data analysis task, is a multidisciplinary field that includes Data Mining, machine learning (ML), artificial neural network (ANN), information retrieval, and statistical analysis [7]. The field of big data analysis is shown in Fig. 1.

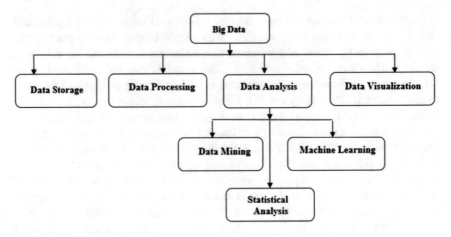

Fig. 1 Field of big data [8]

Data Mining is used as a process to extract useful information from data using different techniques like clustering and classification after putting it into a suitable format. Clustering is one of the data mining techniques that consider unsupervised learning in which data are classified according to their similar characteristics into various groups. Then the groups or clusters are addressed. Clusters are collections of data items similar to one another within the same cluster or group but dissimilar from items in other clusters or groups. The most common clustering methods include K-means, DBSCAN, Hierarchical, and FCM algorithms.

4 Proposed Method

Figure 2 shows the methodology we followed which consists of two phases which are Pre-processing and Patterns discovery and analysis.

4.1 Web Server Log File

Several phases are involved in processing the log file dataset from the web server, as shown in Fig. 2. It contains 293,236 cases that satisfy big data characteristics.

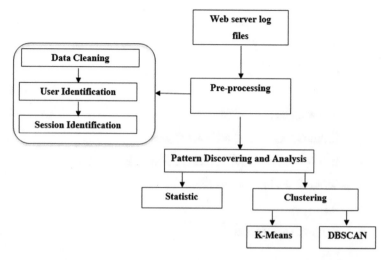

Fig. 2 Methodology

4.2 *Data Preprocessing*

This step involves processing the data into a suitable format with an aim of data size reduction and improving its quality. This consists of splitting text into smaller, valuable parts and identifying each one. This process comprises data cleaning, identifying users and their sessions.

- Data cleaning: In this step, we eliminate missing data or irrelevant information in weblog files such as entries with failure status code and public access to the website's home page because it is familiar to all users, entries that include some other files such as JavaScript, icons, CSS, Png, etc. only demanded entries contents will be considered for implementing next step successfully. Algorithm 1is used to perform data cleaning.

ALGORITHM 1: Data cleaning

Input: Server Log File.

Output: Cleaned log file

1. Begin

2. for each entry in log file

3. If status code >=200 and <=299 then

4. If user agent <> {bot, spider, yandex, crawler} then

5. Save entries in new text file.

6. Else

7. Remove such entries.

8. End if

9. Else

10. Read next entry

11. End if

12. End

Algorithm 1: Data Cleaning

- User identification: Here, the users are identified by their IPs address. Identify unique visitors by matching IP addresses. The IP address that matches the existing one in the IP list is considered an old user, and that doesn't match the one in the IP list is regarded as a new user. It combines agent and referred to identify the unique users. Algorithm 2 is used to perform this task.

ALGORITHM 2: User identification

1. Begin

2. For each record in log file

3. Repeat for each IP address

4. If IP address is found in IPList and user_agent is not the same then

5. Assign new user_id to that entry

6. Else

7. Assign old user_id to that entry

8. Increment user id.

9. End if

10. End

Algorithm 2: User Identification

- Session Identification: After the user identification step, the next step is the session identification, to identify the session we set the time frame to 30 min (The default threshold value). When the duration exceeds the threshold then starts the new session. Algorithm 3 is used to perform this task.

ALGORITHM 3: session identification

1. Start

2. For each entry in log table

3. If IP address is found in IPList and

datetime_difference <= 30 minutes then

4. Assign old session_id

5. Else

6. Assign the new session_id

7. Increment session_id

8. End if

9. End

Algorithm 3: Session Identification

4.3 Pattern Discovery

This phase includes techniques taken from a variety of domains, such as statistics, machine learning, mostly using Association Rules, data mining, and so on, since these approaches are used to find interesting patterns [9]. For this phase there are various methods are developed. Association rules and clustering are commonly used.

- Statistics: The statistical information discovered from weblogs is generated cyclically in a report for the use of website administrators to improve the system's performance, facilitating the task of website modification.
- Clustering: The clustering process is applied to group similar data items into clusters and different data items into different clusters [10]. To do so, we have chosen K-MEANS and DBSCAN clustering algorithms to extract useful information and group elements that have the same content or density.

K-Means Clustering

K-means clustering algorithm applied to find groups of the cluster in a web sites data elements and found every element to which cluster will be assigned. K-means start by initializing centroid to K clusters when a user of website assigned K value, effective use of K-means clustering algorithm relies upon the parameter K. K-means starts by allocating data point to its nearest cluster after that calculating cluster centroid again and change it every time its calculated, and the previous processes persist until no change occurs with the latest result and no transition of data point from its current cluster to any other clusters has happened. In K-Means clustering Euclidean Distance is the distance of the point from the specified mean, where there are many different techniques can be used for calculating the distance, for points' comparison, we use the most commonly used metric. Suppose that there are two points P and Q are defined as [11]:

$$\text{Point } P = (x1(P)), (x2(P)), (x3(P)) \, \& $$
$$\text{Point } Q = (x1(Q)), (x2(Q)), (x3(Q))$$

The calculation of the distance is given below by the following equation:

$$d(P, Q) = \sqrt{(x1(P) - x1(Q))^2 + (x2(P) - x2(Q))^2 \cdots}$$
$$= \sqrt{\left(\sum_{J=1}^{P} (xj(P) - xj(Q))\right)^{-2}} \tag{1}$$

The cluster centroid is the most important parameter; the point whose coordinates match the mean of all the points inside the cluster. The dataset has specific items that may or may not be related to any cluster, and according to that, it cannot be classified under them, such points are called outliers. The following equation's main goal is to get a minimal squared difference between the cluster's centroid and the item contained in the dataset (D) [11].

$$D = \left| X_i^{(J)} - C_j \right|^2 \tag{2}$$

whereas x_i is the item's value, C_j is the centroid of the cluster's value. When we use the k-means clustering technique, we must ensure that the correct or appropriate number of clusters is used. This is done using the elbow point method to determine the suitable number of clusters. The elbow technique works by doing k-means clustering on a dataset for a range of k values (say, 1–10) and calculating the sum of squared errors for each value of k.

DBSCAN Clustering

DBSCAN is an algorithm used for data clustering and it's proposed in 1996 by Martin Ester, Hans- Peter Kriegel, Jörg Sander, and Xiaowei Xu [12]. The technique is a density-based clustering algorithm, which groups together points that are tightly packed together (points with many adjacent neighbors) and identifies as outliers those that stand alone in low-density regions given a collection of points in some space (where their closest neighbors are too far). The basic idea of DBCSAN is to group points that have a high density, marking as outliers the points that lie alone in low-density regions. Local point density at a point p is defined by two parameters required for DBSCN: Calculating the slope between points (ε) or Epsilon and the minimum number of points required to form a cluster (mints). DBSCAN algorithm workflow is described in Fig. 3.

5 Results and Discussion

The result concluded for preprocessing and clustering SUST log file with two clustering algorithms (K-Means, DBSCAN) are summarized in the following sections:

5.1 Data Pre-processing

Tables 1 and 2 describe the result obtained from data cleaning.

After data cleaning, in step 2, only 42,503 records out of 142,345 are remained in the log file. Table 1 shows the comparison between several records before and after data cleaning. The result is presented in Fig. 4. After the cleaning steps, unique visitors' identification by matching IP addresses is done. Table 3 and Fig. 5 shows

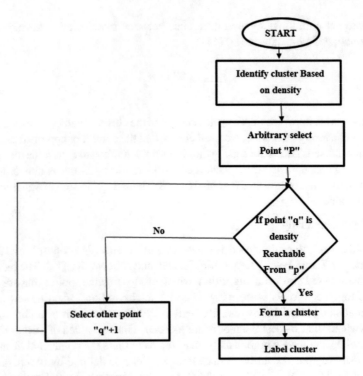

Fig. 3 DBSCAN algorithm workflow

Table1 Statistics of data cleaning step 1

Number of records	
293,236	Original log data
142,345	After cleaning step 1
150,891	Reduced
51.45%	Percentage in reduction

Table 2 Statistics of data cleaning step 2

Number of records	
142,345	Data of cleaning step 1
42,503	After cleaning step 2
99,842	Reduced
70.14%	Percentage in reduction

unique users extracted from the total records at the log file. There are a total of 12,252 unique IP addresses.

The time frame is set to 30 min. Table 4 and Fig. 6 show the session details which present the number of unique sessions extracted from the total record after cleaning,

Fig. 4 Data after cleaning
(step 1)

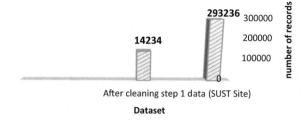

After cleaning step 1 data (SUST Site)

Dataset

Table 3 The number of
records and unique users

Records' number	42,503
Unique Users' number	12,252

Fig. 5 Result of user
identification

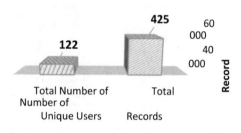

Total Number of
Number of

Unique Users

Total

Records

Table 4 The number of
records, number of sessions

Total number of records	42,503
Number of sessions	16,584

it also showed a substantial sessions' number consisting only of one or two requests.
As shown in Fig. 6 the maximum number of extracted unique web user sessions
didn't extend over 5 visits.

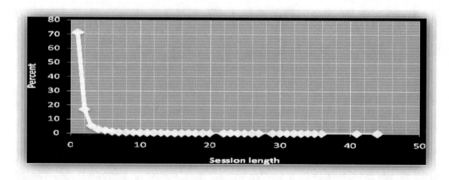

Fig. 6 The session length distribution of dataset

Table 5 Invalid records in
SUST log file

	Error	Error-description	Hits-count
1	302	http moved temporarily	8155
2	304	http not modified	29,274
3	404	http not found	22,587
4	400	http bad request	86
5	500	http internal server error	8566

Table 6 Information about
hits and bandwidth

Summary	
Hits	
Total hits	239,236
Visitor hits	142,345
Spider or crawler hits	150,891
Average hits per day	450
Bandwidth	
Total bandwidth	2550.409 MB
Visitor bandwidth	1449.401 MB

5.2 Pattern Discovery

In this stage, we have implemented mining algorithms depending on our interests. For mining web-logs usage records, a variety of data mining approaches have been used, however, we will focus on clustering and some generalized statistics as described in the following section.

5.2.1 Statistics Results

In this phase, some statistics are illustrated to understand such events of user and data traffic of SUST web site. Table 5, shows the invalid records in the SUST log file, and Table 6, the information about Hits and Bandwidth.

5.2.2 Clustering

The results of clustering techniques are presented in this section.

- **K-MEANS clustering:**

In the K-means algorithm first, we determined the clusters' number for the data based on elbow point. Figure 7 shows the result of the Elbow point method for the data.

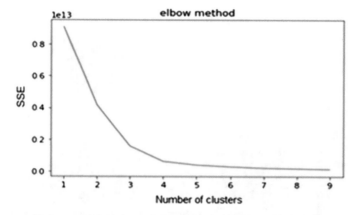

Fig. 7 Elbow point algorithm

From the graph result, we found that the suitable K value is =3. Figure 8 presents the result generated by the K- means on our Data after using optimal K value to produce clusters:

Clustering result shows that there are three clusters as shown in Fig. 8, Cluster1 combines a group of users browsing, downloading images content (.jpg, gif,.png) while Cluster 2 grouping users whose most events is access event as browsing page of our website that written or created using PHP. Finally, Cluster 3 group of users whose most event is searching for specific content in our website.

- DBSCAN Clustering:

After putting suitable values for eps and minPS the clusters of DBSCAN are plotted as follow:

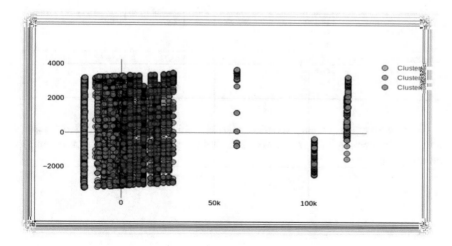

Fig. 8 Result of K-means clustering algorithm

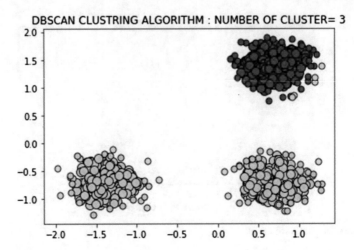

Fig. 9 Result of DBSCAN clustering algorithm

- Estimated number of clusters: 3
- Estimated number of noise points: 14

From Fig. 9 we found that there are three clusters in which each point within a cluster has high density to fall into specific clusters, several cores and border points = 24,205, also there is 9 noise point that is not a core point nor a border point and they are highlighted by yellow. The two clustering methods are compared based on these factors, according to past implementations of data clustering approaches

Algorithm	Parameters	Geometry (metric used)	No. of clusters	Cluster instances	Time talen to build model	Adjusted-rand index	Homogeneity score	Completeness score
K-Means	Number of cluster	Distance between points	3	26,214	0.28	0.25	0.669	0.710
DBSCAN	Neighbourhood size	Distance between neare points	3	24,227	0.12	0.978	0.986	0.934

We conclude that density-based clustering with three clusters gives fair results and greater accuracy than the K-Means technique with three clusters based on this comparison.

6 Conclusion

To improve the quality of websites, the big data generated must be analyzed effectively to support website management and decision-making at the right time. Weblog server file is one of the most important sources of data that gives meaningful value when analyzed. The data have been obtained from the Sudan University of Science and Technology (SUST) website in our study. A variety of data pre-processing steps are applied to the dataset (data cleaning, user and session identification) to prepare it for the clustering process. The results obtained after pre-processing phase on the SUST log file contained valuable information. It shown that data quality will increase when the number of records reduced by (51.45%).K-Means and Density-Based Spatial Clustering of Applications with Noise (DBSCAN) algorithms are then employed to make users' group based on similar browsing content. A high-density element is determined in addition to the core, border, and noise points. The results of the two algorithms are comprised and showed that Density-based clustering with three-cluster yields fair and accurate results than the K-Means with three-cluster.

7 Recommendations

The paper recommends using a fuzzy C-Means algorithm to categorize user sessions to derive groups of users displaying similar access patterns, particularly for heterogeneous and large amounts of data resulting from an exponential growth in the Resources of the web. Large volumes of data stored in servers and organizations increase data and, accordingly, the size of the log files.

8 Dedication

I dedicate this work to the soul of my supervisor Dr. Mohamed Hamed El-hebir who passed away to paradise if Allah says yes, for his patience and generous time spent to guide me through the entire process and also for his continuous support, suggestions, guidance, thorough knowledge and expertise helped me immensely in understanding and developing this dissertation.

References

1. J. Chen et al., "Big data challenge: a data management perspective, Frontiers of Computer Science, Springer" vol.7, no. 2, pp. 157–164, 2013.
2. O.Y. Al-jarrah, P.D. Yoo, S. Muhaidat, G.K. Karagiannidis, Efficient Machine Learning for Big Data: A Review. Big Data Res. 2(3), 87–93 (2015)

3. T.A. Al-asdi, A.J. Obaid, AN EFFICIENT WEB USAGE MINING ALGORITHM BASED. J. Theor. Appl. Inf. Technol. **92**(2), 215–224 (2016)
4. G. K. Lekeas, "Data mining the web: the case of City University's Log Files," 2000.
5. O. CU and P. Bhargavi, "Analysis of Web Server Log by Web usage Mining for Extracting users Patterns," http://www.Tjprc.org/view_archives. Php, vol. 3, no. 2, pp. 123–136, 2013.
6. N. Elgendy and A. Elragal, "Big Data Analytics: A Literature Review Paper," pp. 214–227, 2014.
7. Sandhya, Mala chaturvedi, 2013. "A Survey on Web Mining Algorithms", International Journal of Engineering and Science, vol.2, no.3, pp. 25–30, 2013.
8. S. Abraham, "BIG DATA CHALLENGES AND ISSUES: REVIEW ON ANALYTIC TECHNIQUES", Indian Journal of Computer Science and Engineering (IJCSE), ISSN: 0976- 5166, Vol. 8 No. 3 Jun-Jul 2017375, vol. 8, no. 3, pp. 375–382, 2017.
9. P. S. R. Iswarya, "Predictive Analysis of Users Behaviour in Web Browsing and Pattern Discovery Networks," vol. 4, no. 1, pp. 239–245.
10. S. H. Ganesh, K. Shanmugavadivu, and H. H. T. Rajah, "A Robust Density-Based Clustering Approach Using DBCURE – MapReduce Techniques," vol. 2, no. 3, pp. 215–219, 2017.
11. P. P. Anchalia, A. K. Koundinya, and N. K. Srinath, "Mapreduce design of K-means clustering algorithm," 2013 Int. Conf. Inf. Sci. Appl. ICISA 2013, 2013.
12. V. S. Bawane and S. M. Kale, "Clustering Algorithms in MapReduce: A Review," International Journal of Computer Applications (0975 – 8887), vol. 1, no. 1, pp. 15–18, 2015.

IoT Security in Smart University Systems

**Zahra Oughannou, Amine Atmani, Ibtissame Kandrouch,
Nour el Houda Chaoui, and Habiba Chaoui**

Abstract Through new intelligent, interactive, and most importantly, interconnected services, the connected campus would therefore allow for better control of building operating costs, especially in Smart Universities. Smart University campuses are the subject of current research, movements and advances in the nation. Therefore, transforming the university into a smart campus can provide several benefits to students, faculty, and staff by enhancing their experiences. Indeed, for a student, a campus should primarily facilitate meetings, socialization, and collaboration, before being a technological showcase. It is about optimizing the use of facilities according to the real needs of students, teachers, and visitors, to enhance the quality of life on campus on campus. Thus, students expect to have Internet access everywhere as they would at home, including on campus and in the residence halls. To support tech-savvy, always-connected students, universities must provide modern campus experiences that are convenient, secure, and personalized. However, transforming the university into a Smart Campus presents several challenges, such as security being a crucial challenge. In our article, we have studied the different technologies to make the university smart, made a detailed study of various types of attacks, and the solutions to deal with these attacks. The analysis of the proposed solutions showed us that the

Z. Oughannou (✉) · H. Chaoui
Advanced Systems Engineering Laboratory, National School of Applied Sciences, Ibn Tofail University, Kenitra, Morocco
e-mail: zahra.oughannou@uit.ac.ma

H. Chaoui
e-mail: mejhed90@gmail.com

A. Atmani · I. Kandrouch · N. el Houda Chaoui
Artificial Intelligence & Data Science & Emerging Systems Laboratory, National School of Applied Sciences, Sidi Mohamed Ben Abdellah University, Fes, Morocco
e-mail: amine.atmn@gmail.com

I. Kandrouch
e-mail: ibtissame.Kandrouch@uit.ac.ma

N. el Houda Chaoui
e-mail: nourelhouda.chaoui@usmba.ac.ma

© The Author(s), under exclusive license to Springer Nature Switzerland AG 2022
M. Ouaissa et al. (eds.), *Big Data Analytics and Computational Intelligence for Cybersecurity*, Studies in Big Data 111,
https://doi.org/10.1007/978-3-031-05752-6_16

security mechanisms of traditional computer systems, are not effective for intelligent systems based on IoT. So the combination of Blockchain (BC) and Explainable Artificial Intelligence will be a solid security solution to secure IoT systems within Smart Universities.

Keywords IoT · Smart University · Wireless communication · Security · Blockchain

1   Introduction

While the smart city of tomorrow is being prepared, the university campus is also getting up to date. Under study for several years, the first real smart campus models have now entered the active phase. Just as the smart city concept is not just about a few well thought-out applications and a handful of solar panels, the smart campus is not just about rolling out fiber and offering online courses. It's a much larger project with countless implications, the ins, and outs of which are currently being widely debated around the world. We are dealing with education, and moreover with universities; the subject is therefore highly strategic. A smart campus connects people, devices, and applications, using the Internet of Things. This makes it easier for universities to make decisions, in addition, to improve security and optimizing resources. So a smart campus makes the university safer and universities are becoming connected, using Internet of Things (IoT) technologies, they are connecting everything and anything to create new applications that enhance the campus experience. And they're realizing huge savings in the process. And it's transforming campus life for students, faculty, and visitors. A smart campus connects multiple systems such as door locks, lighting, etc. It also relies on data analytics to ensure a seamless and interconnected experience for all campus components.

Higher education institutions around the world are prioritizing smart campus design to improve student outcomes, reduce costs, and prepare their schools for the future. Once a university has laid the groundwork for wireless and wired services, the possibilities are endless. However, A Smart Campus is not created overnight. Universities need to define a clear vision for leveraging smart technologies on campus. The university as a whole needs to be thought of before designing a smart campus, taking into account several requirements, i.e., energy savings, data analytics, security, which is a critical point, etc.

Smart Universities may face several security issues. Educational institutions are at risk from both physical and cyber-attacks. Due to an open environment either in libraries or community centers that usually have easy access, in addition to the unawareness of students by phishing and other cyber threats. And to maintain their reputation and the trust of students, parents and alumni, these institutions must invest in the security field to ensure comprehensive security of the entire campus. In this context between our work, in which we tried to make an inventory of vulnerabilities and attacks of different types that can infect smart universities, as well as the proposed solutions to deal with these attacks. After our study, we can confirm that the proposed solutions have several limitations. Hence the need for more advanced solutions. So the combination of Blockchain technology (BC) and Explicit Machine Learning (XAI) could be a solid security solution against present and future attacks.

2 Smart Campus

The concept of a smart campus was born since the rise of new domains of IoT. It is a broad concept that includes several physical and electronic objects that communicate and interact with each other. It uses an infrastructure and technologies that allow it to improve its processes [1].

2.1 Smart Universities

The smart university concept is derived from the smart campus concept. Smart campuses are built using the integration of cloud computing and IoT, which help teaching, management, and research in universities [1].

The aspects of smart campus services are many names, environmental, financial, and social aspects. But in our article, we focus on the educational aspect in the context of smart universities. The objective of making universities smart is to offer a smart learning system, which refers to the increasing use of digital tools that allow teaching to progress gradually, in order to satisfy and meet the needs of the new generation of students who are connected to the digital age to inform themselves, learn and stay in touch with the outside world [2, 3]. This leads to the adoption of a variety of smart technologies in academic environments allowing the improvement of the quality of life and performance of teachers and students [3].

The IoT intervenes in all areas and influences the processes of a university, in order to improve them in terms of supporting academic learning, technological development, and finally decision making. The Smart University provides several benefits. It allows for the control of academic flow, i.e. faculties, classrooms and class times, etc., so it allows for knowing the traffic of people attending the university, systematizing all processes, reducing energy consumption, analyzing risks, and decision making based on statistics [1]. For example, a smart school tool that has been developed allows to perform in intelligent way actions that are related to the control of the presence of students, and subsequently sends emergency messages to parents and students, as well as downloading the results of students using school systems [4]. There is also a green campus that has been proposed using IoT concepts, based on an architecture to efficiently manage application systems. These systems provide data to the IT labs to aid in the decision-making process, thereby reducing energy consumption. Other examples include the Smart Library, which is an RFID-based architecture to standardize and protect book selection at Donghua University, and an open-source IoT system to improve environmental sustainability, educating people on their attitudes toward energy consumption. In the design of the system, open hardware units and protocols were used to meet several educational requirements to make the system flexible and scalable [1, 4].

2.2 Smart Universities Technologies and Connectivity

To connect various objects of Smart Universities different technologies are used that ensure communication and offer several high quality services. In this section, we mention the most used technologies in Smart Universities.

2.2.1 Internet of Things (IoT)

IoT is defined as a network of wired and wireless communication technologies. These technologies are interconnected and distributed and can communicate through integrated systems [5]. The internet is not only a system of computers, but it has also turned into a network of devices of all kinds and sizes making any system intelligent [6]. Currently, IoT and ubiquitous technology represent a new communication paradigm that is based on a fundamental architecture designed for the future generation in which all objects and devices are interconnected [7], such as digital machines, sensors, actuators, cell phones, people and other objects that communicate with each other via the internet to create an intelligent system. In smart universities, this technology is designed to perform intelligent actions namely identification, monitoring, and intelligent tracking of any personnel attending the university. It is also used for data management, automation, etc. [2]. IoT devices are inexpensive due to their minimal resources for processing and communication and their limited storage. Therefore, they can be easily employed on a large scale to measure pressure, humidity, temperature, and other environmental factors in smart classrooms. They are naturally heterogeneous as they work with different software platforms and hardware entities [8]. Figure 1 shows an IoT network within Smart Classroom that is used to interconnect the different devices (sensors, cell phones, cameras, etc.).

2.2.2 Cloud Computing and Big Data

The cloud layer is made up of several technologies that enable the processing of big data. At this layer, the data collected by the sensors is stored, which is then analyzed using algorithms designed to make decisions, and finally the results are sent to the service layer for their visualization [4]. The cloud layer in the IoT domain does not have the same processes as cloud or mobile computing. Indeed, the IoT collects a large amount of data from sensors that must be stored and processed. Depending on the amount of data compiled, machine learning techniques are even applicable to the

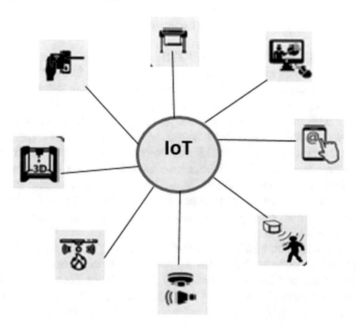

Fig. 1 IoT in smart classroom

cloud environment. Based on the data from the smart campus, a machine learning model can be trained and perform the necessary actions. In a smart system, emergency service needs can be monitored by training and validating student and faculty data [1]. Cloud storage offers several advantages such as scalability, flexibility in addition to its ability to process large data without losing much data [4]. But many cloud users are concerned about access because of the high sensitivity of their data. To solve this kind of problem, a cryptography-based efficient, and sensitive distributed storage system is designed to realize massive and efficient distributed storage service that provides higher security, encrypting all data before distributing it to a variety of servers in the cloud [1].

2.2.3 Wireless Sensor Networks

Wireless sensor networks (WSNs) are used to capture the required data [9]. In intelligent systems, data is easily collected with the help of sensors. There are two types of sensors: dedicated sensors, which are intended for a specific application and generally are not shared with other applications. And the non-dedicated sensors mean they are not dedicated to a specific application, but they are integrated into the smartphones of users who are free to use different applications [10].

2.2.4 Automatic Identification Technologies

- **Radio Frequency Identification (RFID)**: An RFID system consists of a host computer, a reader, and a tag. This technology uses radio frequency to communicate with objects or people who are identified at Smart University. Typically, the tags are given a specific address and then embedded in the objects. They assure an automatic control in real-time without the real presence of a human being, with the help of an RFID reader that can read the information that is stored electronically in the tags. In the context of Smart Universities, this technology can be used to provide access control, attendance management, library management, etc. [2].
- **Near Field Communication (NFC)**: can be considered as an advance of RFID allowing devices such as smartphones to communicate using radio signals. The operation is based on radio waves allowing cell phones to get closer or by making them touch [2].
- **Biometrics**: is the automatic detection of a person using several measures linked to unique human features. We can distinguish between two types of characteristics: behavioral and physiological or biological characteristics. Both types are obtained by using appropriate sensors and the typical characteristics are used to build a biometric model that can be used in the authentication process [2].

2.2.5 Wireless Technology

- **WIFI**: is an IEEE 802.11 standard that is increasingly used to provide Internet access to many devices in businesses and campuses. WIFI offers several advantages namely performance, simple implementation, in addition to a low-cost network, etc. Indeed, thanks to these advantages WIFI can be deployed throughout the smart campus to facilitate the support of a high number of WIFI connections and it also allows users to access the Internet for free [2].
- **WiMAX**: is based on the IEEE 802.16 standard. It can operate in two types of frequencies; frequency division multiplexing (FDD) or time-division multiplexing (TDM).WiMAX data transmission rates range from 1.5 Mbps to 1 Gbps with higher frequency bandwidths ranging from 2 to 66 GHz; however, the vast majority of applications use the 3.5 and 5.8 GHz bands. WiMAX uses microwave radio technology to allow computing devices to be connected to the Internet, so that data can be transferred over long distances of up to 50 km. Therefore, WiMAX has a longer transmission range than Wi-Fi, which reduces interference [4].
- **ZigBee**: It is defined by the IEEE 802.15.4 standard and is typically used to create personal networks with devices and applications that require secure networking, long battery life, and lower data rates. It is therefore often used in control and monitoring applications that are characterized by data reliability, energy efficiency, and cost. ZigBee differs from other personal area networks in its simplicity and low cost [2].
- **Z-wave**: is a WPAN/WLAN solution supported by the Z-Wave Alliance, which is characterized by low cost and low power. It is primarily intended for smart homes

and building automation. This technology is typically used for home automation, heating, ventilation and air conditioning, energy management, security surveillance, etc. [10].

– **Bluetooth Low Energy (BLE)**: This solution is designed to provide low-power, short-range, low-delay, low-rate connectivity. BLE technology operates in the 2.4 GHz frequency band, making it susceptible to interference with many existing wireless protocols like ZigBee and WiFi [10].

– **LoRaWAN**: a proprietary technology developed by Semtech. This technology is based on the physical layer of LoRa. LoRaWAN's PHY layer enables bi-directional communication, making it applicable to both sensors and actuators. To extend its communication coverage by negotiating the data flow, it uses frequency shift keying (FSK) and spread spectrum modulation (CSS) [10].

– **Sigfox**: Ultra Narrow Band (UNB) Sigfox was first released in 2009. It is a low-power proprietary RAT with very limited data rates and a long-nsurange. The main advantages of Sigfox are huge coverage in urban and rural areas, low power consumption, and low deployment costs. In particular, the use of a client–server model reduces costs and improves energy efficiency, while eliminating timing overhead (as opposed to cellular RATs, where devices must wake up periodically for paging) [10].

2.2.6 Cellular Networks

– **NB-IoT**: First introduced in 3GPP Release 13 (R-13), Narrowband-IoT (NB-IoT) can be thought of as simplified LTE for low-power wide-area network (LPWAN) applications. NB-IoT operates in the licensed frequency band from 7 to 900 MHz. It uses two access methods; Orthogonal Frequency Division Multiple Access (OFDMA) to provide downlink and Single Carrier Frequency Division Multiple Access (SC-FDMA) to provide uplink [10].

– **5G**: 5G uses more advanced communication techniques such as small cells, millimeter-wave deployment (mmWaves), spatial division multiple access (SDMA), and cloud radio access network (C-RAN) to meet the size demands [10].

– **6G**: 5G has already started deployment, but currently 6G technology is also appealing to the research community. The 5G network has limitations; it will not be able to meet the futuristic demands of massive connectivity for data-heavy applications such as gaming, telesurgery, holographic communication. Therefore, certainly, the potentialities of 6G communication network will get the interest of application developers soon [8].

2.2.7 Mobile Technology

Laptops, tablets, cell phones, and similar equipment are becoming some of the most dominant devices employed on campus, generally, in everyday life. This technology can be deployed in education to meet the needs of students on campus and lower costs, as well as to improve the quality of education. It offers several capabilities to provide smart campus services, such as door access to enter or exit the building using touchless technology, and the development of mobile options that enable anytime, anywhere education, etc. [2].

2.3   Smart University Vulnerabilities

A vulnerability is a flaw that can be found in a system, application, or service that can be exploited by an attacker to bypass security controls and exploit the system in a way that the developer never anticipated [11]. The unique and specific characteristics of IoT create serious data security issues. If IoT devices are installed in low-traffic areas or left unsupervised, it gets difficult to monitor and secure this vast number of IoT devices that are susceptible to a wide variety of dangers [8]. IoT devices are easily compromised using the default dur code or credentials, via the remote access interface that enables the attacker to use command and control and turn these vulnerable units into a collective of the massive botnet and use them as distributed denial of service (DDOS) to attack sensitive systems [12].

In smart universities, there are several smart devices and connected equipment to provide various services. This larger number of smart devices is a vulnerability that can be exploited by an attacker as an attack entry point into the network. This affects the security of the smart university. For example, enemies can take physical control and capture these IoT units to attack IoT networks. In addition, data from sensors can be captured, distributed, and processed by multiple intermediary systems, enhancing the threat of tampering or alteration [8]. In addition, the complexity of the smart university network as well as the vulnerability of IoT devices to numerous internal and external threats, make network management and monitoring more difficult. Since most devices operate with the help of a human, technical vulnerabilities usually stem from human failures, and also the actual lack of skilled cybersecurity personnel, lack of ongoing training programs and security strategy would render all systems susceptible and make it easier for an attacker to gain access to devices, control systems and steal data [2]. In general, IoT malware has several characteristics. Indeed, IoT malware is used to perform DDoS attacks by scanning open ports of IoT services such as FTP, SSH, or Telnet protocol [13].

2.4   Smart University Susceptible Attacks

Smart University uses several technologies and equipment, which are not secure. Indeed, several devices, systems, and applications use weaker protocols like telnet, HTTP, FTP, etc. And transmit information via unsecured media. [2] This gives rise to several attacks especially attacks that can be applied to wireless and are not specific to a communication protocol, e.g. Jamming Attack can harm the communication of the Smart University system in several modes of communication and is a specialized denial of service attack [14].

Attackers can exploit these vulnerabilities to gain access to systems, recover confidential data for later use [2], they can also target a particular wireless communication (e.g., ZigBee) based on a little technical knowledge and widely available low-cost devices [14]. The main possible attacks against smart universities are cited below:

2.4.1 Physical Attacks (PA)

Physical attacks affect the hardware devices of the Smart University system, such as sensors, controllers, RFID readers, etc. [2]. A physical attack is often the first step in a significant attack. In effect, a damaged device is exploited as an entry point to launch further attacks and affect the rest of the Smart University's network [15]. Physical attacks include:

- **Device theft:** Smart Universities aggregate a set of private and public devices such as laptops, tablets, cameras, etc. that are distributed in Smart Universities with what may be weak protection. Therefore, the theft of these physical objects allows the attacker to obtain physical access to the equipment, and then he can perform several actions such as attacks to violate the privacy of individuals and disrupt the privacy and availability of systems [2].
- **Social engineering:** This attack is based on manipulating individuals to disclose sensitive and confidential data. This is a physical attack, as it exploits human interactions, meaning that social engineers interact physically with campus users to obtain valuable information, which can be used for malicious purposes. This type of attack is applied with the help of campaign emails, phone calls, or direct contact with users [2].
- **Sleep deprivation attack:** Smart University uses many IoT equipment that can be powered by replaceable batteries which provides a certain extension of their life and guarantees a higher performance. The attacker exploits devices that are battery-powered while awake by powering them with erroneous inputs. The batteries then discharge, causing them to shut down [2, 16].
- **Malicious node injection:** In this case, the attacker places a false node between two legitimate nodes on the network to gain full control of the data stream between those nodes. The goal of this attack is to generate anomalous behavior on campus functions and services or to provide the attacker with full control of the affected system [2, 16].
- **Tampering:** action consisting in modifying the physics of a device (for example, RFID) or a communication liaison [16].
- **Malicious code injection**: This action allows the attacker to install malicious code on a physical component, compromising it, which allows him to execute additional attacks [16].

- **RF interference/Jamming:** This attack is based on sending noise signals across the RF/WSN signals which allows the attacker to execute DoS attacks on the RFID tags or systems. On RFID tags and sensor nodes, which impedes communication [16].
- **Permanent denial of service (PDoS):** is a type of DoS attack, also known as phlashing, in which an IoT equipement is affected via physical tampering. The attack is executed by downloading a corrupted BIOS with malware or by destroying the firmware [16].
- **Device spoofing attack:** regardless of the length or combination of camera passwords, attackers can obtain them. Similarly, by listing all possible MAC addresses, the attacker can deploy a tampering attack to search for all online cameras [16].

2.4.2 Software Attacks (SA)

The smart campus system can be affected by software attacks using malware, viruses, and scripts that can steal information, deny services, alter data and even cause damage to smart devices. Software attacks exploit vulnerabilities in any computer system, among these attacks, are: [2]

- **Malware, viruses, worms, Trojan horses, and spyware:** are malicious software used by an attacker to damage the campus system. The sources of this malware can be downloading malicious files from the Internet, emails, etc. Some of these attacks can alter user data without the user's knowledge, and even replicate themselves with no human action [2].
- **Malware:** malware can affect data in IoT devices, which can contaminate data centers or the cloud [16].

2.4.3 Data Attacks (DA)

Currently, confidentiality is a critical aspect because threats to individual privacy limit the success of the IoT [16]. At the Smart University level, there is a wide range of devices and applications producing various types of data that present the most essential resources of the Smart University campus. Therefore, protecting the confidentiality of this data has become a critical necessity due to the sheer volume of data that remote access tools can easily access. In general, attacks against data are those threats that can affect data in different phases namely collection, suppression, use, and storage [2]. Some of these attacks include:

- **Data inconsistency:** In IoT, the attack that targets data integrity that causes inconsistency in data in transit or data that is stored in a central database is called data inconsistency [16].
- **Data Breach:** Data breach (memory leak) is the disclosure of data that is confidential, sensitive, and personal in an unauthorized way [16].
- **Data loss:** It can occur if the data does not have sufficient security, and therefore the data can be very vulnerable to compromise. Some of the most common reasons

for data loss include power outages, data corruption, storage system faults, and data deletion [2].

- **Account or Service Hijacking:** This attack can occur when the criminal obtains the user's credentials that allow access to their account, data, or other confidential information. The attacker relies on these stolen credentials to modify data and damage Smart University services, etc. [2].
- **Unauthorized access:** The role of access control is to allow authorized users to access and deny access to unauthorized users. By exploiting unauthorized access, unauthorized users can appropriate data or access confidential data [16].

2.4.4 Network Attacks (NA)

An attack against the availability of the network mainly takes the form of a denial of service (DoS) [15]. In Smart University, different objects are linked together using many communication technologies namely RFID, WIFI, 5G, and others. Due to network mapping or unauthorized access to gather vulnerability information [2], the attacker may use or overload the network's communication and computing resources, resulting in failed or delayed communications [15], thus these network attacks could disclose the privacy of individuals. The network attacks include:

- **Traffic analysis attacks**: they are based on the interception and examination of network traffic to deduce and obtain essential information from communication patterns. Indeed, the attacker can obtain confidential information on the network, even without approaching it [2, 16].
- **Replay attack**: The attacker can apply this attack by capturing a signed packet and then sending it back to the destination several times. This makes the network busy, which causes a DoS attack [16].
- **Clandestine Eavesdropping**: This attack violates the privacy of Smart Universities systems. It involves an attacker targeting the insecure communication link to transmit and receive data. It does not interfere with the network transmission [8].
- **Denial of service**: The denial of service attack consists of inundating a system's network with huge volumes of spam traffic and data. The objective is to overload it and make it very slow due to a large number of demands [15].
- **Denial of Service/Distributed Denial of Service (DoS/DDoS) attacks**: In the DDoS attack, multiple impacted nodes attack a particular target by inundating connection requests or messages to slow down or even bring down the network system server [16].
- **Malware injection**: This attack involves the installation of malicious software (malware, viruses, spyware, rootkits, ransomware, Trojans, etc....) in cyberspace which causes destruction or disables computers and networks. Among the malware is the WannaCry ransomware that has been exploited before to prevent people from accessing critical services except after paying at least a ransom [15].
- **Phishing**: This is a request for data from a seemingly trusted source. The goal of this attack is to make users think they are trustworthy, so they can perform certain actions, such as clicking on a malicious link or providing sensitive information.

Some of the most popular phishing attacks are smishing (SMS phishing) and spear-phishing [15].

- **SQL injections**: These are injection attacks used by an attacker to compromise data-driven applications. The attacker executes malware SQL queries to the database server of a web application and inserts them into an entry field on the client-side of the application. The goal is to capture, modify or delete the information in the database [15].
- **Man-in-the-Middle Attack (MITM)**: This attack attempts to hack communications between devices implanting false information or distorting messages between them. So far, solutions rely mainly on secure authentication, strong cryptography, and data integrity checks. However, these mechanisms are not sufficient to prevent this type of attack [17].
- **An Advanced Persistent Threat (APT)**: This is a stealthy cyberattack in which an attacker can obtain non-authorized access to a network and remain without detection for a long period [15].
- **RFID spoofing**: Here, the attacker uses an RFID signal that allows him to access the information printed on the RFID tag. Using the original identifier of the tag, the attacker can thus transmit his data by making them pass as valid [16].
- **RFID unauthorized access**: this attack allows the attacker to perform several actions on the data present on the RFID nodes such as reading, modifying, or deleting, due to the absence of appropriate authentication mechanisms [16].
- **Routing information attacks**: These are direct attacks allowing the hacker to spoof or modify routing data and add nuances through actions such as sending error messages, generating routing loops, etc. [16].
- **Selective forwarding**: In this attack, the data transmitted to the destination is incomplete because of a malicious node that may modify, delete or forward certain messages to other nodes in the network [16].
- **Wormhole attack**: In this attack, an attacker must control at least two nodes that are separated from each other and have a high-speed connection between them to tunnel packets from one location to another [17].
- **Sybil attack**: To apply this attack the attacker uses a malicious node claiming multiple identities called Sybil nodes and situated at various locations in the network. This causes a huge allocation of resources unfairly [16].
- **Jamming attack**: This attack focuses on disrupting the communication of smart universities by using data packet flooding, data collision, and external exhaustive noise. This attack is simpler to perform, as it does not need detailed knowledge of the target system [8].
- **Hijacking attack**: The attacker aims to gain full control over the IoT applications of Smart universities, thus updating the stored data. For example, in the context of smart meters, an attacker can manipulate the data while consuming energy [8].
- **Sinkhole and wormhole attacks**: Manhole attacks are used to execute malicious actions on the network. The attacking node passes packets and replies to routine demands. Attacks on the network can further degrade 6LoWPAN operations because of wormhole attacks, which are based on creating a tunnel between

two nodes such that packets arriving at one node immediately reach the other node [18].

2.4.5 Encryption Attacks (EA)

This type of attack poses a serious threat to the confidentiality of cryptographic packages and also to security. The attacker aims to discover the encryption key used in the encryption and decryption of modules, protocols, data using specific mechanisms like [2]

– **Cryptanalysis attacks**: This attack is used to affect the cryptographic security system by discovering the encryption key used which allows the attacker to access the encrypted messages. Various cryptanalysis attack techniques are available, namely differential cryptanalysis attacks, dictionary attacks, full cryptanalysis attacks, etc. [2].
– **Side-channel attack**: Here, the attacker gathers information about what devices are doing when performing cryptographic operations, such as the time it takes operate, power consumption, electromagnetic radiation, frequency of flaws, and then uses this side-channel information to detect the encryption key [16].

2.5 Development of IoT Security Mechanisms in Smart Universities Environments

Security is a crucial point in the development of any IT system. Therefore, it is necessary to apply security mitigation measures that secure IoT infrastructures, users, data, and devices, thus preserving privacy and confidentiality and also ensuring the accessibility of services provided by an IoT ecosystem [19]. Therefore, integrating security from the beginning in the conception and implementation of IoT devices as well as in the development of smart applications is a crucial step to ensure adequate security in IoT [20].

2.5.1 Access Controls

It is a set of rules that are determined in advance to define to a user or device which resources, data files, or devices can be accessed and which cannot be accessed permanently. Using these predetermined access rules to network devices and system functionality reduces the probability of malicious access to network devices. Access controls namely Mandatory Access Control (MAC), Discretionary Access Control (DAC), and Role-Based Access Control (RBAC) can enhance system reliability and remove possible security risks [15].

2.5.2 Authentication

Authentication is a mechanism to identify users and devices in a network to grant access to authorized individuals and unmanipulated devices. The authentication mechanism helps mitigate several attacks that can infect IoT systems. Among these attacks are the impersonation attack, Sybil attack, response attack, and man-in-the-middle attack [19]. To ensure the authentication and encryption of communications, there is the Transport layer Security (TLS). TLS proposes TLS-PSK which uses pre-shared keys, and the TLS-DHE-RSA authentication method using public key cryptographic protocols such as RSA and Diffie-Hellman (DH) [19].

2.5.3 Encryption

To realize effective end-to-end communication with weak power consumption, IoT encryption is used. Thus, to provide end-to-end security in IoT systems, there are light symmetric and asymmetric algorithms [19]. Using strong encryption tools, the integrity and confidentiality of data communications can be guaranteed by ensuring that communications are encrypted. This mechanism reduces the ability of an attacker to generate valid data or hijack communications and subsequently can deceive the system [15]. However, some nodes may be able to integrate general purpose microprocessors, because of the heterogeneity of IoT systems. Low-resource devices may have constraints, as they only integrate ICs specific to a certain application. Therefore, classical cryptographic solutions are not suitable for low-resource smart devices. These devices have limited battery life, low computational power, limited memory, and power supply. Thus, for these devices, light cryptography can be an effective means of encryption [19].

2.5.4 Trust Management

Trust management in the IoT enables secure access control by eliminating malicious nodes. Research on the topic of trust management has shown dynamic and automated trust calculations to assess the trust metrics of the contributing nodes in an IoT network, but most of this research concentrates on detecting malware nodes [19].

2.5.5 Secure Routing

Secure routing can be provided by the RPL protocol, which is a proactive routing protocol that uses the Destination Oriented Directed Acyclic Graph (DODAG). To build the DODAG, the nodes in the network transmit the data packet to the border root via the parent node, then the border root finds out the path to send the packet to the target node. Data packets are forwarded either in the upstream or downstream direction. The DODAG construction process allows for multiple RPL instances in the network. Each RPL instance forms multiple DODAG networks, in which multiple sensor nodes are linked to the root. The internet is used to connect these roots using a transmission link. The main concern of RPL is to construct the routing topology to self-optimize and eliminate loops in the network [21].

2.5.6 Security Patches and Regular Updates

Security updates in IoT devices are necessary. Therefore, IoT devices must be easily upgradable. However, most manufacturers today construct devices without considering future updates to the firmware. Yet, deploying updates can address several issues such as threats and vulnerabilities that may arise as a result of evolving technology, operational systems, and application code. Therefore, implementing a consistent and flexible firmware deployment process will allow organizations to mitigate potential threats and close security gaps in the network. And in this way, enterprises can address cybersecurity challenges that are especially magnified due to the integration of new and legacy systems [15].

2.5.7 Physical Security

Security mechanisms must be used and built into network devices to protect them from unauthorized physical access, as the physical security of devices connected to the network is paramount. Data stored in the devices could be damaged by unauthorized access by personnel. This data may contain critical information, namely identification, authentication, etc. Therefore, to protect private and sensitive data from leakage that can be maliciously exploited by intruders, mechanisms must be in place to remotely wipe and lock devices from the network. The physical security of the rooms where the servers and control areas are situated must also be ensured. These rooms must be secured because they represent a security risk to the entire network, as well as a central access point to the network, and hackers can exploit this point to harm the network [15].

2.5.8 Intrusion Detection Systems (IDS)

To build intrusion detection systems, firewalls and antivirus software can be used. The major intrusion detection methods are:

(a) **Signature-based IDS**

Signature-based detection is based on comparing a possible threat with the type of attack that is stored in the IDS database. Therefore, only intrusions whose signature is in the database can be discovered, and thus new threats that are not stored in the IDS can appear as real attacks that can infect the system [15].

(b)   **Anomaly-based IDS**

Anomaly-based IDS is a dynamic detection technique. This technique uses the behavior by which the IDS searches for weaknesses based on user-defined rules and not based on signatures stored in the IDS. This type of detection relies on artificial intelligence to distinguish between normal and non-normal traffic, making it capable of detecting new or unknown attacks. The advantage of this technique is its ability to distinguish whether the abnormal traffic is malicious or not. Therefore, it is most effective when used to evaluate each reported event to determine the appropriate action to take, thus generating alerts [15].

(c)   **Host-based IDS**

The Host Intrusion Detection System (HIDS) is deployed on a host in the network, allowing it to collect and analyze traffic coming from or going to that host. To monitor specific elements of a host that are not easily available to other systems, HIDS uses its privileged access. This intrusion detection mechanism has only a restricted view of the total network topology, so it can only identify malicious activity on a particular host [15].

(d)   **Network IDS**

Network IDSes (NIDS) operate by surveilling traffic as it passes by the network infrastructure. Contrary to IDSs, NIDSs can to monitor the network and detect malicious activity intended for that network. To remain effective, NIDS must be able to analyze a wide range of network traffic in real-time. Network IDS (NIDS) monitors traffic as it passes through the network environment. NIDS can to monitor the network and detect malicious activity intended for that network, unlike IDS. For NIDS to remain effective, they must be able to analyze a large amount of traffic on the network in real-time [15].

(e)   **Stack-based IDS**

To monitor packets as they move up the OSI layers, the stack-based IDS operates by tightly integrating with the TCP/IP stack. This monitoring technique is more efficient because before the operating system or application has an opportunity to process the packet, the IDS removes it from the stack [15].

2.5.9   Blockchain Technology

To establish trust, standard IoT security solutions rely on a third party. Indeed, unauthorized digital certificates issued by third parties are reasons to report several incidents [22]. Blockchain implementation in IoT offers several benefits and can address several challenges related to managing massive volumes of data, securing IoT devices, protecting users' privacy, assuring trust, confidentiality, integrity, etc. [16]. The operation of blockchain is more complex, but the fundamental idea of distributed ledger technology (DLT) is simple. It is the decentralization of data storage, this technique ensures that the stored data cannot be controlled or manipulated by a central

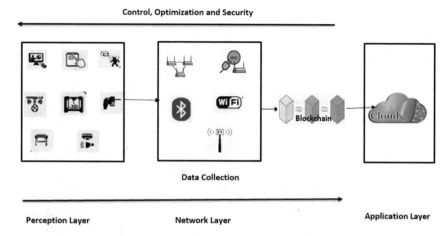

Fig. 2 Integration of the blockchain in the security of smart classroom

entity [8]. In the context of Smart Universities, the integration of Blockchain could have many benefits, mainly the improvement and securing of IoT functionalities, as shown in Fig. 2. And also, storing sensor data, managing device configuration, and enabling micro-payments [23]. Thus, the use of Blockchain can ensure security, indeed Blockchain technology is based on the notion of a chain where blocks are linked to each other and located in a distributive and inflexible way where each blocks stocks specific information. The structure of the Blockchain allows for easy detection of any disappearance or alteration of one or more blocks without the need for an intermediate or central entity The chain of blocks structure can easily detect any disappearance or alteration of one or more blocks without the need for an intermediate or central entity [24].

3 Discussion

To make any system intelligent we rely on the Internet of Things (IoT). However, the diversity of IoT devices with specific characteristics makes them face several challenges namely security, which is the main challenge of different IoT systems. Today, privacy is a critical issue, as privacy challenges can infect the success of IoT. As shown in Tables 1, 2, and 3, several security solutions are addressed to deal with different attacks in different layers of IoT. However, the research done in the field of computer security has shown that the conventional security methods such as authentication, encryption, secure routing, etc., which have been applied in traditional computer systems, have several limitations at the levels of IoT-based smart systems.

Recent research has focused on using Blockchain technology which is based on decentralized data storage to enhance the security of various IoT systems within

Table 1 Overview of attacks and mitigation measures related to the application layer of IoT devices

IoT layers	Security attack	Type of attack	Proposed solutions
Application Layer (Data)	Data breaches	PA + DA	– Server-side; strong access control; accepted authentication practices; storage encryption [14]
	Shared technology vulnerabilities	PA + DA	– Automatic vulnerability assessment (Nessus, Qualys, Burp Suite) [11]
	Cloud computing data security attacks	DA	– Lightweight authentication; encryption [14]
	Driver-based attacks	DA	– Avoid unverified drivers; treat drivers as untrusted software; user awareness [14]
	Software vulnerabilities	DA	– Legacy equipment considerations; frequent updates; vulnerability awareness [14]
	Improper encryption	DA	– Software verification; accepted encryption methodology [14]
	Remote hijacking	DA	– VPNs; strong passwords; configuration best practices; avoid port forwarding [14]
	Data inconsistency	DA	– Blockchain architecture; chaos-based scheme [16]
	Unauthorized access	DA	– Privacy; Preserving ABE; Blockchain-based ABE [16]

Smart Campuses, especially Smart Universities which is our research area. Thus, to our knowledge, until now no research work has combined Blockchain technology (BC) and Explainable Artificial Intelligence (XAI); Which is a method of explicating the results given by artificial intelligence. So we're going to build on these two techniques to propose a robust and credible solution to secure IoT systems in Smart Universities. And this will be among our future works.

4 Conclusion

The concept of Smart Universities was born with the emergence of IoT as an application domain. Universities need to adopt smart campus technologies. Smart campus design is paramount for the successful implementation of Smart Universities, thanks to IoT which has several advantages to solve a wide range of problems. However, there are still challenges to be addressed, these challenges are mainly related to security

Table 2 Overview of attacks and mitigation measures related to the network layer of IoT devices

IoT layers	Security attack	Type of attack	Proposed solutions
Network Layer (Routers)	RFID spoofing	NA	− SRAM based PUF [16]
	Routing information attacks	NA	− Hash Chain Authentication [16]
	Proprietary protocol attacks (serial-based protocols attack)	NA	− Access control; vendor updates and honeypots [14]
	Modbus & BACNet attacks(serial-based protocols)	NA	− Network security; product updates; protocol-specific defenses; protocol improvements [14]
	WEP attacks (WiFi attack)	NA	− Update to WPA2 or WPA3; Update the WiFi security protocol [14]
	WPA/WPA2/WPA3 Attacks (WiFi attack)	NA	− Network security; wireless equipment patching, custom SSID names, WiFi best practices, strong passwords, disabling WPS functionality; firewalls [14]
	ZLL attacks, Ghost-in-Zigbee Zigbee DoS, KillerBee	NA	− Configuration; software/firmware updates; disable nonce reuse; protocol [14]
	Jamming	NA	− Defenses depend on the wireless protocol used [14]
	Sinkhole attack	NA	− Intrusion Detection; Hash Chain Authentication [16]
	Wormhole attack	NA	− Clustering-based Intrusion Detection System [16]
	Sybil attack	NA	− Trust aware Protocol [16]
	Man in the middle attack	NA	− Inter-device Authentication; Secure MQTT [16]
	Dos/DDos attack	NA	− SDN based IoT framework; EDos Server [16]

and privacy. In this work, we studied the concept of Smart Campus in the framework of Smart Universities, namely IoT and technologies to make universities Smart with emphasis on the challenge of security which remains a research focus. Vulnerabilities and different types of attacks that threaten the systems of Smart Universities were studied, as well as the proposed solutions to overcome these attacks. After the analysis of the different solutions, it was deduced that the security mechanisms of traditional computer systems are not effective for the security of smart systems, due to the characteristics of the IoT that requires more advanced techniques such

Table 3 Overview of attacks and mitigation measures related to the physical layer of IoT devices

IoT layers	Security attack	Type of attack	Proposed solutions
Perception layer (devices)	Malicious scripts	PA	– Malware image classification [16]
	Spyware, virus, and worms	PA + SA	– Lightweight framework; high-level synthesis [16]
	Supply chain attacks	PA	– Novel secure supply change architectures; RFID tagging during manufacturing and transport [14]
	Reset sequence attacks	PA	– Access control; limiting device access; device monitoring [14]
	Node capture attacks	PA	– Loss prevention practices; limiting device access; device monitoring; access control [14]
	EM attacks	PA	– Directional antennas; best installation practices [14]
	Video side-channel	PA	– Software-based solutions may help mitigate several video-based side-channel attacks [14]
	Sensory side-channel attacks	PA	– Solutions to sensory attacks as a dedicated machine learning framework [14]
	Battery exhaustion attacks	PA	– Rate limiting solutions; IDS frameworks such as MVP-IDS and B-SIPS; IDS-based solutions designed to protect against battery exhaustion [14]

as Blockchain and XAI. As perspectives, we will base ourselves on these two techniques to propose a solid and credible solution to secure IoT systems, so we will implement the proposed solution in our Smart university model.

References

1. B. Sánchez-Torres, J. Rodríguez-Rodríguez, D. Rico-Bautista, C. Guerrero, Smart Campus: Trends in cybersecurity and future development. Revista Facultad de Ingeniería **27** (2018). https://doi.org/10.19053/01211129.v27.n47.2018.7807
2. G. Ikrissi, T. Mazri, A study of smart campus environment and its security attacks. ISPRS-International Archives of the Photogrammetry, Remote Sensing and Spatial Information Sciences XLIV-4/W3-2020:255–261 (2020). https://doi.org/10.5194/isprs-archives-XLIV-4-W3-2020-255-2020
3. D. Rico-Bautista, G. Maestre-Gongora, C.D. Guerrero, Smart University:IoT adoption model, in *2020 Fourth World Conference on Smart Trends in Systems, Security and Sustainability (WorldS4)* (2020). pp. 821–826
4. K. Lawal, H.N. Rafsanjani, Trends, benefits, risks, and challenges of IoT implementation in residential and commercial buildings. Energy Built Environ. (2021). https://doi.org/10.1016/j.enbenv.2021.01.009
5. S. Haque, S. Zeba, Md Alimul Haque, et al., An IoT model for securing examinations from malpractices. Mater. Today: Proc. https://doi.org/10.1016/j.matpr.2021.03.413
6. D. Al-Malah, H. Jinah, H. Alrikabi, Enhancement of educational services by using the internet of things applications for talent and intelligent schools. Period. Eng. Nat. Sci. (PEN) **8**, 2358–2366 (2020)
7. R. Bdiwi, C. de Runz, S. Faiz, A.A. Cherif, A blockchain based decentralized platform for ubiquitous learning environment, in *2018 IEEE 18th International Conference on Advanced Learning Technologies (ICALT)* (2018). pp. 90–92
8. A. Kumari, R. Gupta, S. Tanwar, Amalgamation of blockchain and IoT for smart cities underlying 6G communication: A comprehensive review. Comput. Commun. **172**, 102–118 (2021). https://doi.org/10.1016/j.comcom.2021.03.005
9. L. Hong-tan, K. Cui-hua, B. Muthu, C.B. Sivaparthipan, Big data and ambient intelligence in IoT-based wireless student health monitoring system. Aggress. Violent. Beh. **101601**(2021). https://doi.org/10.1016/j.avb.2021.101601
10. H. Habibzadeh, T. Soyata, B. Kantarci et al., Sensing, communication and security planes: A new challenge for a smart city system design. Comput. Netw. **144**, 163–200 (2018). https://doi.org/10.1016/j.comnet.2018.08.001
11. Williams R, McMahon E, Samtani S et al., Identifying vulnerabilities of consumer Internet of Things (IoT) devices: A scalable approach, in *2017 IEEE International Conference on Intelligence and Security Informatics (ISI)* (2017). pp. 179–181
12. Sok K, Colin J-N, Po K (2018) Blockchain and Internet of Things Opportunities and Challenges
13. Q.-D. Ngo, H.-T. Nguyen, V.-H. Le, D.-H. Nguyen, A survey of IoT malware and detection methods based on static features. ICT Express **6**, 280–286 (2020). https://doi.org/10.1016/j.icte.2020.04.005
14. L.P. Rondon, L. Babun, A. Aris et al., Survey on enterprise Internet-of-Things systems (E-IoT): a security perspective. Ad Hoc Netw. **125**, 102728 (2022). https://doi.org/10.1016/j.adhoc.2021.102728
15. K. Kimani, V. Oduol, K. Langat, Cyber security challenges for IoT-based smart grid networks. Int. J. Crit. Infrastruct. Prot. **25**, 36–49 (2019). https://doi.org/10.1016/j.ijcip.2019.01.001
16. J. Sengupta, S. Ruj, S. Das Bit, A Comprehensive survey on attacks, security issues and blockchain solutions for IoT and IIoT. J. Netw. Comput. Appl. **149**, 102481 (2020). https://doi.org/10.1016/j.jnca.2019.102481
17. F. Al-Turjman, J.P. Lemayian, Intelligence, security, and vehicular sensor networks in internet of things (IoT)-enabled smart-cities: an overview. Comput. Electr. Eng. **87**, 106776 (2020). https://doi.org/10.1016/j.compeleceng.2020.106776
18. M.A. Khan, K. Salah, IoT security: Review, blockchain solutions, and open challenges. Futur. Gener. Comput. Syst. **82**, 395–411 (2018). https://doi.org/10.1016/j.future.2017.11.022
19. M. Mohamad Noor, binti, Hassan WH, Current research on Internet of Things (IoT) security: a survey. Comput. Netw. **148**, 283–294 (2019). https://doi.org/10.1016/j.comnet.2018.11.025

20. M.G. Samaila, C. Lopes, É. Aires et al., Performance evaluation of the SRE and SBPG components of the IoT hardware platform security advisor framework. Comput. Netw. **199**, 108496 (2021). https://doi.org/10.1016/j.comnet.2021.108496
21. K. Prathapchandran, T. Janani, A trust aware security mechanism to detect sinkhole attack in RPL-based IoT environment using random forest—RFTRUST. Comput. Netw. **198**, 108413 (2021). https://doi.org/10.1016/j.comnet.2021.108413
22. R. Paul, N. Ghosh, S. Sau et al., Blockchain based secure smart city architecture using low resource IoTs. Comput. Netw. **196**, 108234 (2021). https://doi.org/10.1016/j.comnet.2021.108234
23. E. Corradini, S. Nicolazzo, A. Nocera et al., A two-tier Blockchain framework to increase protection and autonomy of smart objects in the IoT. Comput. Commun. **181**, 338–356 (2022). https://doi.org/10.1016/j.comcom.2021.10.028
24. S.A. Latif, F.B.X. Wen, C. Iwendi et al., AI-empowered, blockchain and SDN integrated security architecture for IoT network of cyber physical systems. Comput. Commun. **181**, 274–283 (2022). https://doi.org/10.1016/j.comcom.2021.09.029

The Impact of Big Data and IoT for Computational Smarter Education System

Asma Jahangeer, Ahthasham Sajid⬥, and Afia Zafar

Abstract In modern era of cyber net and latest smart technology i.e., Internet of things (IOT) has played an essential role in every field of life. IOT is known as one of the pillars of 4th industrial revolution (4th IR). Connectivity using different sensor and objects used in daily mankind life are connected to internet that makes communication easy for human to human, human to machine and machine to machine now a day. Many applications of Big data and IOT including smart home, smart hospitals, smart shopping mall, and smart education system have been introduced in past few years. With the passage of time education system and learning methods are massively transformed due to latest technology specifically by use of IOT in education. Using IOT in education is an enhance level of learning opportunity for students. Use of Big data and IOT helps teachers to ass's students in a better way and can take better decisions to fulfill students learning, from data collected by IOT devices to store large amount of data using Big data tools. Administrative staff of educational institutes can take benefit from IOT devices to manage their task more efficiently. Along with advantages there are some challenges to fully apply IOT in field of education but no doubt that the adoption of IOT has completely changed the model of education system in all over the world.

Keywords Smart education · IOT · Big data · Cyber

A. Jahangeer
Faculty of Information and Communication Technology, Department of Information Technology, BUITEMS, Quetta, Pakistan

A. Sajid (✉)
Faculty of Information and Communication Technology, Department of Computer Science, BUITEMS, Quetta, Pakistan
e-mail: ahthasham.sajid@buitms.edu.pk

A. Zafar
Department of Computer Science, NUTECH University, Islamabad, Pakistan

M. Ouaissa et al. (eds.), *Big Data Analytics and Computational Intelligence for Cybersecurity*, Studies in Big Data 111,
https://doi.org/10.1007/978-3-031-05752-6_17

269

1 Introduction

In the era of twenty first century everyone is part of fast-moving technology over the cyber net. People are surrounded by IOT everyday even if they do not notice it. What comes to mind first when they think of IOT? Possibly wearables or smart home devices like smart refrigerators, But IOT is not only limited to these devices. People are likely to not even notice the presence of Internet of Things solutions in their life, but they are nearly everywhere. The Internet of Things provides connectivity for people at any place. With the progress in technology, people are moving towards a society, where everything will be connected through internet. The Big Data and IOT is considered as the future evaluation of the Internet. The question is that what is M2M learning? The answer to this question is that M2M learning stands for machine-to-machine learning. Machine to machine defines to direct communication between devices using any communications channel, like wired and wireless [1].

The basic idea of IOT is to permit autonomous and secure connectivity and exchange of data and applications. The IOT links with real life works and physical devices with the computer-generated world. There are many viruses, hackers, malicious programs which effected the IOT devices so security of IOT is very important. Without security IOT is not powerful. Because all malicious programs, viruses, hackers attack on devices which are connected through internet. In past, no one was paying attention on security issue. Day by day everybody knew and notice that security of IOT devices is very important. Fourth industrial revolution has digitalized the technology. Along with artificial intelligence, block chain, IOT has an important role in this digitalization [2].

Like the other fields of life IOT is an emerging technology in the field of education. With the help of IOT devices in education system, students can participate and interact with other people like teachers, others students and with admin staff in their institutes. IOT brings a lot of possibilities, it helps to develop better Infrastructure of educational institutes as well as improve learning process of students. By using IOT based smart devices higher education institutes can be smarter. Today IOT is a ubiquitous technology, as per estimation by the end of 2020 more than 26.66 billion IOT devices have been active and this number is not stopped here. By the end of 2025 almost 152,200 IOT devices will be able to connect to internet every single minute. Figure 1 has given a representation that how in terms of devices vs. people over the internet has changed drastically after the innovation of internet of things. In 2003 there were allot of human beings' user of the internet whereas there are only few numbers of devices connected to internet; after 2008 with the introduction of IOT this ratio started to changed and we have a good number of devices connected to the internet as human started to carry smart handled devices with them. Right now, in 2020 scenario of connectivity is totally changes; as we now have over 50 billion devices connected to internet which is a huge number if we compare it with the people who are using internet now a days [11].

Basic purpose of IOT is to bring automation in every task in major industries such as IOT is enabled in home automation, in clothing industry, in health care, in

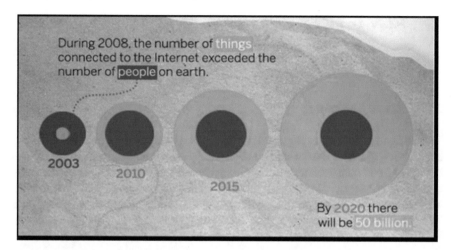

During 2008, the number of things connected to the Internet exceeded the number of people on earth.

2003

2010

2015

By 2020 there will be 50 billion.

Fig. 1 IOT growth over time [20]

agriculture and in education. IOT devices communicate with each other with the help of sensors which is heart of these all devices. Number of small sensors are embedded in single device. All the data from these devices is gathered and collected on cloud. After analytics these devices become actionable. So far there are many IOT devices are introduced such as smart glasses, smart watches, smart door lock, smart fire alarm system, medical sensors and smart security systems. Scanmarker, LOCOROBO, KALTURA, Blackboard are IOT based devices which are specifically used in Education sector. Use of IOT in education brings a valuable change [14].

2   Architecture of Smart Learning Environment

Twenty first century is known as an era of technology, where technology providers have ambition to provide cheap affordable technology to every individual. Now a day's smart devices are becoming smaller and smarter, easy to use and have seamless network. Its living example is our smart phones which are smaller in size but greater in work. Architecture of smart computing is rapidly changed. It has 3-tier architecture, by name cloud computing, fog computing, and swarm computing. The 3-tiers work as companions. Learning applications may have different components that run in 3-tier (Cloud, fog and swarm). All 3 tiers depend on each other to manage and control the resources and to analyze learning content [12].

2.1   3-Tier Architecture

The three-tier architecture is presented below.

2.1.1   Cloud Computing

The 1st layer or inner most layer is known as cloud computing. It provides software as a service (SaaS). It has centralized data storage system along with many remote servers and software networks. Its centralized server makes accessibility of resources and computer services possible. In smart learning environment cloud computing provide centralized data storage, platform visualization, and software services such as smart content, smart pull and push and smart prospect.

2.1.2   Fog Computing

Fog computing have very important role in IOT. Diverse type of services can be produced by it. It has virtualized platform that provide storage, networking services and compute services to data centers in traditional cloud computing and to end devices. Its platform is highly virtualized. In smart learning environment fog computing facilitates with real time interaction, provide support for mobility, location awareness, large scale sensors networks and so on.

2.1.3   Swarm Computing

In 3 tier architecture outer most layer is known as swarm computing. It is also now as environment a wear computing. It executes on swarm of smart devices. As the technology is becoming ubiquities and pervasive, swarm computing envisions to develop an environment that can sense activities on learning devices. Collect data from devices and transfer the data for analysis to data management system.

3   Key Functions of Smart Computing 5 A's

There are five intelligent functions that make computing technology smarter. These smart functions include 5 A's which are actions, audibility, Alternativeness, analysis and awareness. Actions and audibility are functions of cloud computing, Alternativeness and analysis are functions of fog computing and awareness is key function of swarm computing. The all functions are discussed as below:

3.1 Action

Actions can be taken by executing process. These actions can be used by the learners such as to locate places for example campus direction.

3.2 Auditability

Audit report is generated to determine that either learning actions are correct or not. In smart education there is need to make learning more efficient which is possible by analyzing data of learning activities, monitoring, capturing and tracking it at each stage and evaluate. The purpose of audit is to bring improvement in actions.

3.3 Alternativeness

Alternatives can be made to identify that learning flow either through human review or by automatic tasks. Learning pattern depends on alternative course of action.

3.4 Analysis

Different data mining, learning analysis and big data tools are used to analyze real time data received from learner's device. After analysis resources and learning patterns are recommended to learner.

3.5 Awareness

We should be aware of this thing that learning process can be taken place anywhere and anytime. Students' identity, location, condition, status can be captured by swarm computing. For example, by learning analytics, data mining, pattern recognition. This data then can be transfer to central server from learner's device.

4 Smart Environment

The computing paradigm earlier was not that much effective in terms of computational and connectivity prospective (Wireless Communication) capabilities, but

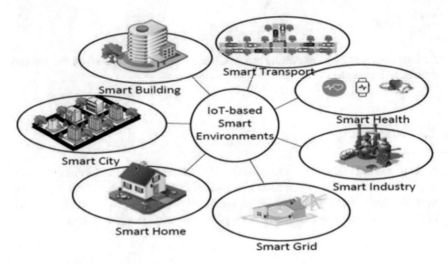

Fig. 2 IOT based smart environments [21]

with the introduction of new paradigm "Connected" everywhere, everyone at any time with objects carried out by the human being with built-in sensing capabilities has tremendously increases an environment to become smarter indeed. IOT systems currently providing such smarter environment with various effective performance characteristics on the other hand also facing some key challenges.

Wireless connectivity is one of the major ingredients in almost every IOT-enables connectivity providing smart features. Challenge is in the selection of right wireless technology for the system so that smart features can be delivered 24/7 (everywhere, everyone at any time) in true spirit [15].

Figure 2 illustrate that with the use of sensor technology and handheld smart devices usage among human being; now a-days almost every environment connected with society (Cities, Offices, Internet of Things Transportation Social Life & Environment Home Automation Cities Office Automation Energy Conservation Health Education Home, Buildings, Transportation, Education) etc. are possibly connected in smart environment. Smart environment could be controlled remotely with smart handheld devices [16].

5 IOT Applications

After the invention of sensing technology with various sensors available now a days in market for almost every field, applications of IOT has also attracted almost every industry belonging to society and human being. Figure 3 bellow has given an insight to this in detail.

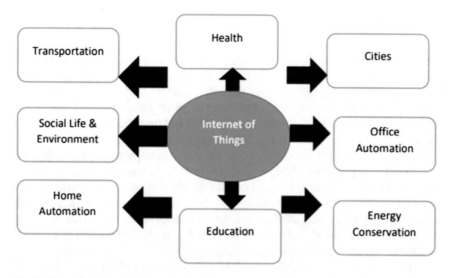

Fig. 3 IOT applications

5.1 Significance of IOT in Education

Use of IOT in education means creating smarter education system which provide structured and an advance value to the system. People get familiar with use of IOT at smart home by using same technology we can get better infrastructure of school buildings. IOT base devices help to reduce men power and save cost. For example, by using smart lightening system in campuses, HR do not need to hire a person for this job instead lights can be turn on and off automatically by sensing a human body in surroundings by this way less electricity will be used. similarly, during the lectures instructor can set the light system to be bright or dim by change of light in classroom students will be more attentive [8]. There are many institutes that have already implemented smart light system and they got better results and found students more productive. IOT devices can be used for security purpose such as smart alarm system that can alert everyone in the building about misfortune or gun shooting can be sense by special purpose sensors. Significance of IOT based devices in education provide secure learning environment. Student performance can be better by use of IOT objects inside the classes especially in circumstances like COVID-19 when learning take place remotely. According to student needs smart devices in the class can be personalized and facilitate students. This application recommends helping material to students so an improve learning environment is able to develop [9].

5.2 Smart Education

Smart education is a term that is being used for the process of applying modern technology in the field of education such as converting teaching into smart teaching, managing educational process with the help of smart devices and use of technology in research area. By implementation of smart education can be a way to improve quality of education, bringing new ideas and innovations in the field and a chance to change traditional way of education. Smart education can change historical process. Like there are eight planets there are eight parts of smart education system has been proposed which are personalized and diverse learning path, culture, skills, resources, global integration of knowledge, economy, technical immersion and system [7]. Online intelligent tutor system, learning analytics, e-learning, knowledge graphs and big data are few directions in smart education. There are many universities which already have developed IOT based smart education system, few of the universities are as following: University of North Carolina, Stanford University, University of Southern California and they have implemented smart INSTRUCT system, MMAP collaborative teaching model teaching system and RIDES Intelligent Tutoring System respectively. Educational Robot, smart campus along with virtual teachers and parallel education are also few examples of already system [17].

5.3 Challenges of Smarter Education

There are few challenges in smart education system such as security, Data visualization, connectivity and prediction system, these are discussed below in detail.

5.3.1 Security

IOT applications in smart education collect data from teaching staff, students and other faculty members, this data is vulnerable to attacks and cause serious security and privacy issue for individuals. Therefore, more research is needed in this aspect [18].

5.3.2 Data Visualization

Data generated from the smart device is very big in size it is very important to take insight from the data collected from smart educational devices, although there are many research papers that have presented data visualization techniques but their need to be correlate already manually collected data and huge amount of data collected from the smart devices. The purpose of smart devices is to provide ease and comfort to users not to create hap hazard with big size of data [18].

5.3.3 Connectivity

Speed of internet depends on the traffic, as there are number of students connected to internet in smart campuses at the same time, so it is discovered that the speed of communication is slow. There is need of a quick protocol such as combination of AR or VR with RFID or any other sensor. These types of protocols introduced more elements and able to respond more quickly as compare to traditional protocols [18].

5.3.4 Prediction Systems

Purpose of smart system is to provide automation, analysis and collect better results, for this purpose there is need to build better prediction modeling algorithm, by this way better results can be collected. For example, by seeing students fee submission it can be predicted that number of student's width drawl or by analyzing student's performance their grades can be predicted. By this analysis correct measures can be taken and resources can be provided for student's improvement [18].

5.4 Effect of Smart Classroom on Student Achievement at Higher Education

A classroom can be an ideal classroom if educators can deliver an engaging session. Students' capabilities can be improved by tech-based learning which is only possible by providing infrastructure of smart classroom. A smart classroom provides better learning environment because this novel atmosphere is more attractive and interactive for the learners [4].

A smart classroom impact student learning but how can we actually define a smart classroom? A smart classroom is defined as room having sensors that can collect data on real time, a computer and an audiovisual system that gives response to instructions. With the help of audio-visual equipment, variety of media can be used by the teachers. By use of internet in the classroom students can get vast knowledge about subject. It increases student interest and bring dynamic perspective to the education. Online learning makes students independent as they can explore about their topics. They can find answers without help of instructors. It makes them more confident and enhance their curiosity for learning. By use of smart devices for learning students can think out of the box. They can analyze, develop better understanding and collect results about tools used in the class [5].

Every student sitting in the classroom have different IQ. They have different way of learning and understanding things. With the help of different sources and media they are not bound to one book. Student's academic performance can be improved by use of technology as it is challenging for students to visualize the concept without technology. A better learning environment that can develop interest and engage students can be achieved by smart classrooms in higher education [19].

5.5   Existing Applications of IOT in Education

There are some of the existing applications of IOT based technology that has implemented and running successfully across the globe.

5.5.1   Smart Lightening System in the Campus

This smart lightening system in the campus help to reduce energy consumption and by saving this energy campus can easily implement the ideal environment of green campus [3]. Smart wearables can help to collect data. Most of these wearables such as hand band or foot wear are connected to web apps and after processing of data feedback is provided to students [8].

5.5.2   Smart Entrance System

The IOT based smart education system have upgraded the learning environment. Manual system of student entrance in campus gates take a lot of time due to security purpose but IOT based RFID ((Radio-frequency Identification) tags can be applied on student's ID cards by that they can enter the campus with ease, these RFID's help to mark an automated attendance. Universities administration can also manage access control of different entrance gates and doors. IT can increase security and enable a comfortable environment for students. Students can use smart devices to reserve their seats in campus library, gymnasium and at auditorium [6].

5.5.3   Student's Healthcare

Fitness band are used to monitor the temperature of students during the school time. It also gathers data for e healthcare and monitor students blood pressure and rate of heart beat. Students having high or low blood pressure or any other illness can be treated in the campus immediately [10].

5.5.4   Smart and an Enhanced Learning Process

Sony smartwatches and google glasses are used to monitor understanding level of students during classes. By collecting data from these smart device's instructor can facilitate students with personalized explanation about any topic. Student behavior can be measured by smart camera.

Let's discuss some of smart devices which are already applied in classrooms.

5.5.5 Scanmarker

Scanmarker is amazing invention of IOT; Its name presents it scan text on the documents, book, magazine and scan it on word or excel document. It works as highlighter all you have to do is scan the text on the book. It has features of text to speech, image reader, and multiple language translation. While you scan you can listen and read the text at same time that prepares students to read fast and absorb quickly. Just imagine transfer of text from physical book to your computer system by just one scan [14].

5.5.6 LOCOROBO MY Loopy

My loopy is first programmable robot for students of early years established by LOCOROBO company. It is designed for learning purpose. My loopy is an interactive way to teach STEM courses to young learners. It develops code writing skills in children because loopy love to response when it is touched by human. Similarly, it response to light, temperature and sound [14].

5.5.7 Kaltura

KALTURA is a video editing tool for kids. If any child wants to edit video KALTURA helps to edit and produce line and on demand video. By this way student can learn tool of video editing and image processing that develops a skill which can be very beneficial for them in future. Integrating IOT in education makes system much easier than ever before, institutes that have established digitized technology have higher value proposition such personalized learning, collaboration, safety and lower cost as compare to traditional systems [14].

5.6 General Challenges of IOT in Education

Along with many benefits there are some challenges that are faced while implementing IOT based infrastructure at educational environment.

IOT based education system have challenges as well some are discussed below:

- Converting traditional system to an IOT based smart system required support from teachers, students and other staff members for that they need to be little tech savvy and should know how to use these smart device and applications. Utilization and managing these resources in the classroom is very important otherwise all effort is useless.

- Making an ordinary system a smart IOT based infrastructure is very costly especially for institutes that have low revenue. In this regard government support is required.
- Security of IOT devices is vulnerable from the beginning although a lot of work has been done for achieving high security but there is still security issue with IOT devices.
- Always required fast and reliable internet connection.
- IOT based data has issue of scalability as data collected from all the devices used in school is so big in volume.
- IOT devices increasing health issue in students and making them less social with in their community. That impact on student's social well-being [13].

6   Conclusion

The role of information technology to improved educational academics activities in society internationally or nationally could not be ignored now days, especially from last 10 years. Internet has shifted from IOP (Internet of People) to IOT (Internet of Things) nowadays. Education could be provided with ease and accessibility on timely and efficient manner with use of smart handheld devices connected over Internet of Things in twenty-first century. Recently an epidemic of COVID 19 has given rise to online education activities in the world; unfortunately, those countries which did not have good internet connectivity and infrastructure suffers allot in delivery of quality education to (School, Colleges and Universities even). A smart educational environment is therefore encouraged with the use of various information technology devices and infrastructure for the quality of education in twenty-first century and in future.

References

1. D.D. Ramlowat, B.K. Pattanayak, Exploring the internet of things (IoT) in education: a review. Information Systems Design and Intelligent Applications (Springer, Singapore, 2019), pp. 245–255
2. M. Bagheri, S.H. Movahed, The effect of the Internet of Things (IoT) on education business model. 2016 12th International Conference on Signal-Image Technology & InternetBased Systems (SITIS). IEEE (2016)
3. M. Maksimović, Transforming educational environment through Green Internet of Things (GIoT). Trend **2017**(23), 32–35 (2017)
4. L. Banica, E. Burtescu, F. Enescu, The impact of internet-of-things in higher education. Sci. Bull. Econ. Sci. **16**(1), 53–59 (2017)
5. M. Abdel-Basset et al., Internet of things in smart education environment: supportive framework in the decision-making process. Concurr. Comput. Pract. Exp. **31.10**, e4515 (2019)

6. J. Marquez et al., IoT in education: Integration of objects with virtual academic communities. New Advances in Information Systems and Technologies. Springer, Cham, pp. 201–212 (2016)
7. J. Bhatt, A. Bhatt, *IoT techniques to nurture education industry: scope & opportunities* (International Journal On Emerging Technologies. Uttarakhand, India, 2017), pp. 128–132
8. A. Zhamanov, Z. Sakhiyeva, M. Zhaparov, Implementation and evaluation of flipped classroom as IoT element into learning process of computer network education. Int. J. Inf. Commun. Technol. Educ. (IJICTE) **14**(2), 30–47 (2018)
9. M.B. Abbasy, E.V. Quesada, Predictable influence of IoT (Internet of Things) in the higher education. Int. J. Infor. Educ. Technol. **7.12**, 914–920 (2017)
10. H. Aldowah et al., Internet of things in higher education: a study on future learning. J. Phy. Conf. Series. **892**(1) (2017)
11. A. Oke, F.A.P. Fernandes, Innovations in teaching and learning: exploring the perceptions of the education sector on the 4th industrial revolution (4IR). J. Open Innov. Technol. Market Compl. **6**(2), 31 (2020)
12. S. Gul et al., A survey on role of internet of things in education. Int. J. Comp. Sci. Net. Sec. **17.5**, 159–165 (2017)
13. A. Alghamdi, S. Shetty, Survey toward a smart campus using the internet of things. 2016 IEEE 4th international conference on future internet of things and cloud (FiCloud). IEEE, (2016)
14. M. Thomas, The Connected Classroom: 9 Examples of IOT in Education. Bulitin. (2020) https://builtin.com/internet-things/iot-education-examples
15. C. Gomez et al., Internet of things for enabling smart environments: a technology-centric perspective. J. Amb. Intell. Smart Environ. **11.1**, 23–43 (2019)
16. E. Ahmed et al., Internet-of-things-based smart environments: state of the art, taxonomy, and open research challenges. IEEE Wirel. Communic. **23.5**, 10–16 (2016)
17. C. Heinemann, V.L. Uskov, Smart university: literature review and creative analysis. International Conference on Smart Education and Smart E-Learning (Springer, Cham, 2017)
18. A.C. Martín, C. Alario-Hoyos, C.D. Kloos, Smart education: a review and future research directions. Multidisc. Dig. Publ. Inst. Proce. **31**(1) (2019)
19. S.Y. Phoong et al., Effect of smart classroom on student achievement at higher education. J. Educ. Technol. Syst. **48.2**, 291–304 (2019)
20. https://www.social-iot.org/d/Tutorial_IC2E2016%20-%20final.pdf
21. E. Ahmed, I. Yaqoob, A. Gani, M. Imran, M. Guizani, Internet-of-things-based smart environments: state of the art, taxonomy, and open research challenges. IEEE Wirel. Commun. **23**(5), 10–16 (2016). https://doi.org/10.1109/MWC.2016.7721736

Transformation in Health-Care Services Using Internet of Things (IoT): Review

Safia Lateef, Muhammad Rizwan, and Muhammad Abul Hassan

Abstract In the twenty-first century, there is a significant change in every aspect of life, such as socially, cultural-wise, economically and health-care-wise, etc., due to many advancements in technology. The most important IoT (Internet of Things) is a new technological paradigm that allows the devices to connect in a network consisting of physical objects or things embedded with electronics, software, sensors, and connectivity to enable objects/devices to exchange information with other connected objects/devices. The healthcare industry gains the most benefits from the uproar of IoT. Therefore, it is hard to underestimate the place of the Internet of Things in healthcare services. This paper addresses the use of the Internet of Things (IoT) in healthcare, its applications, and challenges and indicates long-term benefits in healthcare services provided for human beings.

Keywords Internet of things · Big data · Health care

1 Introduction

The Internet of Things (IoT) is transforming the way of living and working. As the field is growing rapidly and is more relying on smart devices and technology for day-to-day lives. IoT with smart devices/objects are connected and exchange a large amount of data. IoT is once a very small and metier field, but connected devices have outnumbered humans. According to the reports more than 20 billion devices

S. Lateef (✉) · M. Rizwan
Department of Computer Science, Kinnaird College for Women, Lahore, Pakistan
e-mail: safialateef0609@gmail.com

M. Rizwan
e-mail: muhammad.rizwan@kinnaird.edu.pk

M. A. Hassan
Department of Computing and Technology, Abasyn University, Peshawar 25000, Pakistan
e-mail: abulhassan900@gmail.com

© The Author(s), under exclusive license to Springer Nature Switzerland AG 2022
M. Ouaissa et al. (eds.), *Big Data Analytics and Computational Intelligence for Cybersecurity*, Studies in Big Data 111,
https://doi.org/10.1007/978-3-031-05752-6_18

Fig. 1 Smart system and IoT

are connected to IoT and the business has exceeded by 300 billion dollars till 2020 [1, 2].

Internet of Things can be defined as a network of physical objects/devices that can interact with each other to share information and take the required action. The term was introduced by Kevin Ashton in 1999 and the concept of IoT become popular at the Auto-ID Center, MIT [2]. IoT is a network collection of physical devices like home appliances, vehicles, sensors, trackers, and network connectivity that enables the devices to exchange data/ information. It provides organized and accumulated data and resources to people to utilize according to their requirements. On a grand scale, IoT consists of billions of devices and sensors connected over the internet to exchange data continuously. Smart systems and IoT are driven by a combination of the following shown in Fig. 1:

1. Sensors/actuator
2. Connectivity
3. People rocesses

The interaction between these entities is creating new types of smart applications and services. Every industry needs better access, more accurate and real-time data to enhance decision-making. This premise also applies to health-care IoT, due to which patients can remain in touch with the doctor even after leaving hospital premises. Companies that specialize in healthcare or technology tend to heavily invest in IoT. Normally, doctors diagnose the disease and give treatment to a patient in person with help of medical staff in a clinic or hospital. Internet of Things (IoT) enables the health practitioner to remotely monitor the patients and can deliver long-term care. Patients not only are satisfied by doctors as they can contact them easily and efficiently but also, they engage with their care [3].

IoT helps patients to take appropriate measures in case of any emergency. IoT is revolutionizing healthcare services by broadening the horizon for doctors to provide their help and minimizing the medical expenditure for check-ups and medical devices. Nowadays, people use wearable devices that are a part of IoT. These devices help in monitoring physical and physiological activities like a heartbeat, stress level, and many more things [4].

The core focus of this paper lies in the application of IoT in health care and the challenges by different papers are portrayed in the following sections. The main objective of this paper is to provide a collective analysis of several research and review-based studies and explain the core concepts of each one of them.

2   Overview of IoT in Health-Care

The Internet of Things (IoT) is not just limited to the conventional usage of laptops, smartphones, and other gadgets. It can connect a vast set of devices by utilizing embedded-based automation. The devices are connected over internet connectivity and make remote monitoring and exchange of data/information possible [5].

The use of the Internet of Things in health care presents an exciting opportunity to help a large number of people. IoT can be used to enable people with cognitive, sensory, and motor impairments, help people with an illness or injured by accident, support the doctors and medical staff. Machines will not only help in tasks, but also, they act as a substitute for doctors and medical staff for diagnosing and providing the circumstantial treatment [6]. The Fig. 2 shows glimpse of IoT in healthcare:

Internet of Things (IoT) in health care can be referred to as medical devices introduced in the IoT framework. The devices integrated with IoT being used in health care are the Internet of Medical Things (IoMT). IoMT includes intelligent monitoring devices, sensors, electronics, and many more.

The providence of IoT in traditional medical devices empower the doctors to observe efficiently and regularize many things for patient care. IoMT is the key to the future of precision medicine. It enabled the doctor to remotely track patient health, achieving goals like streamlined care, fewer hospital visits and admissions, cost, and efficiency. In addition, patients having chronic illnesses like cardiac problems, high blood pressure, diabetes, and other chronic diseases can benefit from long-term monitoring and increase their life span.

Fig. 2 IoT in health-care

Table 1 Areas covered by IoT

Surgeons	Hospitals	Patients
Improved patient care	Increased efficiency	Shorter hospitalization
Enable complex tasks	Reduce costs	Reduced pain
Improved patient care	Increased efficiency	Shorter hospitalization
Less blood loss	Reduce litigation	Faster recovery time
Smaller incisions care	Marketing tool	Smaller incisions, resulting in re-duced risk of infection, blood loss and scarring
Elminiation of tremor		No Pre-surgery personal blood donation

According to the well-known saying "Health is wealth," there is a need to improve the healthcare facilities and provide sustainable treatments. The researchers are developing and implementing more efficient IoT frameworks to provide early diagnosis and immediate first aid. Apart from providing the best treatment and care for patients, IoT also helps in reducing health expenditure. As per reports, healthcare integrated with the Internet of Things (IoT) market is estimated to be worth 158.1 billion dollars in 2022 [7]. After the studies of the impact of IoT in different fields of life economically, healthcare has the highest percentage of investment with 40 percent of total expenses done in IoT-related subjects [8] The Table 1 shows the important areas covered by IoT technology.

The traditional medical treatment procedure is time-consuming and costly. Thus, by integrating the IoT technology, these two factors can be controlled efficiently, and the hospital system will be upgraded to an automated system like taking appointments, consultation, paying bills, medicine purchasing, collection of reports, and many more [9].

The ways in which Internet of Things is being used in health-care service are shown in Fig. 3 and listed below:

1. Keep record of patient medical history, personal data and medical staff inventory.
2. Manage chronic diseases.
3. Decrease the medicinal procedure cost and time.
4. Remote monitoring of patients using wearable devices.
5. Monitoring the use of drugs and medicine.

Record of Patients and Medical Staff   Medicine & Drug Management   Decrease the Time & Cost of Treatment   Remote Monitoring   Manage Chronic Diseases

Fig. 3 Uses of IoT in health-care

3 Literature Review

Over the years, many state-of-the-art publications have been made on healthcare IoT systems as a whole. Several researchers studied the mechanism of IoT healthcare systems, and others studied their usefulness to society. Studies such as Selvaraj Sureshkumar and Suresh Sundaravaradhan in (Challenges and opportunities in IoT healthcare systems: a systematic review) [11–14], Pranati Rakshit, Ira Nath and Souvik Pal in (Application of IoT in Healthcare) [15], Amit Banerjee, Chinmay Chakraborty, Anand Kumar, Debabrata Biswa in (Emerging trends in IoT and big data analytics for biomedical and health care technologies) [16–20], Shashank Kumar, Arjit Kaur Arora, Parth Gupta, and Baljit Singh Saini in (A Review of Applications, Security and Challenges of Internet of Medical Things) [21], G. Arun Sampaul Thomas and Y. Harold Robinson in (Real-Time Health System (RTHS) Centered Internet of Things (IoT) in Healthcare Industry: Benefits, Use Cases and Advancements in 2020) [22], Matthew N. O. Sadiku, Shumon Alam, and Sarhan M. Musa in (Internet of Things in Healthcare) [26], Pijush Kanti Dutta Pramanik, Bijoy Kumar Upadhyay, Saurabh Pal, Tanmoy Pal in (Internet of things, smart sensors, and pervasive systems: Enabling connected and pervasive healthcare) [33], and many more have focused their studies on the application of IoT in health-care.

Internet of Things (IoT) provides efficient monitoring of patients from the area and provides emergency services. It changes the way facilities are provided and delivered by the healthcare industry. These technologies improved the product by bringing minor to more significant changes. It introduces new tools and devices and fills in the gaps of the healthcare industry. Many devices are invented to monitor the basic vitals at the ease of home. It is evident that the future of medicine and healthcare solely depends on the advancement of IoT. The benefits of integration of IoT in healthcare revolves around better treatments, ease of remote monitoring, cost and time saving, better outcomes for the patients, and timely diagnosis. Some of the factors that are identified by researches are as follows:

- Remote Health Monitoring
- Wearable Devices
- Elderly Care
- Digital Hospitals

IoT enabled the healthcare system to provide and deliver better health facilities compared to the traditional ways and anywhere. It has opened up whole new possibilities of a world where the applications are ten-folds from consultation to diagnosis. It has revolutionized health facilities with intelligent devices like wearable devices, Bluetooth-enabled wearable, mood enhancers, glucose monitors, movement detectors, and many other things. For individuals with chronic diseases like cardiac, cancer, and high blood pressure, wearable devices help them keep updated about their vitals and the need to go to the hospital. The primary goal is to enhance the quality and efficiency of healthcare [23, 24].

4   Architecture of Health-Care IoT

The IoT healthcare system can be defined as a wireless network that provides advanced services by connecting multiple devices in the same domain of the network [10]. The healthcare IoT system is integrated with different IoT functionalities to gain a common healthcare goal. Apart from monitoring, diagnosing, and performing simple tasks, the integrated systems are designed to quickly and efficiently perform complex tasks. The architecture of the IoT healthcare system is divided into three parts, as shown in Fig. 4:

- Physical Layer
- Network Layer
- Application Layer

4.1   Physical Layer

The physical layer is the foundation of the healthcare IoT system. It consists of sensors, wearable devices, actuators, and many more to collect the patient's vital health information. They are connected with other medical devices that store the data of patients working individually or simultaneously. The information collected can be the blood pressure, stress level, heartbeat, temperature, and many other things. This layer uses the technology of RFID, GPS, cameras, smart device sensors, and infrared sensors [3, 10, 11, 14].

4.2   Network Layer

The network layer is responsible for the communication of devices. Its stores and processes the data from the physical layer. Communication mainly uses short and

Fig. 4 Architecture of IoT
in health

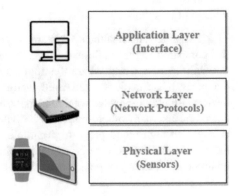

long-range technology like RFID, Bluetooth, ZigBee, WiFi, and GPS. It includes a router, controllers, and network protocols. The information can be stored locally or remote storage like a cloud [25, 27, 28].

4.3 Application Layer

The application performs tasks after interpreting the data from the physical and network layers. This layer indulges in continuous monitoring and can be seen using a smartphone/tablet, smartwatches and others [11, 14, 29].

5 Technologies of IoT in Health-Care

When building an integrated healthcare system using the Internet of Things (IoT), the use of technologies is crucial as it enhances the ability of the IoT health system [3, 30]. The technologies are categorized into following and shown in Fig. 5:

- Identification Technology
- Communication Technology
- Location Technology

5.1 Identification Technology

This technology considers the authentication of patients' data coming from remote areas. For identifying nodes (sensors), a unique Id is assigned to an entity to authorize the nodes and lower the chance of ambiguous data exchange. The unique Id is

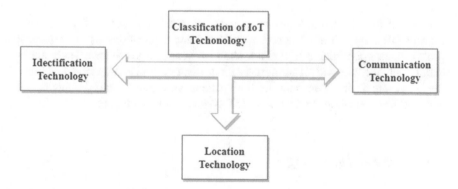

Fig. 5 Classification of technology in health IoT

assigned to every entity related to the healthcare system like doctors, medical staff, medical vehicles, and many more. It ensures the authorized exchange of data and identification among the digital domain. Even the sensors and actuators are also given a unique Id which helps in their proper functioning [31].

5.2 Communication Technology

This technology ensures the flow of communication among the entities present in the network. They are highly dependent on the communication ranges of two types between devices and servers:

- Short range technology (Bluetooth, ZigBee, RFID, WiFi and Body Area Network)
- Long range technology (LTE, WiMAX)

In most health IoT system short range technologies are preferred [7, 32].

5.3 Location Technology

This technology is used to track the location of ambulances, medical staff, and patients within the healthcare network. The most commonly used technology is the global positioning system GPS. It uses satellites to track its position. In case of environmental obstruction between satellites and objects, a local positioning system LPS can be used. The LPS uses short-range technology like RFID (Radio Frequency Identification), Bluetooth, ZigBee [34].

6 Applications of IoT in Health-Care

The rise of IoT is exciting for everybody due to its different scope of use in every aspect of life. In the healthcare sector, IoT has many applications that give medical professionals opportunities to monitor and give treatment to patients. These applications help increase patient comfort, facilitate doctors' decision-making, and make the healthcare environment safer for both patients and medical staff. Following are some of the applications of healthcare IoT systems shown in Fig. 6:

6.1 Remote Monitoring

It is one of the most common applications of healthcare IoT. IoT devices collect the data of patients like temperature, blood pressure, heartbeat, stress level, and more

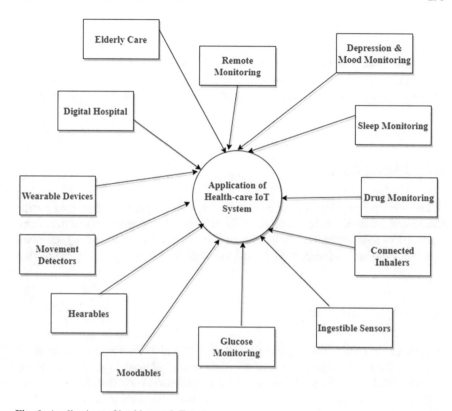

Fig. 6 Applications of health-care IoT system

who are not physically present in the hospital's vicinity. It eliminates the need to travel to the hospital for the patient.

The IoT device collects the data and forwards it to the respective medical staff with the help of an application so that they can view it. If there is any problem with the patient's vitals, an alert will be issued that medical help is needed for the patient. Remote monitoring can also be known as telehealth. If mobility is a problem for the patient, telehealth makes much difference [10, 15, 19, 35].

6.2 Depression and Mood Monitoring

Traditionally it is difficult to tell the symptoms of depression in a person only by a one-on-one session with the healthcare professional. Thus, Mood-Aware IoT devices are introduced to collect the data about heartbeat and blood pressure; even some devices tell the movement of eyes, which helps infer the mental state of a person.

6.3 Sleep Monitoring

Sleep monitoring includes different parameters like temperature, blood pressure, body movement, pulse, and others. Good sleep is a sign of good health; thus, by monitoring the sleep, one can maintain the routine and sleep comfortably and naturally [36].

6.4 Elderly Care

When older people live on their own, there are higher chances of declining health, chances of accidents, and other aspects. The RFID sensor system in IoT devices helps them to live independently. The IoT devices help in keeping track of daily activities, and in any case of a mishap, an alert can be sent to the hospital and relatives [28, 42].

6.5 Digital Hospital

Traditional hospitals are becoming a point of dissatisfaction and challenging to manage. The lengthy medical procedure, paperwork, and time-consuming treatment overload medical staff's work and tire down the patients. Due to this, IoT integration with health care turned everything around. The whole medical process has become intelligent, efficient, and more accessible. The process of making appointments and consultation to treatment has become meticulous [19, 37].

6.6 Wearable Devices

Wearable devices can collect detailed information about a patient by monitoring the vitals throughout the day while being just a smartwatch, smartphone, or any other attachment. The data gathered from the devices can be analyzed to predict one's health. They support fitness, disease management, medication management, and many other things. The wearable devices are best for chronic health patients like cardiac, cancer, diabetes, and Alzheimer [4].

6.7 Drug Monitoring

One of the exciting things about the healthcare IoT system is drug monitoring. The pills have microscopic sensors that send signals to an external device to ensure proper usage and dosage. Also, a label is attached to the medicine that has a unique Id which is difficult to forge and can stop medical fraud and helps as a countermeasure to prevent deaths a result of wrong medication [8, 38].

6.8 Connected Inhalers

Connected inhalers are the best IoT devices for asthma patients. It helps them to keep track of asthma attacks and also alert them to take the inhaler with them in need of any attack.

6.9 Ingestible Sensors

These sensors are also a breakthrough in the healthcare IoT system as no one wants to have a camera or probe being stuck in the body, and also it is very messy to collect data from the body. Ingestible sensors are small enough that they can be swallowed easily. These sensors help monitor the symptoms and provide an early warning for any mishap. The main advantage of these sensors is that they are less invasive [8, 28, 39].

6.9.1 Hearables

Hearables are hearing aids for people who suffer from hearing problems. They are compatible with Bluetooth/WiFi that syncs with smartphones/tablets and smart-watches. It helps filter, equalize, and add layered features to natural world sound.

6.9.2 Mooadables

Mooadables are the new IoT devices that help enhance mood throughout the day. It sounds like movie fiction, and all about it is far from reality [10, 15, 40, 41].

6.9.3 Glucose Monitoring

Glucose monitoring helps diabetic patients to keep a record of their glucose levels throughout the day. It eliminates the traditional record-keeping and alerts the patient or doctor if the glucose level rises. Another IoT device for diabetic patients is an insulin pen which records the time, amount, and type of insulin taken by them [8].

6.9.4   Movement Detectors

Movement detectors are primarily used for physically disabled people to trace their movements. For this purpose, the sensors and devices are attached to their clothes, bed, or body to track the movement. It also assists in monitoring involuntary gestures and gives deeper insight for better medical treatment [42].

Smart healthcare services results in the creation of applications, so that medical staff and patients can easily use the applications. The Table 1 shows different IoT bases healthcare applications:

7   Challenges of IoT in Health-Care

Healthcare IoT system can be revolutionary and highly efficient and has their share of challenges. Health facilities must recognize these challenges to pave down the road of advancement more efficiently. Following are some of the challenges faced by the healthcare IoT system as shown in Table 2.

Table 2 IoT healthcare applications

Condition	Sensors	IoT Operation
Heartbeat monitoring	Capacitive electrodes on electric circuit	Information is transmitted to wireless transmitter using Bluetooth and WiFi
Blood pressure monitoring	Wearable sensors of blood pressure	Measurement, automatic inflation and oscillometric
Diabetes monitoring	Smart phone camera and Opto physiological sensor	Segmentation and measurement of diabetic level
Temperature monitoring	Wearable Sensors	Measurement of skin-based temperature
Medication management	Wireless biomedical sensors suit	Diagnosis and prognosis of essential records which are recorded using GPS, RFIDs and multimedia transmission
Oxygen monitoring	Pulse oximeter wrist	Intelligent detection of pulse time by time
Distant surgery	Surgical robot sensors, Augmented reality	Robotic arms, master controller, information management

7.1 Security

Internet of Things (IoT) is a network of interconnected devices, so like every other network, it is also prone to security threats and attacks. In addition, the hospital and other medical facilities have medical data related to patients, which is very sensitive. Thus, any hospital or medical facility must adhere to robust security policies and guidelines.

7.2 Acceptability and Adoption

The acceptability of advanced healthcare is a significant concern as there is a huge gap in awareness and understanding of data collecting of patients in the generations. People still believe in the traditional way of treatment and medical procedure. The advantages of IoT are somehow clear to some people, but still, a certain amount of people does not understand the value of IoT in health care. On the other side, the transition of the traditional medical procedures to advanced integrated IoT healthcare systems is time-consuming and costly, especially for small health facilities [43].

7.3 Data Management

Data management is one of the biggest challenges in healthcare IoT as the devices generate many data. Some devices store the data for a certain amount of time or send it to other network parts. The data coming from devices is primarily unstructured as it comes from multiple devices. Removing redundancy, process data, and identification of authorized data can be time-consuming and affect the whole network.

7.4 Data Overload

The medical IoT devices generated a large amount of data to give deeper insight into patients' vitals, helping the doctor make a better decision. The rate at which data is produced is enormous, and it burdens the network and lowers down efficiency and effectiveness of IoT. In addition, it delays the processing and feedback of processed data making it difficult for the medical staff to proceed with treatment.

7.5 Integration

Integration of IoT devices is intimidating as there are many potential and complex medical devices. Moreover, device manufacturers have not standardized any rules or policies on integrating the devices to make a robust platform.

8 Conclusion

Like in other aspects of life, the Internet of Things (IoT) also changes the way facilities, procedures, and treatment are delivered by health care. The evolution of IoT in healthcare has a better outcome for patients, a better work environment for medical professionals, and a better patient experience. It gives better record-keeping, patient monitoring, diagnosis, and treatment solutions. Doctors can easily detect the disease, and treatment can be delivered as fast as possible by using this service. This service plays a huge role in time-saving and reducing the cost of the whole medical procedure. The healthcare IoT system opens many opportunities as it moves forward, enabling the medical facilitates to give the best treatments. It is redesigning healthcare with many benefits.

References

1. R.S. Batth, A. Nayyar, A. Nagpal, Internet of robotic things: driving intelligent robotics of future-concept, architecture, applications and technologies. 2018 4th International Conference on Computing Sciences (ICCS). IEEE (2018)
2. B. Chander, G. Kumaravelan, Internet of things: foundation. Princi- ples of Internet of Things (IoT) Ecosystem: Insight Paradigm (Springer, Cham, 2020), pp. 3–33
3. B. Pradhan, S. Bhattacharyya, K. Pal, IoT-based applications in healthcare devices. J. Healthcare Eng. (2021)
4. S.Y. Tun, S.M. Yint, F. Mirza, Internet of things (IoT) appli- cations for elderly care: a reflective review. Aging Clin. Exp. Res. 33(4), 855–867 (2021)
5. R. Mahajan, P. Gupta, Implementation of IoT in Healthcare. Handbook of Research on the Internet of Things Applications in Robotics and Automation (IGI Global, 2020), pp. 190–212
6. V. Kavidha, N. Gayathri, S. Rakesh Kumar, AI, IoT and robotics in the medical and healthcare field. AI and IoT-Based Intelligent Automation in Robotics (2021), pp. 165–187
7. A. Ahad et al., Technologies trend towards 5G network for smart health-care using IoT: A review. Sensors 20.14, 4047 (2020)
8. D. Castro et al., Survey on IoT solutions applied to Healthcare. Dyna 84.203, 192–200 (2017)
9. A. Banerjee, C. Chinmay, M. Rathi Sr., Medical imaging, artificial in- telligence, internet of things, wearable devices in terahertz healthcare technologies. Terahertz Biomedical and Healthcare Technologies (Elsevier, 2020) pp. 145–165
10. B. Pradhan et al., Internet of things and robotics in transforming current-day health- care services. J. Healthcare Eng. 2021 (2021)
11. J.T. Kelly et al., The internet of things: impact and implications for health care delivery. J. Med. Int. Res. 22.11, e20135 (2020)

12. K. Yang et al., Federated machine learning for intelligent IoT via reconfigurable intelligent surface. IEEE Network **34.5**, 16–22 (2020)
13. J. Azeta et al., A review on humanoid robotics in healthcare (2017)
14. S. Selvaraj, S. Sundaravaradhan, Challenges and opportunities in IoT healthcare systems: a systematic review. SN Appl. Sci. **2**(1), 1–8 (2020)
15. P. Rakshit, I. Nath, S. Pal, Application of IoT in healthcare. Principles of Internet of Things (IoT) Ecosystem: Insight Paradigm (Springer, Cham, 2020), pp. 263–277
16. R. Dhaya, R. Kanthavel, F. Algarni, Research perspectives on applications of internet-of-things technology in healthcare WIBSN (wearable and implantable body sen- sor network). Principles of Internet of Things (IoT) Ecosystem: Insight Paradigm (Springer, Cham, 2020), pp. 279–304
17. G. Bodur, S. Gumus, N.G. Gursoy, Perceptions of Turkish health profes- sional students toward the effects of the internet of things (IOT) technology in the future. Nurse Educ. Today **79**, 98–104 (2019)
18. R. Shah, A. Chircu, IoT and AI in healthcare: a systematic literature review. Issues Inf Syst. **19.3** (2018)
19. A. Banerjee et al., Emerging trends in IoT and big data analytics for biomedical and health care technologies. Handbook of data science approaches for biomedical engineering (Academic Press, 2020), pp. 121–152
20. An Approach Towards IoT-Based Healthcare Management System Khushboo Singla, Rudra Arora, and Sakshi Kaushal
21. S. Kumar et al., A review of applications, security and challenges of internet of medical things. Cognitive Internet of Medical Things for Smart Healthcare (2021), pp. 1–23
22. G.A.S. Thomas, Y. Harold Robinson, Real-Time Health System (RTHS) Centered Internet of Things (IoT) in Healthcare Industry: Benefits, Use Cases and Advance- ments in 2020. Multimedia Technologies in the Internet of Things Environment (Springer, Singapore, 2021), pp. 83–93
23. M.K. Slifka, J.L. Whitton, Clinical implications of dysregulated cytokine production. J. Mol. Med. (2000). https://doi.org/10.1007/s001090000086
24. J. Smith, M. Jones Jr., L. Houghton et al., Future of health insurance. N. Engl. J. Med. **965**, 325–329 (1999)
25. J. South, B. Blass, The future of modern genomics (Blackwell, London, 2001)
26. M.N.O. Sadiku, S. Alam, S.M. Musa, Internet of things in healthcare. IoT and ICT for Healthcare Applications (Springer, Cham, 2020), pp. 21–32
27. M.N.O. Sadiku et al., Emerging IoT technologies in smart healthcare. IoT and ICT for Healthcare Applications (Springer, Cham, 2020), pp. 3–10
28. S. Anmulwar, A.K. Gupta, M. Derawi, Challenges of IoT in health-care. IoT and ICT for Healthcare Applications (Springer, Cham, 2020), pp. 11–20
29. S.R. Guntur, R.R. Gorrepati, V.R. Dirisala, Robotics in healthcare: an internet of medical robotic things (IoMRT) perspective. Machine learning in bio-signal analysis and diagnostic imaging (Academic Press, 2019), pp. 293–318
30. M.A. Azzawi, R. Hassan, K.A.A. Bakar, A review on Internet of Things (IoT) in healthcare. Int. J. Appl. Eng. Res. **11.20**, 10216–10221 (2016)
31. J. Kaivo-Oja, S. Roth, L. Westerlund, Futures of robotics. Human work in digital transformation. Int. J. Technol. Manage. **73**(4), 176–205 (2017)
32. P. SDixit et al., Robotics, AI and IoT in medical and healthcare applications. AI and IoT-Based Intelligent Automation in Robotics (2021), pp. 53–73
33. P.K.D. Pramanik et al., Internet of things, smart sensors, and pervasive systems: enabling connected and pervasive healthcare. Healthcare data analytics and management (Academic Press, 2019), pp. 1–58
34. D. Castro et al., Survey on IoT solutions applied to Healthcare. Dyna **84.203**, 192–200 (2017)
35. A.R. Patel et al., Vitality of robotics in healthcare industry: an Internet of Things (IoT) perspective. Internet of Things and Big Data Technologies for Next Generation Healthcare (Springer, Cham, 2017), pp. 91–109

36. J. Kim, G.M. Gu, P. Heo, Robotics for healthcare. Biomedical engineering: frontier research and converging technologies (Springer, Cham, 2016), pp. 489–509
37. L.D. Riek, Healthcare robotics. Communications of the ACM **60.11**, 68–78 (2017)
38. B.K. Chae, The evolution of the Internet of Things (IoT): a computational text analysis. Telecommun. Policy **43.10**, 101848 (2019)
39. B. Mohanta, P. Das, S. Patnaik, Healthcare 5.0: a paradigm shift in digital healthcare system using artificial intelligence, IOT and 5G communication. 2019 International Conference on Applied Machine Learning (ICAML) (IEEE, 2019)
40. O. Vermesan et al., Internet of things cognitive transformation technology research trends and applications. Cognitive Hyperconnected Digital Transformation: Internet of Things Intelligence Evolution (2017)
41. S.S. Mishra, A. Rasool, IoT health care monitoring and tracking: A survey. 2019 3rd international conference on trends in electronics and informatics (ICOEI) (IEEE, 2019)
42. A.S. Yeole, D.R. Kalbande, Use of Internet of Things (IoT) in healthcare: a survey. Proceedings of the ACM Symposium on Women in Research 2016 (2016)
43. M.W. Woo, J.W. Lee, K.H. Park, A reliable IoT system for personal healthcare devices. Fut. Generat. Comp. Syst. **78**, 626–640 (2018)
44. K.H. Rahouma, R.H.M. Aly, H.F. Hamed, Challenges and solutions of using the social internet of things in healthcare and medical solutions—a survey. Toward Social Internet of Things (SIoT): Enabling Technologies, Architectures and Applications (Springer, Cham, 2020), pp. 13–30

A Survey of Deep Learning Methods for Fruit and Vegetable Detection and Yield Estimation

Faiza Aslam, Zia Khan, Arsalan Tahir, Kiran Parveen,
Fawzia Omer Albasheer, Said Ul Abrar, and Danish M. Khan

Abstract Computer vision has a great potential to deal with agriculture problems. It is crucial to utilize novel tools and techniques in the agriculture food industry. The focus of current studies is to automate the fruit harvesting, grading of fruits, fruit recognition, and identification of diseases in the agriculture domain using deep learning and computer vision. Integrating deep learning with computer vision facilitates the consistent, speedy and trustworthy classification of fruit and vegetables compared to the traditional machine learning algorithm. However, there are still some challenges, such as the need for expert farmers to develop large-scale datasets to recognize and identify the problems of agriculture production. This survey includes eighty papers relevant to deep learning and computer vision techniques in the agriculture field.

Keywords Deep learning · Object detection · Computer vision · Yield estimation

1 Introduction

With the growing population, it is necessary to increase the supply of fruit or vegetables as with the increasing demand for fruit and vegetables. Fruit and vegetable are essential for a healthy diet. They are a good source of vitamins, fiber, minerals and

F. Aslam · A. Tahir · K. Parveen
Department of Computer Science, COMSATS University, Islamabad, Pakistan

Z. Khan (✉) · D. M. Khan
Department of Electrical and Electronic Engineering, Universiti Teknologi Petronas, Seri
Iskandar, Malaysia
e-mail: zia.aseer@gmail.com

F. O. Albasheer
Department of Computer Sciences, University of Gezira, Gezira, Sudan
e-mail: fawzia.omer@uofg.edu.sd

S. Ul Abrar
Department of Computer Science, University of Peshawar, Peshawar, KPK, Pakistan
e-mail: saidulabrar@aup.edu.pk

nutrients that keep us healthy and prevent diseases. Food production is crucial to preserve their color, taste, texture, and shape in a specific period [1]. The conventional farming system suffers from a lack of labor, which causes increased challenges in farms [2]. In agriculture field various task using object detection method with the support of robot guidance like harvesting and detection of diseases in plants [3].

In traditional research, most of the work is done manually, such as experts are higher to assess the quality for the inspection of food or a crop. However, the manual task has some flaws like human mistakes lack of knowledge about the characteristics of fruit and vegetable. For this reason, an efficient, consistent system requires that is suitable for the recognition task. The agriculture industry uses an automated system for detecting fruit and vegetable, including pre-harvesting and post-harvesting mechanisms of crops, which mainly depends upon computer vision techniques. Computer vision plays a vital role in agriculture, which exploits fruit classification, fruit harvesting, catalogue tools, and fruit supervision in markets [4]. However, it is crucial to discriminate the fruit based on its visual appearance among various lightning conditions with complex backgrounds [5].

To address these problems, computer vision introduces various algorithms and techniques that are proposed by the various researchers for grading the fruits, such as classification, segmentation and feature extraction, which automate the industrial field, remove the manual authentication of food and increases the quality and inspection of fruit using the guidance of robots [6]. Some authors have focused on the individual fruits to classify them accurately. They discussed the 3-category of oranges. Each category has its properties like color, taste, size and cost. Automatic classification of various fruits is a challenging task in [7]. Fruit detection is a crucial task and state-of-the-art challenge. Multi-Task Convolution Neural Network (MTCNN) is the most popular technique which has made progress for object recognition and classification to precisely target the object-like fruits with superior performance in terms of accuracy and time utilization [8].

In another line of research, post-harvest quality measurement is essential for plant phenotyping and ranking fruits, which helps to calculate the grading of better or poor, fresh or damaged fruits. The convolution Neural Network (CNN) approach was adopted to identify the disease and defects in fruits, specifically in peaches [9], lemons [10], pear [11] and blueberries [12]. Faster Region base Convolution Neural Network (Faster-RCNN) with ResNet101 trained on Common Object in Context (COCO) datasets and designed to detect the green tomato plant with high precision and minor error [13]. The outlook features of fruit like color, shape, and size essentially matter among supermarkets' trading, classification and grading. Cherry usually grows in the form of pairs and clusters. The uneven shape of the cherry causes the disorder during the development and less profitability in markets. The fruit becomes damaged after a specific period. Hence, an efficient algorithm is required to preserve the food from damage and increase its selling rate [14]. A semi-supervised approach was utilized with the combination of U-Net and Faster RCNN models for the yield estimation of detection and counting of apples in the orchard. U-Net was employed for the segmentation task while CNN counted fruit on the individual image dataset. The proposed methodology achieved a higher F1-score, which relies on the technique that has been deployed [15]. The list of Abbreviations is shown in Table 1.

Table 1 List of abbreviations

Abbreviation	Full form
AI	Artificial Intelligence
ANN	Artificial Neural Network
ANN-ABC	Artificial Neural Network-Artificial Bee Colony
ANN-HS	Artificial Neural Network-Harmony Search
AP	Average Precision
CNN	Convolutional Neural Network
CHT	Circular Hough Transformation
COCO	Common Object in Context
CycleGAN	Cycle Consistent Generative Adversarial Network
DasNet	Detection and Segmentation Network
FCM	Fuzzy c-Mean clustering algorithm
FDR	Fruit detection and recognition
GLCM	GLCM & Grey Level Co-occurrence Matrix
GMM	Gaussian Mixture Model
ICCSP	International Conference on Communication and Signal Processing
WGISD	Wine Grape Instance Segmentation Dataset
UAV	Unmanned Aerial Vehicle
SSD	Single Shot Multibox Detector
MTCNN	Multi Task Convolution Neural Network
MRE	Mean Residual Error
NIR	Near Infra-red
SSAE	Stacked Sparse Auto-Encoder
ML	Machine learning
DL	Deep Learning

2 Background Study

Computer vision is applicable in various agriculture fields like production, monitoring, and harvesting the crop. However, there are still some issues raised technological issues, farming automation, environmental influences, building scalable datasets. Therefore, it is necessary to develop a public database to overcome the agriculture challenges [16]. In another line of research, some spatial challenges discussed innovative farming land, automated sensors, Robot farming with the help of exploiting computer vision in agriculture [17]. In agriculture, automated tools and techniques facilitate food grading, fruit harvesting, and production rate to strengthen and preserve the fruit prolong time. In this scenario, the researchers concluded that using image processing various filters and techniques [18] with feature selection process [19] performed accurate classification with the help of novel architecture

Voronoi diagram base class and neuro-fuzzy architecture [20] for the recognition of fruits. Accurate classification and recognition of fruits using machine vision and computer vision techniques have been challenging, considering various circumstances such as the choice of accurate sensors, environmental influences, and heterogeneous variation between interclass and intraclass of fruits [21]. One of the drawbacks of computer vision is designing a dataset that is time-consuming and increases the computational cost. From the literature, it has been analyzed that environmental factor directly influences the detection rate of fruit and vegetable, it also identifies the disease present in soybean. The model could generate different results using similar computer vision techniques on a related dataset under variations in environmental conditions [22].

In this research, CNN architecture is based on the sliding convolution, insufficient for the multi-classification labelling. It only deals with binary classification. Hence multi-layer classification problem is still challenging and crucial to extend the dataset in [23]. Automated estimation of fruit harvesting detection of fruit ripeness accurately is still a challenging and laborious task. In previous studies, machine vision was utilized to estimate the fruit ripeness; now, deep learning with multiple features accounts for promising results [24, 25].

2.1 Computer Vision and Agriculture

Precision Horticulture (PH) is the most trending technology utilized to maximize yield estimation, preserve fruits from diseases, and automate fruit harvesting in orchards [26]. In similar research, various algorithms were adopted for the fruit harvesting robots. With the development in agriculture and current imaging technologies, most of the information is visualized in a better way and precisely target the fruits that assist the fruit recognition process also support the growth of fruit picking. The quality of the fruits detection system depends upon the various light conditions, stroke, and the environment in which the robot survives with suitable sensors. Expert farmers are the basic need of the farms. It is crucial to collect precise information about the growth of the crop. Due to the manual system, agriculture industries face many problems such as labor and time cost, lack of knowledge, and less experience of the workers causes the reduction of farming in orchards as discussed in [27].

Most of the focus of this research is to measure fruit quality. In-depth analysis, the computer vision and image processing comparison reported for fruit and vegetable quality assessment in the food industry, looks at the image feature and segmentation problem. The analysis of fruit and vegetables relies on the color, shape, size, texture, and disease identification [28]. Deep learning has made tremendous progress in the past few years in agriculture. CNN architecture used computer vision techniques to identify the potato disease in plants. The performance of the proposed architecture varies with the ratio of training data 90–96 using database images in the tomato field [29]. According to this study [30], Generative Adversarial Network (GAN) and CNN were introduced to recognize diseases in plant leaf with the support of android apps.

In similar research, AlexNet with SequeezNet was deployed to detect 9 various kinds of diseases in tomato farms.

The proposed model is trained over the dataset of the plant village. The results taken over the AlexNet, which achieved an accuracy of 95.65 while SequeezNet attained an accuracy of 94.3 with less computational resources. To the best of this research [23], CNN considerably has better classification accuracy than Support Vector Machine (SVM) for the real-time identification of diseases in plants. The result showed that using a cloud-based system proposed model trained over the 1030 images yielded an accuracy of 93.4 on pomegranate and 88.7 on Firecracker images. In this survey, DensNet with 152 layers was proposed to classify multiple diseases in 14 kinds of plants accurately. DensNet achieved 99.75 accuracy on the plant village dataset using less number of parameters. However, it improved the computational time and performed best compared to the other architectures such as VGG16, ResNet-50, ResNet-101, and ResNet-152 and Inception-v4 in [31]. In similar research, tomato disease was identified using plant village data, and the model AlexNet estimated higher accuracy than VGG16 with feasible computational time in [32]. Transfer learning was utilized to detect tomato and sugar beet plants accurately. It also compared six different kinds of convolutional network architectures such as ResNet-101, ResNet-50, AlexNet, inception-v3, Google Net, and VGG19 under the consideration of various lighting conditions. The experiments declared that AlexNet showed a higher accuracy of 98.0 while VGG19 estimated 98.7 accuracy in [33].

3 Literature Review of Surveys

In this survey, Machine Learning (ML) presented tremendous progress in various agriculture applications such as disease detection, weed detection, preserving the crop from diseases and, most commonly, prediction and yield estimation. Artificial Neural Network (ANN) exploited for this purpose. However, using ML, new methods are proposed to save the agricultural food products, urging in [34]. In another research, deep learning was employed to address the challenge of food manufacturing in the agriculture domain. Besides the approaches mentioned above, deep learning achieved better accuracy and precision in the case of classification and regression problems. It also reduced the regression error. Despite extensive training, deep learning emerged in the agriculture field to solve various problems [35–37]. In similar research, a mixture of ML and image processing techniques was developed to facilitate an automated system for the precise recognition and grading of fruit and used to discriminate the fruit based on fruit appearance, diversity and maturity level. Moreover, image processing facilitates continuous, sterilized rapid growth in the fruit industry [38].

In similar research, Faster-RCNN cope with inception v2 and single-shot multi-box detector cope with Mobile-Net deployed for counting fruit containing 3 categories like Avocado, lemon, and Hass. The experiments showed that Faster-RCNN efficiently performed with 93.1 accuracy compared to MobileNet estimated 90 accuracy while counting fruit [39]. This survey analyzed computer vision techniques to address professional challenges in various fields of agriculture. Unmanned Aerial vehicle technique utilized to keep track of crop development, disease precaution, automate the harvesting and quality evaluation of agriculture products as discussed in [16]. This research concludes that scarcity of datasets is a common problem because newly developed RGB-D sensors have not been utilized to classify fruit [21]. This research briefly analyzed the quality inspection of fruit or vegetable-based on texture, pattern, color, size and shape characteristics. Besides the advancement of computer vision, multi-dimension images were not utilized for the quality evaluation of fruit or vegetable. Only a single image focused on the grading of fruit. A generic framework for classification, segmentation, sorting and grading on multiple fruits is required [6]. In reference [40], various image processing techniques with CNN mainly focus on three approaches of fruit detection, quality assessment control and fruit classification. This study also supports robot harvesting. The complete Literature Review is summarized in Table 2. The results showed that CNN and pretrained network explicitly outperformed for these tasks and achieved almost 100 accuracy. This survey analyzed that computer vision and ML integrated to solve the problem of agriculture domain and performed the brief analysis of seed, crops and fruits also improved their quality in [41]. ML, along with artificial intelligence, performed the agriculture supply chain assessment. Various ML algorithms were utilized to develop the permissible agriculture supply chain, which increased their yield [42].

3.1 Deep Learning Framework for the Detection of Fruit and Vegetable

Deep learning methods have been commonly used in recent research to successfully detect various kind of fruits.

A : Artificial Neural Network (ANN)

Prediction of the vineyard for better yield is a necessary and challenging task to estimate the productivity rate in viticulture at various vineyard zones. ANN, combined with the association of vegetation index and vegetation fraction, was covered using computer vision techniques to address these problems. The proposed methodology is based on remote sensors and Unmanned Aerial vehicles (UAV), facilitating prior pre-diction instead of ground base measurement [67, 68]. Many authors have studied the various aspects of fruit classification on a public dataset on RGB images. In a similar study, the author presented the classification of 18 different categories of fruit using a computer vision algorithm. The proposed scheme showed that 99.8 accuracy were achieved on fruits like strawberries, blueberries, blackberries, pineapples, green grapes, red grape, black grape, and cantaloupes using Feed Forward Neural Network (FNN) with a deep learning algorithm [43]. This study presented a classification of three varieties of oranges using hybrid Artificial Neural Network—Artificial Bee Colony (ANN-ABC) with an accuracy of 97, Artificial Neural Network—Harmony

Table 2 Literature review table

Authors	Article publication	Fruit/ Veg	Datasets	Techniques used	Results
Zhang, Yudong [43]	2016 Wiley	18 fruits	1653 images	FNN, Deep Learning	99.88
Stein [44]	Sensors, 2016	Mangoes	RGB and NIR images	Faster RCNN	Error rate of LIDAR 1.36
Sa and Ge [45]	Sensors, 2016	Sweet pepper	RGB and NIR images	VGG-16	F1-Score 0.838
Cen [46]	ASABE 2016	Cucumber	Hyperspectral images	CNN-SSAE	91.1 and 88.6
Tan [47]	Multimedia tools and application 2016	Melon	Skin lesion images captured by Infrared video	5-layer CNN LeNet-5 B-LeNet-4	Accuracy 97.5 and recall 98.5
Jawale [48]	2017 (ICCSP)	Apple		ANN	94.94
Zaborowicz [49]	Scientia Horticulturae 2017	Tomato		ANN	98.50
Rahnemoonfar [50]	Sensors, 2017	Tomato	Tomato synthetic images	Inception-ResNet	91–93 accuracy
Bargoti [51]	Journal of Field Robotics, 2017	Apple	N/A	RGB images VGG-16	0.791 F1-Score
Chen [52]	IEEE robotics and automation letters 2017	Apple and Orange	Distinct dataset green apple and orange	Two CNN model	Oranges 0.813, Apples 0.838
Cavallo [53]	Journal of food engineering 2018	Ice berge Lettuce	320 images	CNN	86.00
Wajid [54]	(iCoMET) 2018	Orange	335 images	Naive Bayes, ANN, Decision Tree	93.45
Oo [55]	Biosystem engineering 2018	Strawberry	337 strawberry sample	ANN	90.00

(continued)

Table 2 (continued)

Authors	Article publication	Fruit/ Veg	Datasets	Techniques used	Results
Zhang [56]	EURASIP Journal 2018	Banana	17,312 images with different ripening stages	CNN	95.6
Wang [12]	Sensors, 2018	Blueberry	Hyperspectral images	ResNet and ResNet	Accuracy 0.8844 F1-Score 0.8784, Accuracy 0.8952 F1-Score 0.8905
Habaragamuwa [57]	Environment and Food, 2018	Strawberry	RGB-D camera images	DCNN and Mask-RCNN	88.03–77.21
Williams [58]	Biosystem engineering 2019	Kiwi	RGB images modified	VGG-16	Harvesting 51 perc achieved
Yu, Zhang [59]	Computers and Electronics in Agriculture, 2019	Strawberry	2000 images	Mask-RCNN	95.78
Ganesh, [60]	IFAC Papers Online, 2019	Orange	RGB and HSV images	Mask-RCNN	97.53
Liu [61]	IEEE Access 2019	Kiwi	RGB-D and NIR image	VGG-16	90.7
Ge, Yuanyue [62]	IFAC Papers On Line, 2019	Strawberry	RGB-D images	DCNN	94
Altaheri [63]	IEEE Access,2019	Date fruit	8072RGB images	AlexNet and VGG-16	99.01–97.01,98.59
Lin and Tang [64]	Sensors, 2019	Guava	RGB-D images	VGG-16, Google Net	98.3–94.8
Barre [65]	Computers and Electronics in agriculture 2019	Grapevine	N/A	LSL, CNN model of epicuticular waxes	97.3
Munasingha [66], Tran et al. [67]	ICACT 2019	Papaya	Publicly available dataset	CNN	92
Tran [64]	Applied Sciences, 2019	Tomato	RGB images	Inception-ResNet [60], Autoencoder	87.273 and 79.091
Jahanbakhshi [75]	Scientia Horticulturae 2020	Lemons	RGB images	Three CNN with 11–16–18 layers	97.3

(continued)

Table 2 (continued)

Authors	Article publication	Fruit/ Veg	Datasets	Techniques used	Results
Santos [43]	Computers and Electronics 2020	Wine grapes	WGISD Public dataset of 300 RGB images	ResNet	F1-Score 0.91
Ballesteros [68]	Precision agriculture 2020	Vineyard	Multi-spectral and Hyperspectral images	ANN	28.7 RE, 12.8 RMSE

Search (ANN-HS) provided 94 accuracy. From the comparative analysis with the traditional K-Nearest Neighbor (KNN) approach, the proposed method has been a significant advantage over the KNN 70.88 in [7]. The Artificial Neural Network (ANN) architecture can be seen in Fig. 1.

B: DCNN

Recently, deep learning has made significant progress in object detection. In this research, cascaded CNN architecture used with augmentation method for better detecting fruit like apple images is collected in orchards. In addition, the image Net dataset was used to generate the dataset. The model is applied for the other

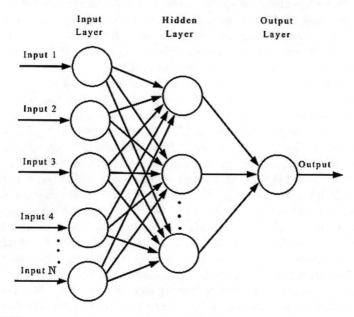

Fig. 1 ANN

type of fruit, such as strawberries and oranges, on the test dataset and achieved remarkable results [8]. Image acquisition has been made from various resources. Therefore, it is a difficult task to identify the object accurately. For this purpose, a novel Fruit Detection and Recognition (FDR) algorithm was proposed. Most of the well-known architecture CNN implemented precisely nominates the classification of fruit. The model performance showed high accuracy results using its dataset containing various images with less computational complexity [69]. In this research, hyperspectral images were exploited to classify fruit and vegetable using a pre-trained network with CNN of RGB data. A dataset of hyperspectral images is captured from the real images. The analysis estimates that Google Net with pretrained pseudo-RGB images is calculated from hyperspectral images achieved an average accuracy of 85.23, which was enhanced by using the compression of kernel module of 92.3 in [70].

The CNN model is efficiently designed to detect and recognize 60 categories of fruits. The model has been trained on the Fruit-360 dataset for early detection of fruits. Experiment results showed that 96.3 accuracy was achieved while training the NN. However, the model was not suitable for real-time application. It was just limited to Fruit-360 in [71]. Automatic identification of a defect in fruits analyzed exploiting CNN architecture such as tuta absoluta defect exists in tomato plants. For this purpose, 3 pre-trained networks like inception v3, VGG16, VGG19 and ResNet module were proposed. Inception v3 well performed with accuracy of 87.2 compared to the other model. These pretrained networks easily calculate the variation among the severity condition of tuta absoluta at low-, high, and no tuta.

The results showed that mango 88 accuracy, lime 83 accuracy, pitya 99 accuracy by the utilization of video streaming efficiently [72]. Fully convolution network developed for automatic detection and semantic segmentation of guava fruit and branches with 3D-pose estimation in the orchard, and accuracy of 0.893 with 0.806 IOU estimated of guava fruit on segmentation. Although on-branch segmentation it is a difficult task. However, their results performed better than the traditional algorithm for the detection of guava of 0.983 precision and 0.948 recall [64]. CNN architecture was implemented for precise on-branch-based fruit recognition using the PH method. The proposed algorithm is designed for real-time applications. For the experiments, data has been collected from six kinds of fruits: apricot, apple, nectarine, sour cherry, peach, and colored plums in orchards using RGB images. The proposed model attained an accuracy of 99.76 with 0.019 cross-entropy loss. Hence, it declared that the proposed technique efficiently works compared to the traditional approaches Yolo, ResNet, and VGG16 in [26]. In this paper, the CNN model is presented to accurately classify fruit and vegetable. In the addition of VGG architecture on a publicly available dataset, the model achieved 95.6 accuracy. This task is accomplished using the data preprocessing step, feature extraction method, and multiple classifiers to classify the images using different performance metrics [73]. With the fusion of two feature learning algorithms such as CNN and multi-scale multi-layered perceptron's, a pixel-wise fruit segmenting was proposed for the fruit detection using watershed segmentation and the Circular Hough Transformation (CHT) for the individual supervision of the fruits while counting the images captured

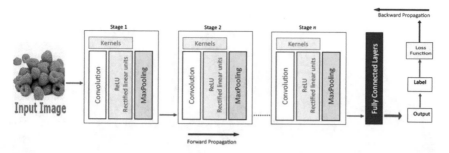

Fig. 2 DCNN

in orchards. The performance of the proposed model achieved the best results with watershed segmentation by the utilization of a squared correlation coefficient of 0.826 in [51].

This paper presented a framework of a deep convolution neural network trained on a small custom dataset pretrained on a large dataset for developing a high-performance fruit detection system. The authors utilized a faster region-based R-CNN network to combine two modules of RGB and Near Infra-red (NIR) images with early and late fusion enhanced the DCNN, and Fig. 2 shows the architecture of DCNN. The results showed that the proposed scheme gave better results than the conventional system [45]. DCNN presented a challenging classification of cherry fruit due to its irregular shape. The performance of the proposed algorithm is enhanced by the addition of hybrid max and average pooling. The results showed 99.4 accuracy using the data augmentation method is higher than traditional ML methods such as KNN, ANN, EDT, and fuzzy logic [14].

C: Mask-R-CNN

Mask R-CNN was adopted on the public dataset to target the three main problems in this research: object recognition, semantic segmentation, and instant segmentation [43]. In another line of research, Mask-RCNN with Feature pyramid architecture was exploited for automated identification of strawberry harvesting under various lighting conditions, occlusions, and complex backgrounds. The model's performance was evaluated over the 100 images and achieved 95.7 precision, 98.4 recalls with 0.89 intersections over union [59]. Mask-RCNN is specifically designed for the object detection and instant segmentation task for the pixel-wise detection of each fruit as shown in Fig. 3. The developed framework performed the experiments over the RGB and HSV data undergoing natural environmental conditions in orange orchards. The output of the model showed 0.89 F1-Score, including RGB and HSV images. Robot harvesting is one of the advantages of the mask segmentation approach [60]. The WGISD dataset is utilized to detect wine grapes using the ResNet architecture for counting and tracking fruits. On the other hand, bounding box techniques were applied for the object recognition task and targeted the grape cluster successfully using the structuring element hit and miss strategy with the precise shape and size

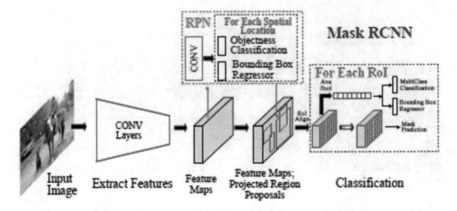

Fig. 3 Mask R-CNN

of the fruit. Instant segmentation maintains the tracking with mask annotation using the CNN architecture. Mask R-CNN was deployed for all three approaches, while the Yolo was employed just for the object recognition task. Hence, the Yolo-v3 approach was utilized for the multi-label classification. The results showed the 3D model employed for grape segmentation with a 0.90 F1-score [43].

D: Faster-RCNN

Faster RCNN is one of the famous frameworks for object recognition in [74]. Multi-function architecture was proposed for the detection and segmentation of fruits for robot harvesting in apple orchards. The proposed technique Detection and Segmentation Network (DasNet) outperformed with 83.6 Average Precision (AP) and 0.832 F1-score rather than three traditional schemes Yolo-v3, Faster-RCNN, and ResNet-101. In addition, the light-weighted network achieved the best results F1-score of 0.827 on the classification of apple and 87.6 and 77.2 on segmentation of apple and branches in the orchard, which increases the model's performance in [75]. In fruit classification, Faster-RCNN was adopted for the parallel detection of various fruits, including mango, apple, and almond. It improved the performance via data augmentation and reduced the labeling cost [76]. The combination of Faster-RCNN with three residual networks of ResNet50, ResNet101, and ResNet inception-v2 accurately detected tomato plants using the COCO dataset the architecture of Faster-RCNN shown in Fig. 4. Experiments take a long training time; hence the proposed technique improved accuracy with an F1-Score of 83.67, AP of 87.83, and IOU greater than equal to 0.5. Therefore, Faster-RCNN with ResNet101 strengthen the fruit counting, robot harvesting and is pertinent for yield prediction [13].

E: Yolo Network

Specifically, this research utilized a supervised-based Yolo-v2 architecture to detect green mango under various lightning postures. A novel method, UAV introduced for visual detection in the orchard. The proposed algorithm showed 96.1 precision

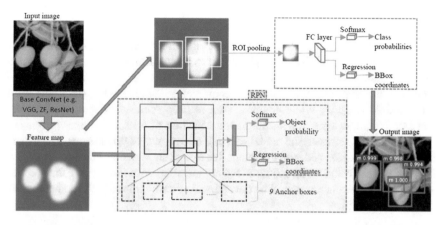

Fig. 4 Faster-RCNN

and 89.0 recall considering illumination effects in [77]. In another research line, DensNet-Yolo-v3 with the fusion of anthracnose method was introduced to detect apple lesions in orchards. DensNet is useful for Yolo-V3 to optimize the feature extraction process and help to minimize their resolution. In addition, Cycle Consistent Generative Adversarial Network (CycleGAN) was deployed to enhance the dataset. The proposed technique efficiently worked compared to the traditional Faster-RCNN and VGG16 model with less detection time in a real-time environment [78]. In this paper, a novel YOLO-V3-dense model is developed to detect fruit to monitor its various growth stages in orchards. For this purpose, real-time data was approximately convenient to prevail. Figure 5 shows the YOLO Network architecture. The Dense-Net method was exploited to substitute a low-resolution feature layer in the Yolo architecture. The results showed that the YOLO-V3 Dense-Net model had been a more significant advantage over YoloV3 and Faster RCNN algorithm in [79].

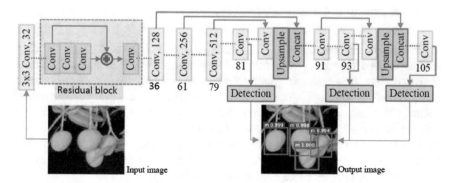

Fig. 5 YOLO network

F: SSD

In similar research, Faster-RCNN cope with inception v2 and single-shot multi-box detector cope with Mobile-Net deployed for counting fruit containing 3 categories like Avocado, lemon, and Hass. The experiments showed that Faster-RCNN efficiently performed with 93.1 accuracy compared to Mobile Net, which has been estimated 90 accuracy while counting fruit in [39]. A novel system Efficient-Net and Mix-Net, developed for the automatic detection of fruit, overcomes prolonged training and testing time. The model was trained over the 48,905 images. The experimental results performed over the ImageNet dataset showed that Mix-Net speeds up the model's performance and reduces the computation cost. However, Efficient-Net achieved better accuracy than the conventional system and reduced the number of parameters in [80].

4   Datasets

The dataset has been crucial to solving various problems such as object recognition, classification, and segmentation in research. Availability of the dataset affects the prediction of the model to achieve the desired outcome using a compatible algorithm. Increasing the training dataset and developing a new dataset for the best performance are needed to solve the critical, challenging problem. The availability of various images assisted in making different kinds of comprehensive datasets over the internet. Therefore, with the presence of millions of images, the scalability of datasets has become covered. Object recognition has made an extraordinary performance with the significant development in datasets. This survey reported various datasets, which are listed in below Table 3.

5   Performance Assessment Metrics

Various assessment metrics have been utilized to evaluate the performance of deep learning models that vary to the corresponding problem. Deep learning used different types of performance evaluation parameters such as True Positive (TP), True Negative (TN), Precision, Recall, F1-Score, Average Precision (AP), and Intersection over Union (IOU), Root Mean Square Error (RMSE), Mean Residual Error (MRE) and Relative Error (RE). These metrics are designed to validate the prediction of various models. Some researchers used individual metrics to measure the performance, and some used a combination of metrics. Table 4 lists these metrics with their symbols and formulas. All Performance metrics are listed in Table 4.

Table 3 Datasets

Dataset name	Type of images	Fruit/ vegetable	References
ImageNet	RGB images	18 fruits	Duong et al. [81], Lu [82]
Fruit 360	RGB images	60 fruits	Steinbrener et al. [71]
Own dataset	Hyperspectral Images	2 fruits	Wang et al. [12], Tan et al. [46], Khan et al. [70], Rubanga et al. [73]
Supermarket dataset	RGB images	15 type of fruit/veg	Zhu et al. [83]
WGISD	RGB images	Grape fruit	Stein et al. [43]
Custom dataset	Hyperspectral images	multiple fruit	Cen et al. [45], Khan et al. [70]
Public dataset	RGB images	Papaya	Munasingha et al. [66]
Leaf Diseases dataset	N/A	Soybean	Shrivastava et al. [22]
Dataset of papaya	RGB images	Papaya	Patino-Saucedo et al. [84]
Citrus Disease Image Gallery Dataset	RGB images	citrus fruit	Lu et al. [82]
Plant village dataset	RGB images	multiple fruit	Gandhi et al. [30], Habib et al. [85], Sharif et al. 86]
Strawberry dataset	RGB-D camera images	Strawberry	Zhang et al. [55], Williams et al. [57], Ganesh et al. [59], Altaheri et al. [62]
Apple dataset	RGB Images	Apple	Zaborowicz et al. [48], Chen et al. [51], Cavallo et al. [52]
Orange dataset	RGB and HSV image	Orange	NZJBE [54], Liu et al. [60]
Tomato image dataset	RGB and Synthetic images	Tomato	Rahnemoonfar and Sheppard [49], Bargoti et al. [50], Tran et al. [67]
COCO datasets	RGB images	Green tomato	Mu et al. [13], Liu et al. [60]

A: P

P stands for precision which measures the TP observations from predicted positive observations.

B: R

R stands for recall which predicted the TP detections from the actual annotations.

Table 4 Performance evaluation metrics

Metric abbreviation	Metric name	Formula
Ac	Accuracy	$\frac{Tp+TN}{TP+FP+FN+TN}$
P	Precision	$\frac{Tp}{TP+FN}$
R	Recall	$\frac{Tp}{TP+FN}$
F1	F1 score	$2.\frac{precision.recall}{precision+recall}$
IOU	Intersection over Union	$\frac{Area of overlap}{Area of union}$
MAP	Mean average precision	$\frac{1}{N}\sum_{i=1}^{n}(AP_i)$
RE	Relative error	$\frac{Absolute error}{Measurement being taken}$
MAE	Mean absolute error	$\frac{1}{n}\sum_{i=1}^{n}|y_i - x)|$
RMSE	Root mean square error	$\sqrt{\frac{1}{n}\sum_{i=1}^{n}(y_i - x_i)^2}$

C : *AP*

AP metric stands for average precision that is most widely used for object detection tasks. Typically, this metric calculates the accuracy of deep learning models.

D: *IOU*

IOU metric accurately performed the object localization and was mainly used for object detection purposes as shown in Fig. 6. A few research studies considered threshold value greater than equal to 0.5, which showed better prediction of the models.

E: F1-Score

F1-score defines the harmonic mean of precision and recall curve, which is widely used to test the performance of ML and deep learning models. F1-Score is the best

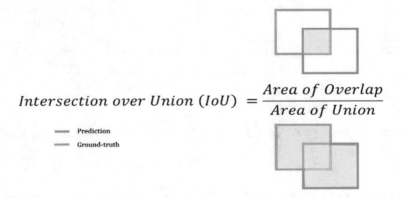

$$\textit{Intersection over Union (IoU)} = \frac{Area\ of\ Overlap}{Area\ of\ Union}$$

— Prediction
— Ground-truth

Fig. 6 IOU

Fig. 7 Tradeoff between the precision and recall curve

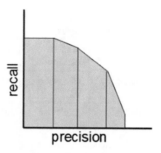

choice of the researcher to optimize the model performance and reduce the FP and FN rates. F1-Score is considered the best case when it's equal to 1, whereas it is considered worse when it's equal to 0. The Tradeoff between the precision and recall curve can be seen in Fig. 7. However, few metrics have not been able to decrease both FP and FN rates simultaneously, which is why this metric is used commonly for specific problems. Table 5 shows the Comparative analysis of Fruit/ Vegetable using machine learning and digital image processing techniques.

6 Conclusion

We have discussed the comparison of state-of-the-art deep learning techniques in detail for detecting and classifying fruit and vegetable under the supervision of computer vision. Computer vision supports the steady, fast and trustworthy fruit and vegetable yield estimation. Hence, our work reported that our proposed deep learning framework copes with various challenges of the agriculture domain compared to the traditional ML approaches. Our study mainly focuses on computer vision techniques with a deep learning algorithm for the accurate detection of fruit or vegetable, which constitute various developed datasets, training models, and performance evaluation parameters. Future researchers focus on state-of-the-art deep learning algorithms, which would be the best approach to automate the farming system that deals with agriculture problems with the help of computer vision.

6.1 Key Challenges

With the development of deep learning algorithms, the researchers propose many techniques; still, there are some challenges to overcome, and computer vision, as a promising technology, will continuously play an essential role in the quality inspection of fruits and vegetables. The demand for large-scale datasets has been increasing to address the state-of-the-art challenges in farming. It would be more interesting to adopt the hybrid approach of computer vision and artificial intelligence with the

Table 5 Comparative analysis of fruit/ vegetable using machine learning and digital image processing techniques

Authors	Problem description	Year	Fruit/Vegetable	Dataset	Technique	Results
Arakeri [87]	Automatic fruit	2016	Tomato	520 images	FNN	96.47 to 100.0
Ortac [88]	Quality of food	2016	Figs	120 hyperspectral images	LDA, SVM	100.0
Samajpati [89]	Detection of diseases	2016	Oranges and Apples	N/A	Random Forecast	N/A
Nandi [90]	Grading of mango	2016	Mango	200 images	SVM	87.00
Rachmawati [4]	Multiclass Fruit	2017	Apple lime, banana, lemon, peach, pear	N/A	Hierarchical multi-feature classification	N/A
Mulyani [91]	Maturity Fuji classification Fuji apples	2017	Fuji Apple	N/A	Fuzzy Logic classification	85.71
Nasirahmadi [92]	Recognition of almond exploiting SIFT descriptor	2017	Almond	2000 images	KNN, L-SVM, Chi-SVM	91.00
Akhter [93]	Eggplant grading	2017	Eggplant	50images taken	KNN	88.00
Perez [94]	Detection of grapevine using SIFT	2017	Grapevine	760	SVM	97.70
Qureshi [95]	Fruit or vegetable yield estimation using machine vision	2017	Mango	2464 images	SVM	98.00
Zeng [73]	Grading of fruit using computer vision	2017	Vege	(26 classes)	KNN, SVM	95.60
Wan [96]	Detection of tomato using computer vision	2018	Tomato	150 sample of tomato for each variety of Roma and Pear	BPNN	100.00

(continued)

Table 5 (continued)

Authors	Problem description	Year	Fruit/Vegetable	Dataset	Technique	Results
Choi [97]	Real-time smart fruits assessment grading system	2018	Pears	19,000 images	ANN	97.4
Sidehabi [98]	Classification on passion fruit	2018	Passion fruit	Use 75 passion fruit videos	ANN and K-mean clustering	90.0
Mim [99]	Maturity recognition of mango	2018	Mango	100 images of mangoes	Decision tree	96.0
Li [100]	Object detection using salience and curve fitting method	2018	Apple	55	K-means, FCM	91.84
Amiryousefi [101]	Classification of seed using image base clustering	2018	Pomegranate	20	Cultivars	100.00
Kuang [102]	Fruit detection using multi-class	2018	Fruit	1778	SVM	98.50
Xiong [103]	Identification of lichi using robot	2018	Litchi	480	FCM	97.50
Pereira [104]	Grading of ripe papaya using image processing	2018	Papaya	114 images	Decision tree	95.98
Hassan [105]	Defect detection of olive fruit	2018	Olive fruit	2969	KNN, FCM, K-means	100.00
Habib [84]	Detection of papaya disease using machine vision		Papaya	129 images	SVM, Decision Tree, Naive bayes	90.15

diversity of scalable datasets [16]. With the growth of un-organized situations plus obstruction, varying lighting conditions and inconsistent clustering have been significant challenges for detecting fruits in orchards [8]. Fruit and vegetable accurate detection is a state-of-the-art challenge. Due to the similar color, size, and shape characteristics, it is difficult to discriminate between the apple and tomato. For this purpose, an efficient algorithm is exploited to distinguish between the similar properties of the items based on feature and texture information [71, 102].

Instant segmentation is still a challenging task because it provides fruit's overall geometry like its color, shape, and size. In the future, it would be best practice to implement the expert technique to obtain the desired results of instant segmentation [75]. Pose estimation is necessary to target the exact region of the object instead of colliding non-interested object or background, which enhances harvesting profit. In [64], the 3D pose estimation is improved, which is still a challenging task. The above research analyzed the numerous aspects of computer vision for fruit and vegetable detection. Hence, it declares that the existing system has some problems, which is still challenging. One of the major flaws of the classification and recognition of the system is the less availability of dataset of fruit and vegetable. Lack of knowledge about the utilization of techniques like A.I, ANN, ML, fuzzy logic, etc., in [73]. Due to the various bad illumination effect background complexity of the fruit detection system, it is difficult to design the automated robot for the fruit and vegetable in yield estimation.

6.2   Future Directions

Several newly developed sensors have still not been utilized for the detection of fruit and vegetable. However, it is crucial to design a significant dataset that would be enough to acquire benefits from RGBD-sensors in the future [21]. In the future, there is a need to explore an expert system for detecting lesions on multiple fruits and focus on identifying the kind of lesion that facilitates the diagnosis process to prevent plant diseases [78]. One of the major drawbacks of computer vision is the extensive computational complexity in large recognition time. Implementing a system that takes less time in recognition would be the best practice. In another way, various descriptors are exploited to acquire the best performance for classifying and detecting fruit or vegetable-based on variant color, shape, size, and texture, as discussed in [105]. Future research is trying to exploit the GAN to generate synthetic images similar to real data, significantly affecting the classification and recognition of an object in computer vision [106–108].

References

1. Food and A. Organization, How to Feed the World in 2050. In Executive Summary-Proceedings of the Expert Meeting on How to Feed the World in 2050. 2009. Food and Agriculture Organization Rome, Italy (2009)
2. C. Hung, J. Underwood, J. Nieto, S. Sukkarieh, A feature learning based approach for automated fruit yield estimation. in Field and service robotics (Springer, 2015), pp. 485−498
3. K.P.J.C. Ferentinos, E.I. Agriculture, Deep learning models for plant disease detection and diagnosis, vol. 145, pp. 311−318 (2018)
4. E. Rachmawati, I. Supriana, M.L. Khodra, Toward a new approach in fruit recognition using hybrid RGBD features and fruit hierarchy property. In 2017 4th International Conference on Electrical Engineering, Computer Science and Informatics (EECSI) (IEEE, 2017)
5. J. Feng, L. Zeng, L.J.S. He, Apple fruit recognition algorithm based on multi-spectral dynamic image analysis, vol. 19, no. 4, pp. 949 (2019)
6. A. Bhargava, A.J.J.O.K.S.U.-C. Bansal, I. Sciences, Fruits and vegetables quality evaluation using computer vision: a review (2018)
7. S. Sabzi, Y. Abbaspour-Gilandeh, G.J.I.P.I.A. Garc´ıa-Mateos, A new approach for visual identification of orange varieties using neural networks and metaheuristic algorithms, vol. 5, no. 1, pp. 162–172, (2018)
8. L. Zhang, G. Gui, A.M. Khattak, M. Wang, W. Gao, J.J.I.A. Jia, Multi-task cascaded convolutional networks based intelligent fruit detection for designing automated robot, vol. 7, pp. 56028–56038 (2019)
9. Y. Sun, R. Lu, Y. Lu, K. Tu, L.J.P.B. Pan, and technology, Detection of early decay in peaches by structured-illumination reflectance imaging, vol. 151, pp. 68–78 (2019)
10. A. Jahanbakhshi, et al., Classification of sour lemons based on apparent defects using stochastic pooling mechanism in deep convolutional neural networks, **263**, 109133 (2020)
11. X. Yu, H. Lu, D.J.P.B. Wu, Technology, Development of deep learning method for predicting firmness and soluble solid content of postharvest Korla fragrant pear using Vis/NIR hyperspectral reflectance imaging, vol. 141, pp. 39–49 (2018)
12. Z. Wang, M. Hu, G.J.S. Zhai, Application of deep learning architectures for accurate and rapid detection of internal mechanical damage of blueberry using hyperspectral transmittance data, vol. 18, no. 4, p. 1126 (2018)
13. Y. Mu, T.-S. Chen, S. Ninomiya, W.J.S. Guo, Intact Detection of Highly Occluded Immature Tomatoes on Plants Using Deep Learning Techniques, vol. 20, no. 10, p. 2984 (2020)
14. M. Momeny, A. Jahanbakhshi, K. Jafarnezhad, Y.-D.J.P.B. Zhang, Technology, Accurate classification of cherry fruit using deep CNN based on hybrid pooling approach, vol. 166, p. 111204 (2020)
15. N. Ha¨ni, P. Roy, VJJOFR. Isler, A comparative study of fruit detection and counting methods for yield mapping in apple orchards, vol. 37, no. 2, pp. 263–282 (2020)
16. H. Tian, T. Wang, Y. Liu, X. Qiao, Y.J.I.P.I.A. Li, Computer vision technology in agricultural automation—a review, vol. 7, no. 1, pp. 1–19 (2020)
17. A. Colantoni, D. Monarca, V. Laurendi, M. Villarini, F. Gambella, M. Cecchini, Smart machines, remote sensing, precision farming, processes, mechatronic, materials and policies for safety and health aspects, ed: Multidisciplinary Digital Publishing Institute (2018)
18. Y. Chen et al., The visual object tracking algorithm research based on adaptive combination kernel, vol. 10, no. 12, pp. 4855–4867 (2019)
19. F. Moslehi, A.J.J.O.A.I. Haeri, H. Computing, An evolutionary computation-based approach for feature selection, pp. 1–13 (2019)
20. S. Misra, R.H.J.J.O.A.I. Laskar, H. Computing, Development of a hierarchical dynamic keyboard character recognition system using trajectory features and scale-invariant holistic modeling of characters, vol. 10, no. 12, pp. 4901–4923 (2019)
21. K. Hameed, D. Chai, A.J.I. Rassau, V. Computing, A compre-hensive review of fruit and vegetable classification techniques, vol. 80, pp. 24–44 (2018)

22. S. Shrivastava, S.K. Singh, D.S.J.M.T. Hooda, and Applications, Soybean plant foliar disease detection using image retrieval approaches, vol. 76, no. 24, pp. 26647–26674 (2017)
23. L. Jain, H. Vardhan, M. Nishanth, S. Shylaja, Cloud-based system for supervised classification of plant diseases using convolutional neural networks. in 2017 IEEE International Conference on Cloud Computing in Emerging Markets (CCEM) (IEEE, 2017), pp. 63–68
24. K. Kangune, V. Kulkarni, P.J.A.J.F.C.I.T. Kosamkar, Automated estimation of grape ripeness (2019)
25. M.I. Al-Hiyali, N. Yahya, I. Faye, Z. Khan, K.A. Laboratoire, Classification of BOLD FMRI signals using wavelet transform and transfer learning for detection of autism spectrum disorder. In *2020 IEEE-EMBS Conference on Biomedical Engineering and Sciences (IECBES)* (IEEE, 2021), pp. 94–98
26. S.I. Saedi, H.J.E.S.W.A. Khosravi, A Deep Neural Network Approach Towards Real-Time On-Branch Fruit Recognition for Precision Horticulture, p. 113594 (2020)
27. Y. Zhao, L. Gong, Y. Huang, C.J.C. Liu, E.I. Agriculture, A review of key techniques of vision-based control for harvesting robot, vol. 127, pp. 311–323 (2016)
28. S.R. Dubey, A.S.J.J.O.I.S. Jalal, Application of image processing in fruit and vegetable analysis: a review, vol. 24, no. 4, pp. 405–424 (2015)
29. D. Oppenheim, G.J.A.I.A.B. Shani, Potato disease classification using convolution neural networks, vol. 8, no. 2, p. 244 (2017)
30. R. Gandhi, S. Nimbalkar, N. Yelamanchili, S. Ponkshe, Plant disease detection using CNNs and GANs as an augmentative approach, in 2018 IEEE International Conference on Innovative Research and Development (ICIRD) (IEEE, 2018), pp. 1–5
31. E.C. Too, L. Yujian, S. Njuki, L.J.C. Yingchun, E.I. Agriculture, A comparative study of fine-tuning deep learning models for plant disease identification, vol. 161, pp. 272–279 (2019)
32. A.K. Rangarajan, R. Purushothaman, A.J.P.C.S. Ramesh, Tomato crop disease classification using pre-trained deep learning algorithm, vol. 133, pp. 1040–1047 (2018)
33. H.K. Suh, J. Ijsselmuiden, J.W. Hofstee, E.J.J.B.E. van Henten, Transfer learning for the classification of sugar beet and volunteer potato under field conditions, vol. 174, pp. 50–65 (2018)
34. K.G. Liakos, P. Busato, D. Moshou, S. Pearson, D.J.S. Bochtis, Machine learning in agriculture: a review, vol. 18, no. 8, p. 2674 (2018)
35. A. Kamilaris, F.X.J.C. Prenafeta-Boldú, E.I. Agriculture, Deep learning in agriculture: a survey, vol. 147, pp. 70–90 (2018)
36. Z. Khan, N. Yahya, K. Alsaih, M.I. Al-Hiyali, F. Meriaudeau, Recent Automatic Segmentation Algorithms of MRI Prostate Regions: A Review. *IEEE Access* (2021)
37. S.K. Behera, A.K. Rath, A. Mahapatra, P.K.J.J.O.A.I. Sethy, H. Computing, Identification, classification and grading of fruits using machine learning and computer intelligence: a review, pp. 1–11 (2020)
38. A. Rafi, Z. Khan, F. Aslam, S. Jawed, A. Shafique, H. Ali, A Review: Recent Automatic Algorithms for the Segmentation of Brain Tumor MRI. *AI and IoT for Sustainable Development in Emerging Countries*, 505–522 (2022)
39. J.P. Vasconez, J. Delpiano, S. Vougioukas, F.A.J.C. Cheein, E.I. Agriculture, Comparison of convolutional neural networks in fruit detection and counting: a comprehensive evaluation, vol. 173, p. 105348 (2020)
40. J. Naranjo-Torres, M. Mora, R. Hernández-García, R.J. Barrientos, C. Fredes, A.J.A.S. Valenzuela, A Review of Convolutional Neural Network Applied to Fruit Image Processing, vol. 10, no. 10, p. 3443 (2020)
41. A. Paul, S. Ghosh, A.K. Das, S. Goswami, S.D. Choudhury, S. Sen, R. Sharma, S.S. Kamble, A. Gunasekaran, V. Kumar, A.J.C. Kumar, O. Research, A systematic literature review on machine learning applications for sustainable agriculture supply chain performance, p. 104926 (2020)
42. Y. Zhang, P. Phillips, S. Wang, G. Ji, J. Yang, J.J.E.S. Wu, Fruit classification by biogeography-based optimization and feedforward neural network, vol. 33, no. 3, pp. 239–253 (2016)

43. M. Stein, S. Bargoti, J.J.S. Underwood, Image based mango fruit detection, localisation and yield estimation using multiple view geometry, vol. 16, no. 11, p. 1915 (2016)
44. I. Sa, Z. Ge, F. Dayoub, B. Upcroft, T. Perez, C.J.S. McCool, Deepfruits: A fruit detection system using deep neural networks, vol. 16, no. 8, p. 1222 (2016)
45. H. Cen, Y. He, R. Lu, Hyperspectral imaging-based surface and internal defects detection of cucumber via stacked sparse auto-encoder and convolutional neural network, in 2016 ASABE Annual International Meeting, p. 1: American Society of Agricultural and Biological Engineers (2016)
46. W. Tan, C. Zhao, H.J.M.T. Wu, and Applications, Intelligent alerting for fruit-melon lesion image based on momentum deep learning, vol. 75, no. 24, pp. 16741–16761 (2016)
47. D. Jawale, M. Deshmukh, Real time automatic bruise detection in (Apple) fruits using thermal camera, in 2017 International Conference on Communication and Signal Processing (ICCSP) (IEEE, 2017), pp. 1080–1085
48. M. Zaborowicz, P. Boniecki, K. Koszela, A. Przybylak, J.J.S.H. Przybył, Application of neural image analysis in evaluating the quality of greenhouse tomatoes, vol. 218, pp. 222–229 (2017)
49. M. Rahnemoonfar, C.J.S. Sheppard, Deep count: fruit counting based on deep simulated learning, vol. 17, no. 4, p. 905 (2017)
50. S. Bargoti, J.P.J.J.O.F.R. Underwood, Image segmentation for fruit detection and yield estimation in apple orchards, vol. 34, no. 6, pp. 1039–1060 (2017)
51. S.W. Chen et al., Counting apples and oranges with deep learning: a data-driven approach, vol. 2, no. 2, pp. 781–788 (2017)
52. D.P. Cavallo, M. Cefola, B. Pace, A.F. Logrieco, G.J.J.O.F.E. Attolico, Non-destructive automatic quality evaluation of fresh-cut iceberg lettuce through packaging material, vol. 223, pp. 46–52 (2018)
53. A. Wajid, N.K. Singh, P. Junjun, M.A. Mughal, Recognition of ripe, unripe and scaled condition of orange citrus based on decision tree classification, in 2018 International Conference on Computing, Mathematics and Engineering Technologies (iCoMET) (IEEE, 2018), pp. 1–4
54. L.M.O. NZJBE. Aung, A simple and efficient method for automatic strawberry shape and size estimation and classification, vol. 170, pp. 96–107 (2018)
55. Y. Zhang, J. Lian, M. Fan, Y.J.E.J.O.I. Zheng, V. Processing, Deep indicator for fine-grained classification of banana's ripening stages, vol. 2018, no. 1, pp. 1–10 (2018)
56. H. Habaragamuwa et al., Detecting greenhouse strawberries (mature and immature), using deep convolutional neural network, vol. 11, no. 3, pp. 127–138 (2018)
57. H.A. Williams et al., Robotic kiwifruit harvesting using machine vision, convolutional neural networks, and robotic arms, vol. 181, pp. 140–156 (2019)
58. Y. Yu, K. Zhang, L. Yang, D.J.C. Zhang, E.I. Agriculture, Fruit detection for strawberry harvesting robot in non-structural environment based on Mask-RCNN, vol. 163, p. 104846 (2019)
59. P. Ganesh, K. Volle, T. Burks, S.J.I.-P. Mehta, Deep Orange: Mask R-CNN based Orange Detection and Segmentation, vol. 52, no. 30, pp. 70–75 (2019)
60. Z. Liu et al., Improved kiwifruit detection using pre-trained VGG16 with RGB and NIR information fusion (2019)
61. Y. Ge, Y. Xiong, P.J.J.I.-P. From, Instance Segmentation and Localization of Strawberries in Farm Conditions for Automatic Fruit Harvesting, vol. 52, no. 30, pp. 294–299 (2019)
62. H. Altaheri, M. Alsulaiman, G.J.I.A. Muhammad, Date fruit classification for robotic harvesting in a natural environment using deep learning, vol. 7, pp. 117115–117133 (2019)
63. G. Lin, Y. Tang, X. Zou, J. Xiong, J.J.S. Li, Guava detection and pose estimation using a low-cost RGB-D sensor in the field, vol. 19, no. 2, p. 428 (2019). A review on agricultural advancement based on computer vision and grapevine berries using light separation and convolutional neural net-works, vol. 156, pp. 263–274 (2019)
64. P. Barre´ et al., Automated phenotyping of epicuticular waxes of machine learning, in Emerging Technology in Modelling and Graphics (Springer, 2020), pp. 567–581
65. R. Sharma, S.S. Kamble, A. Gunasekaran, V. Kumar, A.J.C. Kumar, O. Research, A systematic literature review on machine learning applications for sustainable agriculture supply chain performance, p. 104926 (2020)

66. L. Munasingha, H. Gunasinghe, W. Dhanapala, Identification of Papaya Fruit Diseases using Deep Learning Approach, 2019: 4th International Conference on Advances in Computing and Technology (ICACT)

67. T.-T. Tran, J.-W. Choi, T.-T.H. Le, J.-W.J.A.S. Kim, A Comparative Study of Deep CNN in Forecasting and Classifying the Macronutrient Deficiencies on Development of Tomato Plant, vol. 9, no. 8, p. 1601 (2019)

68. T.T. Santos, L.L. de Souza, A.A. dos Santos, S.J.C. Avila, E. Agriculture, Grape detection, segmentation, and tracking using deep neural networks and three-dimensional association, vol. 170, p. 105247 (2020)

69. R. Ballesteros, D.S. Intrigliolo, J.F. Ortega, J.M. Ram´ırez-Cuesta, I. Buesa, M.A.J.P.A. Moreno, Vineyard yield estimation by combining remote sensing, computer vision and artificial neural network techniques (2020)

70. R. Khan, R.J.I.J.O.I. Debnath, Graphics, S. Processing, Multi class fruit classification using efficient object detection and recognition techniques, vol. 11, no. 8, p. 1 (2019)

71. J. Steinbrener, K. Posch, R.J.C. Leitner, E.I. Agriculture, Hyperspectral fruit and vegetable classification using convolutional neural networks, vol. 162, pp. 364–372 (2019)

72. H. Muresan, M.J.A.U.S. Oltean, Informatica, Fruit recognition from images using deep learning, vol. 10, no. 1, pp. 26–42 (2018)

73. D.P. Rubanga, L.K. Loyani, M. Richard, S.J.A.P.A. Shimada, A Deep Learning Approach for Determining Effects of Tuta Absoluta in Tomato Plants (2020)

74. G. Zeng, Fruit and vegetables classification system using image saliency and convolutional neural network, in 2017 IEEE 3rd Information Technology and Mechatronics Engineering Conference (ITOEC) (IEEE, 2017), pp. 613–617

75. S. Ren, K. He, R. Girshick, J. Sun, Faster r-cnn: Towards real-time object detection with region proposal networks, in Advances in neural information processing systems, pp. 91–99 (2015)

76. H. Kang, C.J.S. Chen, Fruit detection and segmentation for apple harvesting using visual sensor in orchards, vol. 19, no. 20, p. 4599 (2019)

77. S. Bargoti, J. Underwood, Deep fruit detection in orchards, in 2017 IEEE International Conference on Robotics and Automation (ICRA) (2017, IEEEE), pp. 3626–3633

78. J. Xiong et al., Visual detection of green mangoes by an unmanned aerial vehicle in orchards based on a deep learning method, vol. 194, pp. 261–272 (2020)

79. Y. Tian, G. Yang, Z. Wang, E. Li, Z.J.J.O.S. Liang, Detection of apple lesions in orchards based on deep learning methods of cyclegan and yolov3-dense, vol. 2019 (2019)

80. Y. Tian et al., Apple detection during different growth stages in orchards using the improved YOLO-V3 model, vol. 157, pp. 417–426 (2019)

81. L.T. Duong, P.T. Nguyen, C. Di Sipio, D.J.C. Di Ruscio, E.I. Agriculture, Automated fruit recognition using EfficientNet and MixNet, vol. 171, p. 105326 (2020)

82. Y.J.A.P.A. Lu, Food image recognition by using convolutional neural networks (cnns) (2016)

83. L. Zhu, Z. Li, C. Li, J. Wu, J.J.I.J.O.A. Yue, B. Engineering, High performance vegetable classification from images based on alexnet deep learning model, vol. 11, no. 4, pp. 217–223 (2018)

84. A. Patino-Saucedo, H. Rostro-Gonzalez, J. Conradt, Tropical fruits classification using an AlexNet-type convolutional neural network and image augmentation, in International Conference on Neural Information Processing (Springer, 2018), pp. 371–379

85. M.T. Habib et al., Machine vision based papaya disease recognition, vol. 32, no. 3, pp. 300–309 (2020)

86. M. Sharif et al., Detection and classification of citrus diseases in agriculture based on optimized weighted segmentation and feature selection, vol. 150, pp. 220–234 (2018)

87. G. Wang, Y. Sun, J.J.C.I. Wang, and neuroscience, Automatic image-based plant disease severity estimation using deep learning, vol. 2017 (2017)

88. M.P.J.P.C.S. Arakeri, Computer vision based fruit grading system for quality evaluation of tomato in agriculture industry, vol. 79, pp. 426–433 (2016)

89. G. Ortac, A.S. Bilgi, Y.E. Görgülü, A. Günes, H. Kalkan, K. Taşdemir, Classification of black mold contaminated figs by hyper-spectral imaging, in 2015 IEEE International Symposium on Signal Processing and Information Technology (ISSPIT) (IEEE, 2015), pp. 227–230

90. B.J. Samajpati, S.D. Degadwala, Hybrid approach for apple fruit diseases detection and classification using random forest classifier, in 2016 International Conference on Communication and Signal Processing (ICCSP) (IEEE, 2016), pp. 1015–1019

91. C.S. Nandi, B. Tudu, C.J.I.S.J. Koley, A machine vision technique for grading of harvested mangoes based on maturity and quality, vol. 16, no. 16, pp. 6387–6396 (2016)

92. E.D.S. Mulyani, J.P. Susanto, Classification of maturity level of fuji apple fruit with fuzzy logic method, in 2017 5th International Conference on Cyber and IT Service Management (CITSM) (IEEE, 2017), pp. 1–4

93. A. Nasirahmadi, S.-H.M.J.B.E. Ashtiani, Bag-of-Feature model for sweet and bitter almond classification, vol. 156, pp. 51–60 (2017)

94. Y.A. Akter, M.O. Rahman, Development of a computer vision based eggplant grading system, in 2017 4th International Conference on Advances in Electrical Engineering (ICAEE) (IEEE, 2017), pp. 285–290

95. D.S. Pe'rez, F. Bromberg, C.A.J.C. Diaz, e. i. agriculture, Image classification for detection of winter grapevine buds in natural conditions using scale-invariant features transform, bag of features and support vector machines, vol. 135, pp. 81–95 (2017)

96. W. Qureshi, A. Payne, K. Walsh, R. Linker, O. Cohen, M.J.P.A. Dailey, Machine vision for counting fruit on mango tree canopies, vol. 18, no. 2, pp. 224–244 (2017)

97. P. Wan, A. Toudeshki, H. Tan, R.J.C. Ehsani, E.I. Agriculture, A methodology for fresh tomato maturity detection using computer vision, vol. 146, pp. 43–50 (2018)

98. H.S. Choi, J.B. Cho, S.G. Kim, H.S. Choi, A real-time smart fruit quality grading system classifying by external appearance and internal flavor factors, in 2018 IEEE International Conference on Industrial Technology (ICIT) (IEEE, 2018), pp. 2081–2086

99. S.W. Sidehabi, A. Suyuti, I.S. Areni, I. Nurtanio, Classification on passion fruit's ripeness using K-means clustering and artificial neural network, in 2018 International Conference on Information and Communications Technology (ICOIACT) (IEEE, 2018), pp. 304–309

100. F.S. Mim, S.M. Galib, M.F. Hasan, S.A.J.S.H. Jerin, Automatic detection of mango ripening stages–An application of information technology to botany, vol. 237, pp. 156–163 (2018)

101. B. Li, Y. Long, H.J.I.J.O.A. Song, B. Engineering, Detection of green apples in natural scenes based on saliency theory and Gaussian curve fitting, vol. 11, no. 1, pp. 192–198 (2018)

102. M.R. Amiryousefi, M. Mohebbi, A.J.F.S. Tehranifar, and nutrition, Pomegranate seed clustering by machine vision, vol. 6, no. 1, pp. 18–26 (2018)

103. H. Kuang, C. Liu, L.L.H. Chan, H.J.N. Yan, Multi-class fruit detection based on image region selection and improved object proposals, vol. 283, pp. 241–255 (2018)

104. J. Xiong et al., The recognition of litchi clusters and the calculation of picking point in a nocturnal natural environment, vol. 166, pp. 44–57 (2018)

105. L.F.S. Pereira, S. Barbon Jr, N.A. Valous, D.F.J.C. Barbin, E.I. Agriculture, Predicting the ripening of papaya fruit with digital imaging and random forests, vol. 145, pp. 76–82 (2018)

106. N.M.H. Hassan, A.A.J.M.S. Nashat, S. Processing, New effective techniques for automatic detection and classification of external olive fruits defects based on image processing techniques, vol. 30, no. 2, pp. 571–589 (2019)

107. M.K. Tripathi, D.D.J.I.P.I.A. Maktedar, A role of computer vision in fruits and vegetables among various horticulture products of agriculture fields: a survey (2019)

108. I.A. Quiroz, G.H.J.C. Alfe'rez, E.I. Agriculture, Image recognition of Legacy blueberries in a Chilean smart farm through deep learning

Bird Calls Identification in Soundscape Recordings Using Deep Convolutional Neural Network

Muhammad Azeem, Ghulam Ali, Riaz Ul Amin, and Zaheer Ud Din Babar

Abstract Bird species diversity assumes a significant part in giving essential dimensions to people. Many bird species are threatened by climate change. To safeguard them, we must first determine the species to which they belong and then necessary precautions must be taken to ensure their existence. A conventional way to deal with the inspection of bird diversity is through manual review. This methodology depends on experts ecologists to accomplish exact outcomes. Another methodology is bioacoustics observing, which utilizes computerized recorders to gather wildlife vocalizations for helping researchers in bird studies. For researchers, conservation biologists, and birders, identifying bird species accurately in recorded audio files would be a game-changing tool. So an automatic bird detection system by their calls is the need of the hour. In this sense, artificial neural networks have better the recognition excellence of machine learning and deep learning methods for bird species recognition expressively in recent years. Convolutional neural networks (CNNs) are machine learning algorithms that are effective in image processing and sound detection. A CNN system for classifying bird calls is proposed and evaluated in this work using various setups and hyperparameters. The proposed model has achieved 96% of the testing accuracy of the BirdCLEF Kaggle competition 2021 dataset.

Keywords Convolutional Neural Network (CNN) · Artificial Neural Networks (ANN) · Machine Learning (ML) · Deep Learning (DL) · Bird calls · Kaggle BirdCLEF 2021 · Bird species classification · Bio acoustical monitoring · Mel spectrograms · Precision · Recall · F1-Measure · Accuracy · Evaluation matrices · Librosa · Audio detection

M. Azeem (✉) · G. Ali · R. U. Amin · Z. U. D. Babar
Department of Computer Science, University of Okara, Okara, Pakistan
e-mail: azeemchaudharyg@gmail.com

G. Ali
e-mail: ghulamali@uo.edu.pk

R. U. Amin
e-mail: dr.riazulamin@uo.edu.pk

325

1 Introduction

Birds contribute an important part to ecology. Several birds, however, have been endangered as a result of human activities and environmental changes. Therefore, there is a vital need to keep track of species diversity. Bird species diversity monitoring can take two forms: field observation and acoustical monitoring. In contrast to field observation, sound monitoring provides significant benefits in terms of research scalability by gathering massive volumes of bird data. There are around 10,000 bird species in the world. Therefore, there is an urgent need to automatically classify bird species based on their calls. Taking this into account machine learning techniques are playing an important part to classify birds from their voices even from dense and noisy audios. Many state-of-the-art machine learning models have been used for this purpose like support vector machine, decision tree, resnet50, etc. Although, these models perform brilliantly in-plane datasets but losses their accuracies with datasets having noise and dense environment audios. Because birds have resilient connections with species of different kinds in ecology, they are usually recognized as a reliable indicator of animal species diversity. Around two key methods are exist to keep track of bird variety: (1) physical reviews established on explanations on the ground, and (2) audio checking utilizing independent recording units. The physical techniques, for example, the five-minute bird include utilized in New Zealand, depending on expert information on specialists and can accomplish solid outcomes. However, because the majority of species of birds are movable and counted on a spot basis, there is a chance that certain species would be missed. Manual approaches are also limited in their sustainability due to the cost of keeping professionals in remote areas. Acoustical monitoring uses acoustic sensors to support ecologists in bird studies.

Sensors can work consistently for a longer period and the gathered sounds can give an industrious and unquestionable record of the acoustic soundscape. Moreover, sensing technology is these days a modest method to assist ecologists with considering singing species when joined with automatic exploration methods. Over many years and in various locations, audio sensors were used to collect singing animal sounds. It is a difficult assignment for environmentalists to snoop to entire accounts or else to evaluate the conforming audio signals produced from sound data from the outside. There is a critical need for automated tools to manage the recordings that were obtained. Character recognition technologies have recently been used to computerize bird sound detection in acoustic records. Many example acknowledgment approaches have been investigated for programmed bird species acknowledgment. Commercial software, for example, Raven, and Song Scope is presently accessible to section and describe bird calls. While completely automated analysis techniques can be increased to deal with huge volumes of sound information, in practice, their unwavering quality and accuracy stay tricky. Building precise bird call recognizers on these actual birds singing accounts is a difficult task because various ecological rushes and calls vary geologically, occasionally, and then over the entire lifespan of animal classifications.

Consequently, semi-automated methods, where specialists included are needed to work on the accuracy of mechanized methodologies, have been investigated.

In the recent past, much research was undertaken to dissect potential methods to the stated challenge. The yearly BirdCLEF recognition challenge: a biodiversity data review campaign, may be to blame for the increased interest. The BirdCLEF 2017 [1] training dataset includes approximately 36,000 audio files from 1500 distinct species collected from Xeno-canto, with classes having varied numbers of sound samples. Infield recordings, the problem centers on recognizing single auditory species as well as separating several overlaid sounds. In 2017 [2], take up the task, with 60% accuracy for overlaid sounds and 68% for distinguishing the dominating species in their trials. Pick [3] takes a similar method in 2016, with 41.2% accuracy for multi-labeled data and 52.9% for single-labeled data. Networks trained from scratch, melscaled power spectrograms, an upper-frequency cap, and noise filtering are all used in his research. The winner of the 2016 BirdCLEF competition [4], with a single labelling score of 68.6% and a multiple labelling score of 55.5%, discusses categorization using a CNN with five convolutional and one deep layer. After isolating the noise from the genuine bird sound, spectrograms are generated using the audio files as input. Unsupervised learning [5], decision tree-based feature selection [6], recurrent CNNs [7], and Hidden Markov models [8] are all proposed as alternatives or complements to CNNs in the literature. This study looks towards generic feature extraction for a large range of species of birds. The paper makes key contributions to propose a state-of-the-art deep convolutional neural network model for bird calls identification from soundscape recordings having noises in the background. The rest of the chapter is organized as follows in section II related work will be discussed, in section III methodology is described, in section IV experiments results are discussed, and in last the conclusion a future directions will be discussed.

2   Related Work

Convolutional neural network (CNN) based models have become the most well-known methodology in birdcall recognition. All in all, the spectrogram of bird sound is viewed as the info and the model would treat the bird-call recognizable proof assignment as a picture grouping issue. This is natural, since highlights of birds call one of a kind to every animal group, for example, the pitch and the tone, which can be seen in the spectrograms by experienced natural eyes. In [9] Normalization has been applied, the mean and the fluctuation for change were determined from the whole preparing dataset. Tracking down the best CNN design is a tedious assignment and is regularly done simply by instinct. Present status of-the-art approaches attempt to handle this issue with mechanized hyperparameters search. In [10] To lessen the measure of conceivable plan choices and dependent on currently accepted procedures for CNN formats have been chosen. Every weighted layer (aside from input and output layers) use Batch Normalization, exponential linear units for unit initiation. It has not been utilized any of the extra metadata except for the class id of closer view species. The presence of various background species misshapes the training data and makes a single mark preparing especially testing. In [11] the ResNet-50 (a 50 layer deep-CNN architecture), is the preliminary deep CNN design that used residual learning in 2015. ResNet-50 has been fruitful in expanding precision in computer vision seat stamping difficulties, winning first prize in the ImageNet Large Scale Visual Recognition Challenge 2015.

Given the achievement of ResNet-50 architecture in the computer vision space, in this examination, the ResNet-50 design for mechanized bird call recognition has been used. Keras ResNet-50 is applied to other bird sounds, the outcomes might measure up more genuinely. This improved the accuracy to about 72% training precision and 65% validation precision. The accuracy started to level after 400 epochs. The utilization of more limited info spectrograms to improve the precision of the ResNet-50 model in the future. In [12] The research describes a convolutional neural network-based deep learning approach for bird melody order that was utilized in a sound record-based bird ID challenge, called BirdCLEF 2016. The preparation and test set contained about 24k and 8.5k chronicles, having a place with 999 bird species. The recorded waveforms were different regarding length and substance. We changed over the waveforms into a recurrence area and split them into equivalent portions.

In [10] The portions were taken care of into a convolutional neural organization for including realizing, which was trailed by completely associated layers for grouping. In the authority scores, our answer arrived at a MAP score of more than 40% for primary species, and a MAP score of more than 33% for primary species blended in with foundation species. In [13] Deep convolutional neural networks, at first expected for picture portrayal, are changed and adjusted to perceive the presence of birds in strong accounts. Diverse data extension systems are applied to extend model execution and further develop hypotheses to cloud account conditions and new regular environmental elements. In [12] the proposed approach is surveyed on the dataset of the bird sound location task which is fundamental for the IEEE AASP Challenge on Detection and Classification of acoustic scenes and occasions 2018. It beats past bleeding-edge achieving a region under the bend more than 95 % on the public test leaderboard. A pre-prepared Inception-v3 convolutional neural organization has been used in this exploration. The association was tweaked on 36,492 sound records tending to 1,500 bird species in the particular situation of the BirdCLEF 2017 endeavor. In [14] sound records were changed into spectrograms moreover, further took care of by applying bandpass adjusting, clatter modifying, besides, and calm area departure. For data development purposes, time moving, time broadening, pitch moving, and pitch expanding were applied. This paper [15] shows that no-tuning a pre-arranged convolutional neural organization performs better contrasted with setting up a neural association without any planning. Region variety from picture to sound space could be viably applied. The associations' results were surveyed in the BirdCLEF 2017 endeavor besides, cultivated an authority mean typical precision (MAP) score of 0.567 for standard records and a MAP score of 0.496 for records with establishment species on the test dataset.

3   Methodology

The strategies utilized to the appliance and calculate the bird species models are described in the chapter. The research design that was adopted for this study is discussed here and the procedures for data collecting are also outlined. The methodologies utilized to analyze the data are highlighted, as well as the procedures that were followed to carry out this study, are described. Furthermore, a novel CNN model is designed and trained for bird species classification.

3.1   Dataset

The dataset used in this work has been acquired from Kaggle Official BirdCLEF challenge 2021. Dataset comprises of total 6,900 audio files with Ogg format. In this research due to some system limitations, only 1500 files are considered to train and test the proposed model. The training data that has been acquired comprises 6,900 audio files for 397 species. This is too much for this research due to some system and environment limitations, dataset has been limited to species that have at least 200 recordings with a rating of 4 or better. For this purpose train_metadata.csv file is used which plants with 27 species from 8,548 audio files. As the audio file cannot pass directly to the CNN model. Therefore, spectrograms for the audio are generated which are about 4,157 in total for 27 bird species. After that spectrograms are divided into training and testing segments. A total of 1,247 spectrograms are used for testing and 2,883 spectrograms are used for training purposes which are about 70% training and 30% testing of ratio. The species list comprises very communal species such as the House Sparrow (houspa), Blue Jay (blujay), or Song Sparrow (sonspa). This is not a wicked choice at all to start experimenting.

3.2   Mel-Spectrograms

A spectrogram is a visual representation of a signal's frequency spectrum, where the frequency spectrum of a signal refers to the frequency range that the signal covers. Underdone acoustic data is not suitable for neural network input, and consequently, the acoustic pointer is typically converted interested in a time-spectral depiction. A function that extracts spectrograms for a given audio file has been defined. That function requires to contents a file with Librosa (we just use the primary 15 seconds in this work), get Mel spectrograms, and save each spectrogram as a PNG image in an operational directory for future access. Total 1500 audio files that comprise 27 species have been considered final and extracted the spectrograms. A total of 4,157 training spectrograms have been extracted. That's approximately 150 for every species which is not too wicked. To make definite the spectrograms appearance right and display

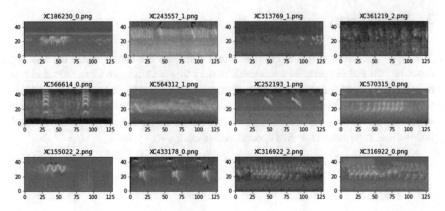

Fig. 1 Extracted spectrograms for training

the first 12. Parse audio files from train_short_audio data source and extract training samples (Fig. 1 shows the first 12 extracted spectrograms for training).

3.3 Proposed CNN Model

The proposed CNN model has comprised of four CNN blocks. Each block has the sequence like a convolutional layer (Conv), rectified linear unit (ReLU) as activation function, batch normalization layer used to overcome the gradient vanishing problem and, in the last max-pooling layer has been used to extract the more prominent features from the data. After CNN blocks global average pooling has been performed and the additional two dense layers. The last layer of the proposed model is the classification layer which is acts as an output layer as well with softmax activation function has been used for this particular layer. The convolutional layer in the first block has comprised of 16 filters with a 3 x 3 filter size a relu activation function and input size of data as hyperparameters. Max pooling layer with a pooling size of 2 x 2 has been used.

The convolutional layer in the second block has comprised of 32 filters with a 3 x 3 filter size a relu activation function and input size of data as hyperparameters. Max pooling layer with a pooling size of 2 x 2 has been used. The convolutional layer in the third block has comprised of 64 filters with a 3 x 3 filter size a relu activation function and input size of data as hyperparameters. Max pooling layer with a pooling size of 2 x 2 has been used. The convolutional layer in the fourth block has comprised of 128 filters with a 3 x 3 filter size a relu activation function and input size of data as hyperparameters. Max pooling layer with a pooling size of 2 x 2 has been used. The dense layers in the dense block have comprised 256 input channels and a relu activation function followed by a dropout layer with a rate of 0.5. Below is the summary of the model. All the detail of the results will be discussed in the next chapter results and analysis (Fig. 2 shows the model diagram).

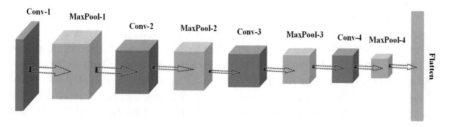

Fig. 2 Proposed CNN model

Table 1 Model training parameters

Sr. No	Parameter	Values
1	Learning rate	0.001
2	Batch size	32
3	Epochs	100
4	Optimizer	Adam
5	Activation	ReLU

3.4 Model Training

Before training the model has been compiled using Adam optimizer with an initial learning rate of 0.001 and accuracy metrics have been used for performance measures. Callbacks have been used to monitor the checkpoints as to where the model starts overfitting training will stop automatically so it makes things so easy. To train the model dataset has been divided into 70% for training and 30% for testing purposes. To fit the model for training batch size of 32, epochs size 100, and validation split of 0.2 has been used. Below is the training verbose of the model (Table 1 depicts important parameters for model training).

4 Experiments and Results

The experiment was done to examine if the system functioned appropriately with the default configuration. For the training, a preliminary LR of 0.001, Adam as an optimizer, and Cross-entropy Loss. BirdCLEF 2021 dataset is used with a partition of 70% & 30% for training and testing respectively (Table 2 shows the experimental results with accuracy).

Table 2 Experimental results

Batch size	Epochs	Mean square error	Val. accuracy
32	100	4.01	96%

It is noticed that validation accuracy does not increase significantly rather than a slight change in it. So an early stopping has occurred after 32 epochs which unloaded the computational burden from the system. The model has been trained on focal recording and soundscape recordings simultaneously. Focal audio recordings as training data can be deceptive and soundscapes have abundant greater noise levels and also comprise very unclear bird calls. The accuracy has been observed at the end of the training about 96%, which is relatively good for novel CNN-based model experiments.

A soundscape from the training data has been picked, however, the whole procedure can simply be automated and then applied to all soundscape files. The file has been loaded with Librosa, spectrograms have been extracted for 5-second chunks, and each chunk has been passed through the model and ultimately consign a label to the 5-second audio chunk. A soundscape is required that contains some of the species that have been trained the proposed model for. The "28933_SSW_20170408.ogg" file has been picked up from the dataset which looks to comprise a plethora of Song Sparrow (sonspa) vocalizations. After analyzing soundscape recording from the selected soundscape audio about 42 birds species have been identified that was encouraging results for a simple model.

Song Sparrow (sonspa) vocalizations have been identified significantly, however missed some others. The Northern Cardinal (norcar) and Red-winged Blackbird (rewbla) have not been detected successfully all though they had been in the training data. This would be a great example of the difficulties faced while analyzing soundscapes. Focal recordings as training data can be deceptive and soundscapes have much higher noise levels and also comprise very unclear bird calls. Therefore, these challenges can be mitigated by further improvement in the model organization and by preprocessing the dataset properly (Table 3 shows the soundscape analyses results of audios).

A predictive model's performance is measured using different evaluation metrics. This usually entails training a model on a dataset, then using the model to generate

Table 3 Soundscape analysis results

Row ID	Site	Audio ID	Seconds	Birds	Predictions
28933_SSW_5	SSW	28,933	5	Sonspa	Sonspa
28933_SSW_10	SSW	28,933	10	Rewbla	Rewbla
28933_SSW_15	SSW	28,933	15	Sonspa	Nocall
28933_SSW_20	SSW	28,933	20	Sonspa	Sonspa
28933_SSW_25	SSW	28,933	25	Houspa	Houspa
28933_SSW_30	SSW	28,933	30	Blujay	Blujay
28933_SSW_35	SSW	28,933	35	Sonspa	Nocall
28933_SSW_40	SSW	28,933	40	Sonspa	Sonspa
28933_SSW_45	SSW	28,933	45	Rewbla	Rewbla
28933_SSW_50	SSW	28,933	50	Houspa	Nocall

Table 4 Model evaluation matrices

Accuracy	Precision	Recall	F1-Measure
96%	100%	100%	100%

predictions on a holdout dataset that was not used during training, and comparing the predictions to the predicted values in the holdout dataset. Metrics used to evaluate the proposed model in this study are accuracy, precision, recall, and F1-measure (Table 4 shows the evaluation metrics to observe the efficiency of the proposed model).

The validation loss (blue) is shown on the right y-axis about 5% pieces of audio categories ordered by the number of training sections within every 5% piece in Fig. 3. On the left y-axis, also it displays the training loss (green) for every piece (green).

To summarize, the proposed model is trained on 4,152 spectrograms for 27 birds species. Model is trained using 32 batch sizes and 100 epochs. To observe the efficiency of the proposed model different evaluation metrics are used such as accuracy, precision, recall, and F1-measure. To make the proposed model much suitable for unseen data and minimize its error rate to enhance its accuracy and efficiency different optimizers are used such as Adam, Adamax, Nadam, Adaleta, Adagrad, SGD, and RMSprop. Adam, SGD, and Adelta show promising results. It is observed that the proposed model has shown 96% of accuracy which is comparatively good with other state-of-the-art models.

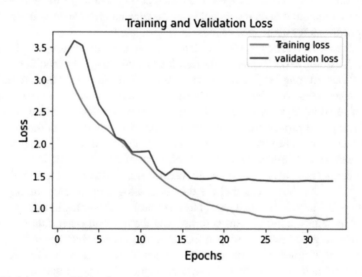

Fig. 3 Training and validation loss graph

5   Conclusion and Future Work

The field of bird calls identification utilizing AI strategies has seen a consistent expansion in recent years, with most works focusing on training different neural networks from the start. In this research a novel CNN model has been proposed by utilizing a visual portrayal of sound; in this case, spectrograms. In this thesis work, the objective of developing a CNN model for the bird calls identification has been achieved, and test it, using a dataset acquired from Kaggle BirdCLEF 2021 with different recording circumstances. Furthermore, a signal pre-processing system has been developed that permits fetching spectrograms, and also allows fetching measurements from the training model. In retrieving the specified bird species, the approach suggested in this thesis shows good results. There are, however, some limits to be aware of. First, the birdcall identification model is evaluated on a set of unique datasets obtained during the Kaggle BirdCLEF-2021 competition. The proposed model must train numerous parameters of the machine learning algorithms utilized to adopt this strategy to different datasets. This means that using this method in different recordings may necessitate more parameter adjustment research. Second, the generated characteristics were destined to be general for the designated classes (27 bird classes), covering previously discovered common birdcall structures. The general feature is insufficiently wide to encompass all of the bird species recorded in the audio records. There are over 390 different bird species in the universe. Due to a lack of annotation data, the model is only evaluated on 27 species in this thesis. The implications for other species must be investigated.

The research's shortcomings, as stated in the last subsection, inspire future research. To improve performance, the developed model might be modified and optimized further. At long last, notice some methods of performance improvement that have been accepted could be a continuation of this research work. More bird species will be studied in the future, and further animal classes might be explored as well. To evaluate the re-productiveness of the constructed procedures, additional tests on diverse datasets must be investigated. Other automatic birdsong analyses, like categorization, can benefit from the datasets utilized in the study. As a result, it is worthwhile to make the datasets available to public researchers. Concerning the signal preprocessing part: it has been seen that the dataset is altogether different among others available publically, only one dataset has been used in this work to observe performance. A potential improvement is to upgrade them for all other datasets as well. The point is to get a system equipped for extricating cleaner bird sound features for an assortment of non-bird or background sounds in the input data. Procedures, for example, commotion decrease, for instance, Wiener filtering, or data enhancement and correlation normalization. As to the training part: a basic model with a few layers and improvement of certain boundaries has been used. A possible improvement is to change the design of this model, mixing different layers and playing with different configurations. Experiments with different analyzers fluctuate their hyperparameters since they would be set just the most basic ones. Some important improvements can be made by using data augmentation and transfer learning techniques. It is assumed that these techniques would be very beneficial in terms of increasing the proposed model accuracy and efficiency.

References

1. A. Joly, et al., Lifeclef 2017 lab overview: multimedia species identification challenges. in International Conference of the Cross-Language Evaluation Forum for European Languages (Springer, 2017)
2. S. Kahl, et al., Large-Scale Bird Sound Classification using Convolutional Neural Networks, in CLEF (working notes) (2017)
3. K.J. Piczak, Recognizing Bird Species in Audio Recordings using Deep Convolutional Neural Networks. in CLEF (working notes) (2016)
4. E. Sprengel et al., Audio based bird species identification using deep learning techniques (2016)
5. D. Stowell, M.D. Plumbley, Audio-only bird classification using unsupervised feature learning (2014)
6. M. Lasseck, Improved Automatic Bird Identification through Decision Tree based Feature Selection and Bagging. CLEF (working notes), **1391** (2015)
7. E. Cakir, et al. Convolutional recurrent neural networks for bird audio detection. In *2017 25th European Signal Processing Conference (EUSIPCO)* (IEEE, 2017)
8. P. Jančovic, et al., Bird species recognition using HMM-based unsupervised modelling of individual syllables with incorporated duration modelling. in 2016 IEEE International Conference on Acoustics, Speech and Signal Processing (ICASSP) (IEEE, 2016)
9. L. Solé Franquesa, Birds Sound Detection Using Convolutional Neural Networks. Universitat Politècnica de Catalunya (2019)
10. D. Stowell et al., Automatic acoustic detection of birds through deep learning: the first bird audio detection challenge. Methods Ecol. Evol. **10**(3), 368–380 (2019)
11. M. Sankupellay, D. Konovalov, Bird call recognition using deep convolutional neural network, ResNet-50. in Proc. ACOUSTICS (2018)
12. M. Lasseck, Acoustic bird detection with deep convolutional neural networks. in Proceedings of the Detection and Classification of Acoustic Scenes and Events 2018 Workshop (DCASE2018) (2018)
13. J. Xie et al., Investigation of different CNN-based models for improved bird sound classification. IEEE Access **7**, 175353–175361 (2019)
14. A. Incze, et al., Bird sound recognition using a convolutional neural network. in 2018 IEEE 16th International Symposium on Intelligent Systems and Informatics (SISY) (IEEE, 2018)
15. A. Joly, et al., LifeCLEF 2015: multimedia life species identification challenges. in International Conference of the Cross-Language Evaluation Forum for European Languages (Springer, 2015)